SOLID STATE ASTROPHYSICS

ASTROPHYSICS AND
SPACE SCIENCE LIBRARY

A SERIES OF BOOKS ON THE RECENT DEVELOPMENTS

OF SPACE SCIENCE AND OF GENERAL GEOPHYSICS AND ASTROPHYSICS

PUBLISHED IN CONNECTION WITH THE JOURNAL

SPACE SCIENCE REVIEWS

VOLUME 55

PROCEEDINGS

SOLID STATE ASTROPHYSICS

PROCEEDINGS OF A SYMPOSIUM
HELD AT THE UNIVERSITY COLLEGE,
CARDIFF, WALES, 9–12 JULY 1974

Edited by

N. C. WICKRAMASINGHE and D. J. MORGAN

University College, Cardiff

With a Preface by Sir Fred Hoyle, F.R.S.

D. REIDEL PUBLISHING COMPANY

DORDRECHT-HOLLAND / BOSTON-U.S.A.

Library of Congress Cataloging in Publication Data

Main entry under title:

Solid state astrophysics.

    (Astrophysics and space science library ; v. 55)
    Bibliography: p.
    Includes indexes.
    1. Cosmic dust—Congresses. 2. Neutron stars—
Congresses. 3. Solids—Congresses. I. Wickramasinghe, Nalin
Chandra. II. Morgan, D. J., 1936–   III. Series.
QB791.S64   523.01   75–34355

ISBN-13: 978-94-010-1886-9   e-ISBN-13: 978-94-010-1884-5
DOI: 10.1007/978-94-010-1884-5

Published by D. Reidel Publishing Company,
P.O. Box 17, Dordrecht, Holland

Sold and distributed in the U.S.A., Canada, and Mexico
by D. Reidel Publishing Company, Inc.
Lincoln Building, 160 Old Derby Street, Hingham,
Mass. 02043, U.S.A.

TABLE OF CONTENTS

PART II / NEUTRON STAR PHYSICS

PREFACE

Over the past decade the study of the formation and properties of interstellar grains has assumed a growing importance, going much beyond what might have been guessed only a few years ago. It has come to be understood that grains play a role in processes other than the simple absorption and scattering of starlight, which was all that the astronomers of a generation ago considered to be their relevance. Grains indeed play a critical role in controlling the temperature, composition, and states of aggregation of the whole interstellar medium. Among the still mysterious problems is the origin of the vast clouds of obscuring material that is observed in radiogalaxies like NGC 5128 and M 82, which may well be associated with the explosions of very massive objects.

It is safe to say that from this growing field of study much still remains to be discovered. The topics discussed in this volume will make clear to the reader the range and versatility of the subjects.

F. HOYLE

FOREWORD

by

THE PRINCIPAL

The Symposium on Solid State Astrophysics held in July 1974 brought to University College Cardiff a large and very distinguished gathering of astronomers. It was the first time that such a collection of scholars, absorbed with the problem of the systems of outer space, had collected together in Wales, and so provided a splendid springboard for the researches of the newly founded group of astronomers in the Department of Applied Mathematics and Astronomy at University College.

I know that Professor Wickramasinghe and his colleagues care deeply that the University of Wales, and, in particular, the Cardiff group, should be one of the world centres for the study of Theoretical Astronomy, and I believe that this volume will make an important contribution to the achievement of that hope.

C. W. L. BEVAN

INTRODUCTION

An International Symposium on Solid State Astrophysics was held at University College, Cardiff, during the period 9–12th July 1974. The symposium was sponsored by the European Physical Society and was financed in part by a grant from the Royal Society.

The recognition of the importance of solid state physics in several branches of astrophysical research has been relatively recent. Solid state physics is relevant to problems associated with the behaviour of interstellar solid particles, the physics of the Moon and inner planets and the crystallization of neutron stars. On account of the strong interest in problems relating to interstellar dust in the Department of Applied Mathematics and Astronomy at Cardiff, it was considered appropriate to concentrate on this topic and to restrict the subject matter of the symposium to interstellar dust and neutron stars. The main aim of the symposium was to bring together astronomers, solid state theorists and laboratory physicists working on problems relating to these two major areas of contemporary astrophysics.

Part I of the proceedings published here contains invited papers and contributions in the general field of interstellar dust and related laboratory astrophysics. Questions relating to the chemical composition and optical properties of interstellar dust have been debated by astronomers for over 40 years. A resurgence of interest in these subjects has taken place in recent years, mainly due to advances in infrared and ultraviolet astronomy which are providing vital clues regarding the composition of these grains. Although the existence of silicates and graphite particles in both circumstellar shells and interstellar space appears to be fairly well established, the question of what other materials, if any, condense on these grains in the interstellar medium remained an open question at the end of the symposium. It was generally accepted that interstellar dust grains play a crucial role in several astrophysical processes – e.g. molecule formation in interstellar space, star formation and infrared radiation from nebulae and galaxies.

Part II of the proceedings contains invited papers and contributions relating to applications of solid state physics to the theory of neutron stars. Important questions relating to the equation of state of neutron star matter, including that of the conditions for crystallization, were discussed by invited speakers and contributors.

In the case of papers which have been published elsewhere than in *Astrophysics and Space Science*, only abstracts are presented here.

The Organizing Committee gratefully acknowledges financial assistance received from the Royal Society and wishes to express its gratitude to University College, Cardiff, for the facilities provided during the period of the symposium.

N. C. WICKRAMASINGHE
D. J. MORGAN

ORGANIZING COMMITTEE

A. R. Beattie (Cardiff)

B. Fitton (E.S.T.E.C., Noordwijk)

S. Hayakawa (Nagoya)

T. Lukes (Cardiff)

P. G. Mezger (Bonn)

D. J. Morgan (Cardiff)

M. Rees (Cambridge)

E. E. Salpeter (Cornell)

K. H. Schmidt (G.D.R.)

N. C. Wickramasinghe (Cardiff)

PART I

INTERSTELLAR DUST

ON THE RATIO OF THE TOTAL TO SELECTIVE ABSORPTION

W. A. SHERWOOD

Astronomisches Institut der Ruhr-Universität, Bochum-Querenburg, W. Germany

Abstract. The ratio of total to selective absorbtion, R, has been found to remain constant as dust is processed in clouds from low to high density, through H II regions and open clusters, and returned to the interstellar medium. R has the same value in dense dust clouds as it has in H II regions of different ages. Variations in R values obtained from stars in H II regions may be due to errors in special type classification. Globular cluster diameters show no tendency to increase with distance from the Sun when $R=3.2$ is used. Large grains evidently do not exist in the interstellar medium. There is no evidence for neutral extinction in the Galaxy at large.

1. Introduction

One of the observable properties of interstellar grains is the ratio of total to selective absorption, $R=A_V/E_{(B-V)}$. It has long been suspected that the value of R is not constant but varies as some function of galactic longitude (Fernie and Marlborough, 1963; Schmidt-Kaler, 1965). This variation is attributed to the alignment of grains in spiral arms. A much larger variation has been reported in the direction of certain dust clouds, open clusters, and H II regions (Johnson, 1968; Crézé, 1972; Racine, 1974). Changes in R reflect changes in the size distribution of the grains which may be related to changes in grain density. The question raised here is: Does the value of R change as the interstellar medium evolves from low density to the high density required for star formation? The answer to this question has been sought among dense dust clouds, H II regions, and open clusters. In addition, the interstellar medium is searched to determine whether there is any evidence for neutral extinction.

2. Dense Dust Clouds

Carrasco *et al.* (1973) found that a quantity $E_{(V-K)}/E_{(B-V)}$ which is comparable with R increases with $E_{(B-V)}$. $E_{(V-K)}$ is equivalent to the optical depth of the cloud at visual wavelengths. The trend for R to increase is marginal if one omits three slightly reddened stars which give a low value of R; the picture certainly does not begin to compare with the large increase in particle density one would expect for the observed total photographic extinction (Bok, 1956). The use of a photometric colour to derive R is open to two criticisms:

(i) K at $2.2\,\mu$ may include some infrared emission which would then distort R (Schmidt-Kaler, 1967), and

(ii) one may be including the effects of faint red companion stars. Of the 13 stars used, all the bright stars (7) and at least 2 of the fainter stars are not single stars.

It is concluded that although the density of the cloud may increase by a factor of 50–100 from the outside of the cloud to the centre, the value of R shows no marked change.

A dust complex in Aquila ($l^{II} = 39°\!.5$, $b^{II} = -1°$) has been studied by means of photoelectric and photographic photometry and star counts in B and V (Sherwood, 1974). The extinction reaches the order of 20 mag. kpc^{-1} with densities of >20 atoms cm^{-3}. At such densities, the clouds should be undergoing collapse and, indeed, there seems to be evidence for star formation. Determination of the total visual and blue absorptions yields an estimate of $R = 3.5 \pm 1$. Observations by Rössiger (1971) indicate that the value of R does not vary significantly from cloud to cloud.

3. H II Regions

Three types of H II regions are studied: radio compact with diameters of <0.5 pc, optical compact with diameters of <5 pc (<10 pc, Georgelin et al., 1973), and others with larger diameters. H II regions not readily associated with star formation are not considered here. The ages of H II regions may be estimated partly by their size and partly by the spectral type of the earliest star (an upper limit for main sequence stars). The ordering by diameter here is one of increasing age; i.e., older H II regions have larger diameters.

Pottasch (1974) has demonstrated that the dust in radio compact H II regions has the same properties as the dust observed in the interstellar medium. He also found that N'_H/A_V, the ratio of the column density of neutral and molecular hydrogen to the visual absorption is essentially constant, implying the constancy of R. This means that the distribution of grain sizes curve is merely shifted up or down with changes in the numbers of grains but the shape is not significantly altered.

Recently Georgelin et al. (1973) have discussed the discrepancy between photometric and kinematic distances to H II regions. The largest disagreement is found among optically compact H II regions (diameter <10 pc). They offered two explanations: large scale streaming motions of 10–15 km s^{-1} and/or $R \sim 4$–7. They found it difficult to account for the discrepancy by streaming since the kinematic distances are always less than the photometric distances and the compact H II regions are the most affected. However, Kraft and Schmidt (1963) and Schmidt-Kaler (1967) found peculiar motions of 10–20 km s^{-1} over regions 1 kpc in diameter. Many of the compact H II regions having discrepant distances lie in regions of this size and so could be subject to motion departing from that of the Schmidt galactic model.

The alternative, that R is between 4 and 7, does not appear to be tenable. In the Puppis region, for example, FitzGerald and Moffat (1974) find that the reddening affecting the H II region NGC 2467 is largely foreground. There is no reason to postulate a large value of R without accompanying reddening; i.e., there is no reason for neutral extinction.

There is a third possibility: these these stars are underluminous. If those compact H II regions with uncertain spectral types are omitted from Figure 2 of Georgelin *et al.* (1973) and a correction of 0″75 is made to the remaining 12 objects having photometric distances much larger than the kinematic distances, the discrepancy is markedly reduced.

As representative of other H II regions, R was determined for 20 stars in 12 H II regions observed spectrophotometrically by Anderson (1970). The results are given in Table I. The data in columns 4 and 5 are the spectral type (SpT) and colour type (sp) of the brightest physical members (Becker and Fenkart, 1971). The distribution of R as a function of spectral type is shown in Figure 1a. The scatter is quite large: ± 0.7 rms for a single star. What is the source of this scatter? The uncertainty due to the photometry and R determination is estimated to be only ± 0.15.

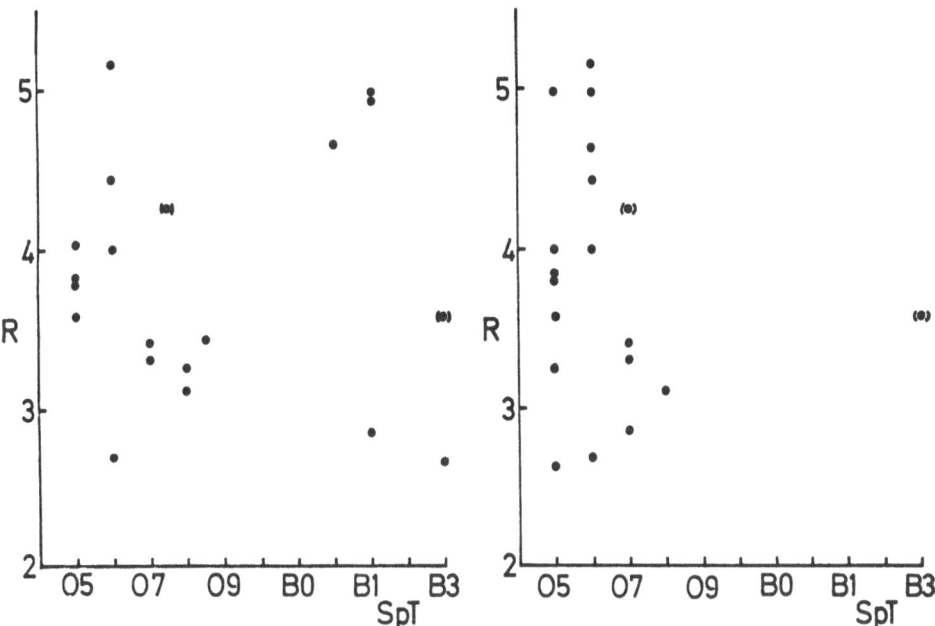

Fig. 1a.   Distribution of R as a function of spectral type for stars in H II regions listed in Table I.

Fig. 1b.   Distribution of R as a function of the earliest spectral type in an H II region. For example, if an H II region contains an O5 and a B2 star, the R value for the B2 star would be plotted as though it were an O5 star since the age of the H II region must be less than or equal to that of the earliest star.

In Figure 1a it is seen that R is not a function of spectral type/temperature since stars of type O6 and B1 have the full range of R values. It is also not a function of the age of the H II region as is shown in Figure 1b where the values of R are plotted against the spectral type of the earliest star in the H II region. R does not appear to

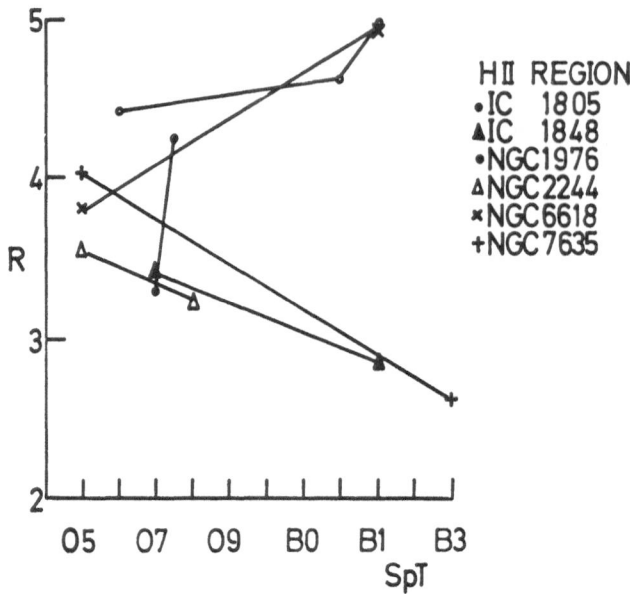

Fig. 1c.   Distribution of R in H ɪɪ regions in which two or more early type stars were observed by Anderson (1970). The stars in a given H ɪɪ region ought to be of the same age and the clouds ought to have similar properties.

be a function of chemical abundance when individual H ɪɪ regions are examined (Figure 1c). If the R values are random, then perhaps the source of scatter lies in small spectral classification errors and the problems of fitting real stars to atmospheric models even when effective temperatures and surface gravity can be determined from the observations. The two stars in IC 1805 may be taken as an example: HD 15 570 is classified as O5f(Ia) by Ishida (1970) and as O7f by Anderson (1970). HD 15 629 is classified as O5(V) by Ishida and O7 by Anderson. Becker and Fenkart (1971) give O6 as the earliest spectral type in the cluster. The former star gives $R=(4.2)$ (the brackets indicate uncertainty in the spectral classification). The percentage error in M_V due to misclassification is, for high luminosity stars ($M_V \sim -7$), of the same order as the dispersion in R. It seems reasonable to attribute the scatter in R to errors in spectral classification, and therefore the mean error of ± 0.1 for the mean value of R in Table I is not statistically meaningful. The value $R=3.7\pm0.6$ does not differ significantly from the value $R=3.2$.

4. Open Clusters and the Interstellar Medium

Most of the large R values in open clusters (Johnson, 1968) have been refuted. They depended mainly on the 'cluster diameter' method which Lynds (1967) and Harris (1973) have shown results in a constant value of $R=3$ being acceptable. Becker (1966) and, more recently, Moffat (1974) have shown that use of the 'variable extinction' method invariably results in large R values through the influence of photometric

TABLE I

Values of R for H II Regions

HD/BD	NGC/IC	Star SpT	Cluster SpT	sp	R
5 005	281	O6			4.00
15 570	1805	O7f	O6	O	(4.24)
15 629	1805	O7			3.30
17 505	1848	O7	O7	O	3.40
237 015	1848	B1V			2.85
37 020	1976	B1			4.97
37 022	1976	O6p			4.43
37 023	1976	B0.5Vp			4.62
42 088	2174	O6			2.68
46 150	2244	O5	O5	O	3.56
46 485	2244	O8			3.24
164 492	6514	O8+O9			3.42
164 492A	6514	O8			3.10
Her 36	6523	O6n			5.15
Anon A	6618	O5			3.80
Anon C	6618	B1Vp			4.94
200 775	7023	B3Vep			(3.58)
215 835	7080	O5	O9	O9	3.84
220 057	7635	B3Vp			2.63
+ 60°2522	7635	O5			4.02

Mean and standard deviation for O stars	3.7±0.6
Mean and standard deviation for all stars	3.8±0.8
Error of the mean	±0.2

errors and field stars; e.g., IC 2581. Garrison (1970) has shown that $R=3$ is better for III Cep than Johnson's value of $R=4.8$.

The 'colour-difference' method based on infrared observations has been criticized (Schmidt-Kaler, 1967, 1971) for ignoring the effects of emission on the determination of R. Abnormal values of R based on IR observations may reflect the existence of circumstellar shells rather than genuinely large values of R.

There remains the possibility that large grains have been removed from the stars in which they were formed or from the sites of star formation. Are large values of R found for the interstellar medium away from regions of star formation?

Neckel (1967) and Shane and Wirtanen (1967) reported evidence for a large R value at the galactic poles. Peterson (1970) and de Vaucouleur and Malik (1969) found normal values. These results can be tested by studying variations in globular cluster diameters as a function of galactic latitude. New distance determinations by Kukarkin and Rusev (1972) and van den Bergh (1967) have been combined with Arp's (1965) values to convert the photoelectrically defined apparent globular cluster diameters (Kron and Mayall, 1960) to linear diameters.

Variations in diameter as a function of Shapley concentration class were first determined. Figure 2 shows that the difference from the mean diameter for each class shows no tendency to increase with distance from the Sun and galactic latitude. $R=3.2$ was used. (With Arp's data alone and $R=3$ ($R_B=4$), there was a significant trend for diameters at low latitude to increase with distance.) The fact that low latitude globular clusters appear to have smaller diameters is a result of the photometric definition of angular diameter containing 90% of the light with respect to a rich background of stars in the Milky Way. The diameters are underestimated by 10% or more and this affects the mean value for the higher latitude diameters. Intermediate latitude clusters are scattered about zero.

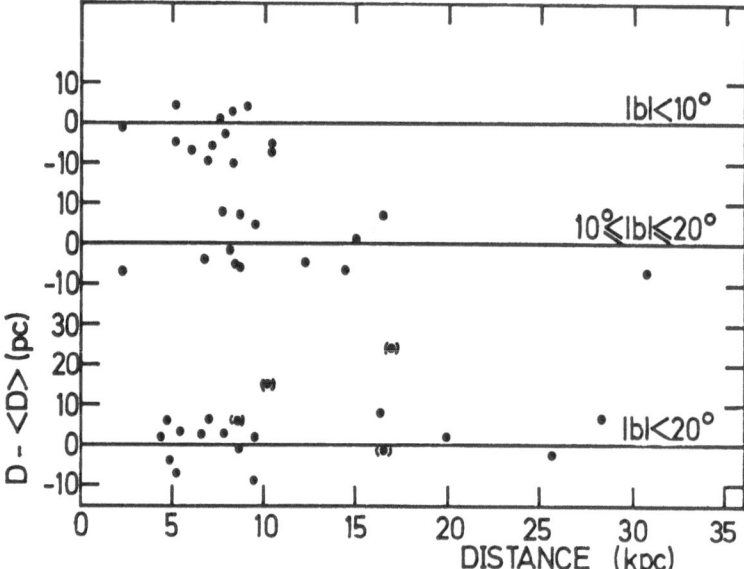

Fig. 2. Distribution of the difference between a globular cluster diameter, D, and the mean value appropriate to its concentration class, $\langle D \rangle$, as a function of distance from the Sun for three galactic latitude zones. The low latitude diameters are systematically smaller than average as a consequence of their photoelectric definition (Kron and Mayall, 1960) which is affected by the rich background of field stars.

5. Summary

The values of R found for the various types of objects are summarized in Table II. Figure 3 shows schematically these values of R as a function of evolutionary time. Clouds increase in density and form stars, but R remains unchanged to the limits of observation. The time between free fall collapse and the formation of radio compact H II regions is very short and the lack of optical observations makes the task of determining R very difficult. However, it seems reasonable to assume from this diagram that the value of R remains constant for all types of objects discussed here. Thus the shape of the distribution of grain sizes remains relatively constant and the curve is simply scaled up or down according to the number of grains present.

Values of R for dense dust complexes, H II regions, and young open clusters do

TABLE II

Values of R

Type of object	Method	Source	R	rms
Dense dust clouds (Aquila)	Star counts	Sherwood	3.5	±1.0
Dense dust clouds (Several)	Star counts	Rössiger	3.7	±0.5
Dense dust clouds (Coal sack)	Star counts	Tapia	3.3	±0.8
Dense dust clouds (ϱ Oph)	$E_{(V-K)}/E_{(B-V)}$	Carrasco *et al.*	3.5	±p.5
H II regions (Several)	Radio N'_H/A_V	Pottasch	3	
H II regions (Several)	$m(\lambda)$ O stars	Sherwood (Anderson)	3.7	±0.6
	$m(\lambda)$O+ B stars	Sherwood (Anderson)	3.8	±0.7
Open clusters (Several)	Cluster diameters	Lynds	3	
Open clusters (Several)	Cluster diameters	Harris	3.15	±0.20
Interstellar medium (globular clusters)	Cluster diameters	Sherwood	3.2	

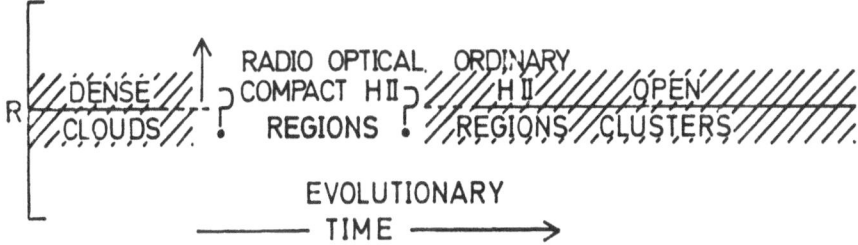

Fig. 3. Schematic diagram of R as a function of evolutionary time for a variety of young objects. No marked variations in the value of R are apparent.

not differ statistically from $R=3.2$. For the interstellar medium in the direction of globular clusters, R is also 3.2. The most accurate values of R are still those summarized by Schmidt-Kaler (1967, 1971): $R=3.25$. No evidence was found for neutral extinction within the galactic halo and large R values in the direction of the galactic poles must be attributed to measurement error.

References

Anderson, C. M.: 1970, *Astrophys. J.* **160**, 507.
Arp, H. C.: 1965, *Stars and Stellar Systems* 5, 401.
Becker, W.: 1966, *Z. Astrophys.* **64**, 77.
Becker, W. and Fenkart, R.: 1971, *Astron. Astrophys. Suppl.* **4**, 241.
Bergh, S. van den: 1967, *J. Roy. Astron. Soc. Com.* **61**, 179.
Bok, B. J.: 1956, *Astron. J.* **61**, 309.
Carrasco, K., Strom, S. E., and Strom, K. M.: 1973, *Astrophys. J.* **182**, 95.
Cohen, M. and Gaustad, J. E.: 1973, *Astrophys. J.* **186**, L131.
Crézé, M.: 1972, *Astron. Astrophys.* **21**, 85.
Fernie, J. D. and Marlborough, J. M.: 1963, *Astrophys. J.* **137**, 700.
FitzGerald, M. P. and Moffat, A. F. J.: 1974, *Astron. J.* **79**, 873.
Garrison, R. F.: 1970, *Astron. J.* **75**, 1001.
Georgelin, Y. M., Georgelin, Y. P., and Roux, S.: 1973, *Astron. Astrophys.* **25**, 337.

Harris, D. H.: 1973, in J. M. Greenberg and H. C. van de Hulst (eds.), 'Interstellar Dust and Related Topics', *IAU Symp.* **52**, 31.

Ishida, K.: 1970, *Publ. Astron. Soc. Japan* **22**, 277.

Johnson, H. L.: 1968, *Stars and Stellar Systems* **7**, 167.

Kraft, R. P. and Schmidt, M.: 1963, *Astrophys. J.* **137**, 249.

Kron, G. E. and Mayall, N. U.: 1960, *Astron. J.* **65**, 581.

Kukarkin, B. V. and Rusev, R. M.: 1972, *Soviet Astron. AJ* **16**, 95.

Lynds, B. T.: 1967, *Publ. Astron. Soc. Pacific* **79**, 448.

Moffat, A. F. J.: 1974, *Astron. Astrophys.* **32**, 103.

Neckel, H.: 1967, in J. M. Greenstein and T. P. Roark (eds.), *Proc. IAU Colloq. on Interstellar Grains*, NASA SP-140, p. 39.

Peterson, B. A.: 1970, *Astron. J.* **75**, 695.

Pottasch, S. R.: 1974, *Astron. Astrophys.* **30**, 371.

Racine, R.: 1974, *Astron. J.* **79**, 945.

Rogerson, J. B., York, D. G., Drake, J. F., Jenkins, E. B., Morton, D. C., and Spitzer, L.: 1973, *Astrophys. J.* **181**, L110.

Rössiger, S.: 1971, *Astron. Nachr.* **293**, 211.

Schmidt-Kaler, Th.: 1965, in H. H. Voigt (ed.), *Landolt-Bornstein*, New Series VI/1, Springer-Verlag, Berlin, p. 284.

Schmidt-Kaler, Th.: 1967, in H. van Woerden (ed.), 'Radio Astronomy and the Galactic System', *IAU Symp.* **31**, 161.

Schmidt-Kaler, Th.: 1971, in L. N. Mavridis (ed.), *Structure and Evolution of the Galaxy*, D. Reidel Publ. Co., Dordrecht-Holland, p. 85.

Shane, C. D. and Wirtanen, C. A.: 1967, *Publ. Lick Obs.* **23**, Pt. 1.

Sherwood, W. A.: 1974, *Publ. Roy. Obs. Edinburgh*, **9**, 85.

Tapia, S.: 1973, in J. M. Greenberg and H. C. van de Hulst (eds.), 'Interstellar Dust and Related Topics', *IAU Symp.* **52**, 43.

Vaucouleurs, G. de and Malik, G. M.: 1969, *Monthly Notices Roy. Astron. Soc.* **142**, 387.

FEATURES OF THE INTERSTELLAR EXTINCTION CURVE

D. H. MORGAN

Royal Observatory, Edinburgh, Scotland

Abstract. The extinction curves for spherical particles are subject to the errors of the particle material's refractive index. Their sensitivity to these errors has been investigated and is found to be dependent upon wavelength. For graphite, significant errors are produced in the far ultraviolet part of the extinction curve; for silicates, in the near ultraviolet; while for iron the error is relatively small.

The wavelength dependence of the 10 μm and 20 μm absorption bands of small silicate spheroids upon their shape and alignment has been studied. It is found that the bands can be displaced by ~ 1 μm towards longer wavelengths from their positions for corresponding spheres: and that a further, though small, displacement can be superimposed upon this by their subsequent alignment.

One of the most important ways of attempting to determine the composition of the interstellar dust, or to place constraints upon its composition, is to compare the observed extinction curves with theoretical ones. In particular, current interest lies in the model fitting of the recent ultraviolet extinction curves obtained from satellite observations. These theoretical curves are usually constructed using the Mie theory to obtain the extinction efficiencies of small homogeneous spherical particles of specified size and complex refractive index, $m = n - ik$. A variety of size and material distributions can, of course, be considered. The refractive indices are determined from experiment and therefore suffer an experimental uncertainty. Here we have investigated the way in which the Mie theory transmits errors in the refractive indices of several materials and have compared this with the results of changes in particle size.

An estimate of the uncertainties in refractive indices can be obtained by comparing the values of Taft and Philipp (1965) and Tosatti and Bassani (1971) for graphite. Their estimates of n generally agree to $\sim 5\%$, but to only 25% near 3.5 μm^{-1}; and of k to 1% near 4 μm^{-1}, 30% near 2 μm^{-1} and 70% near 8 μm^{-1}. We shall take 10% to be representative of the uncertainties involved.

Using the Mie theory we have calculated the extinction efficiency Q_{ext}^* of a graphite sphere of radius $a = 0.02$ μm for an adopted set of refractive indices and have recalculated it, Q_{ext}^i, $i = n$, k or a, with 10% increases and decreases first in n, then in k and finally in a. The adopted refractive indices were taken from Greenaway *et al.* (1969) for $\lambda^{-1} \leqslant 4$ μm^{-1} and Taft and Philipp for $\lambda^{-1} > 4$ μm^{-1}. The transmitted errors

$$\Delta^i = (Q_{\text{ext}}^i - Q_{\text{ext}}^*)/Q_{\text{ext}}^*, \qquad i = n, k \text{ or } a,$$

are shown in Figure 1, the continuous and broken lines referring to positive and negative initial errors respectively. Δ^n and Δ^k are almost complementary, Δ^n being the larger in the visible and far ultraviolet and Δ^k being the larger in the near ultraviolet. In general Δ^k is less than the input error but Δ^n is not. Δ^n and Δ^a are generally comparable in magnitude. The large values of Δ^n in the far ultraviolet are due to a change with n of the wavelength position of a second peak in the extinction efficiency (Morgan and Nandy, 1974a).

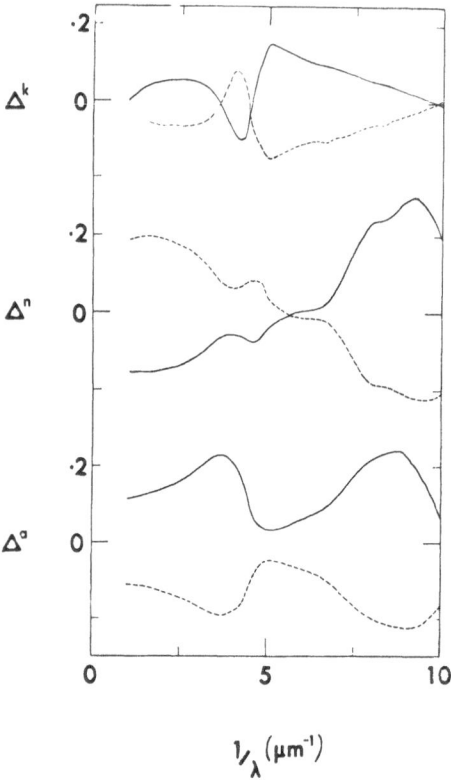

Fig. 1.   The errors transmitted by the Mie theory, Δ^k, Δ^n and Δ^a from 10% errors in k, n and a respectively, in 0.02 μm radius graphite spheres. The continuous lines are for positive input errors and the dashed lines for negative input errors.

Figure 2 shows the same quantities for 0.08 μm radius spheres of the silicate brown enstatite using the refractive indices of Huffman and Staap (1971). The important features of these results are the insensitivity of the particles to both refractive index and size for $\lambda^{-1} > 5\ \mu$m^{-1}, and the very large transmitted errors, greater than 50% on occasions in the visible and near ultraviolet due to both n and a. As with graphite, Δ^n and Δ^a are remarkably similar.

A third material considered as a grain constituent is iron. The transmitted errors are again similar but of a much smaller magnitude. However, the refractive indices of iron are rather uncertain at present.

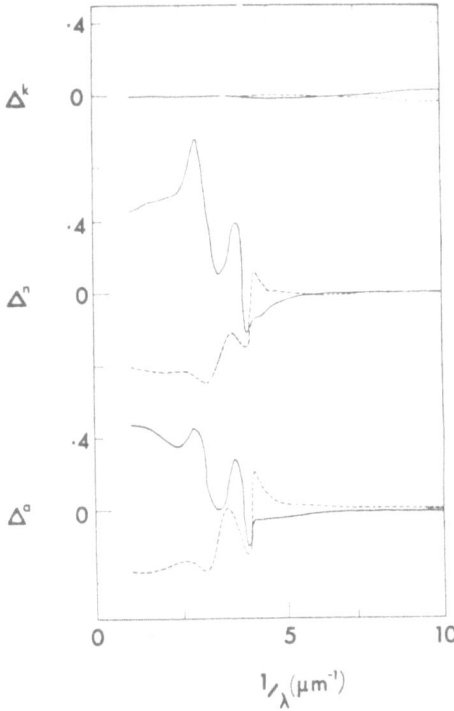

Fig. 2.   As Figure 1 but for 0.08 μm radius silicate spheres.

In addition to recent attention to the ultraviolet extinction curve, considerable interest is shown in the infrared bands seen at 10 μm and 20 μm in emission and absorption by circumstellar and interstellar dust (Woolf and Ney, 1969; Knacke *et al.*, 1969; Hackwell *et al.*, 1970). Many silicates produce extinction peaks at these wavelengths and the refractive indices of several silicates have recently been measured by Pollack *et al.* (1973). We have performed similar calculations to those above for these bands for 0.1 μm radius obsidian spheres, and the results for the 10 μm band are shown in Figure 3. This time the ordinate is Q_{ext}. The continuous curves refer to the adopted refractive indices, and the broken and dashed curves to 10% increases and decreases in the refractive indices. The upper set of curves shows the transmission of uncertainties in n, and the lower set, uncertainties in k. The results for the 20 μm band are qualitatively similar. k is seen to be more influential on the short wavelength side of the band and n more influential on the long.

The foregoing has been concerned with spheres, but it is necessary to consider the effect of particle shape on the extinction curves. Gilra (1971) has shown that the wavelength position of the ultraviolet peak for small graphite spheroids is very sensitive to their shape. We shall consider the extent to which shape affects the wavelength positions of the 10 μm and 20 μm bands of the silicate andesite using the data given by Pollack *et al.* (1973). At these infrared wavelengths the particles' dimensions,

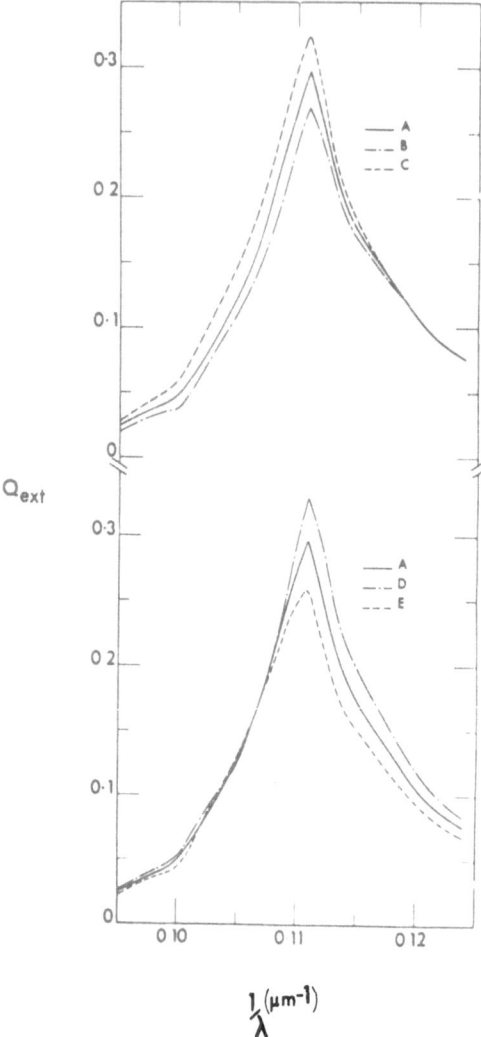

Fig. 3. The extinction efficiency of 0.1 μm radius obsidian spheres. Curve A is for the adopted refractive indices, curves B and C for 10% increases and decreases in n, and D and E for 10% increases and decreases in k.

thought to be ~ 1 μm in circumstellar dust shells, are still small compared with the wavelength. Thus the Rayleigh approximation for small spheroids (see, for example, Wickramasinghe, 1967) is appropriate. Moreover, the albedo for these particles is very small so that the extinction efficiency is virtually independent of particle size but dependent on $x=a/b$, where a, b and c are the lengths of the spheroid axes and $b=c$. For an assembly of spheroidal particles the mean extinction cross-section is

$$C_{\text{ext}} = \frac{1}{n+1} \{C_{\text{ext}}^{\perp} + nC_{\text{ext}}^{\|}\},$$

where $C_{\text{ext}}^{\perp}$ and $C_{\text{ext}}^{\|}$ are the extinction cross-sections with unique axes, a, perpendicular and parallel to the electric vector of the radiation; and n is a parameter determined by the degree of alignment of the particles. For particles in random orientation $n=0.5$.

Figure 4 shows the extinction cross-sections, normalized to unity at $9.0\,\mu$m, of randomly orientated andesite spheroids of size $b=0.1\,\mu$m for values of x ranging from 1.0 (spheres) to 0.01 (discs) – Figure 4(a); and 1.0 to 100.0 (needles) – Figure 4(b). Two distinct maxima exist; one at $9.0\,\mu$m, the other at $10.2\,\mu$m, and particle shape controls the relative importance of the two. The $10.2\,\mu$m feature is absent for oblate and prolate spheroids with $0.5<x<2.0$ and the $9.0\,\mu$m feature is absent only for oblate spheroids with $0.01<x<0.1$. Otherwise both features are present; oblates with $0.1<x<0.5$ and prolates with $0.5<x<100.0$.

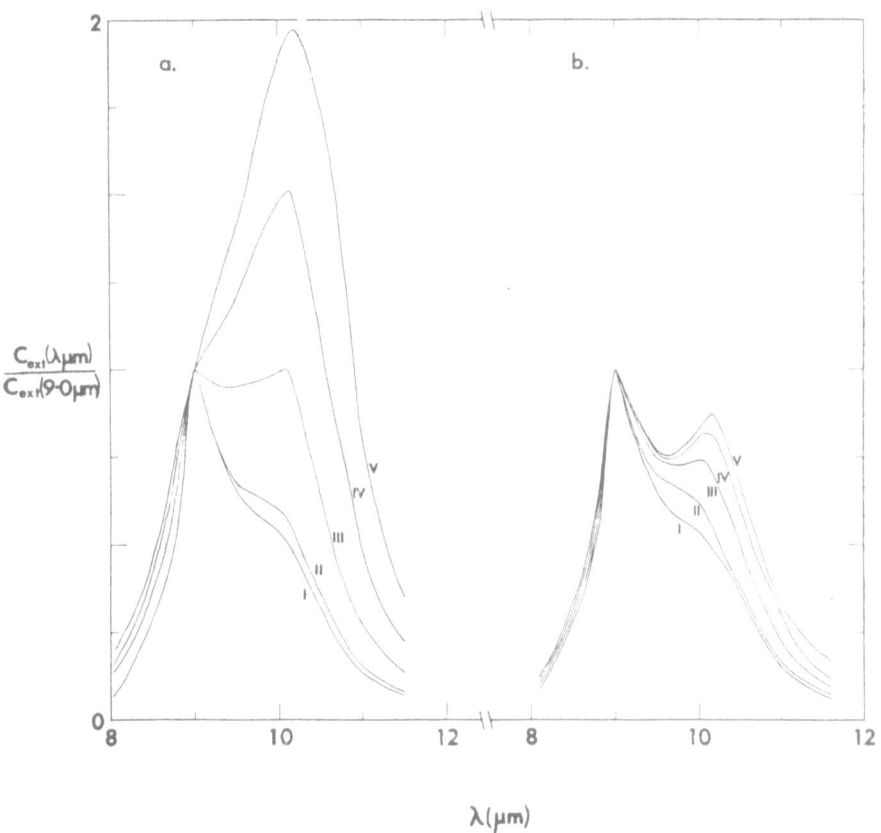

Fig. 4.   (a) The extinction cross-sections of oblate andesite spheroids of several shapes. Curves I to V are for values of x of 1.0, 0.5, 0.25, 0.1 and 0.01 respectively; and $b=0.1\,\mu$m. (b) As (a) but for prolate spheroids. Curves I to V are for values of x of 1, 2, 4, 10 and 100 respectively.

Figures 5 and 6 show the extinction cross-sections of oblate and prolate spheroids of $b=0.1\,\mu$m and $x=0.1$ and $x=10.0$ respectively, for several values of the alignment parameter n. n like x affects the relative importance of the $9.0\,\mu$m and $10.2\,\mu$m peaks. For prolates the peak is at $9.0\,\mu$m for $0.0<n<0.4$; at $10.2\,\mu$m for $2.0<n<10.0$;

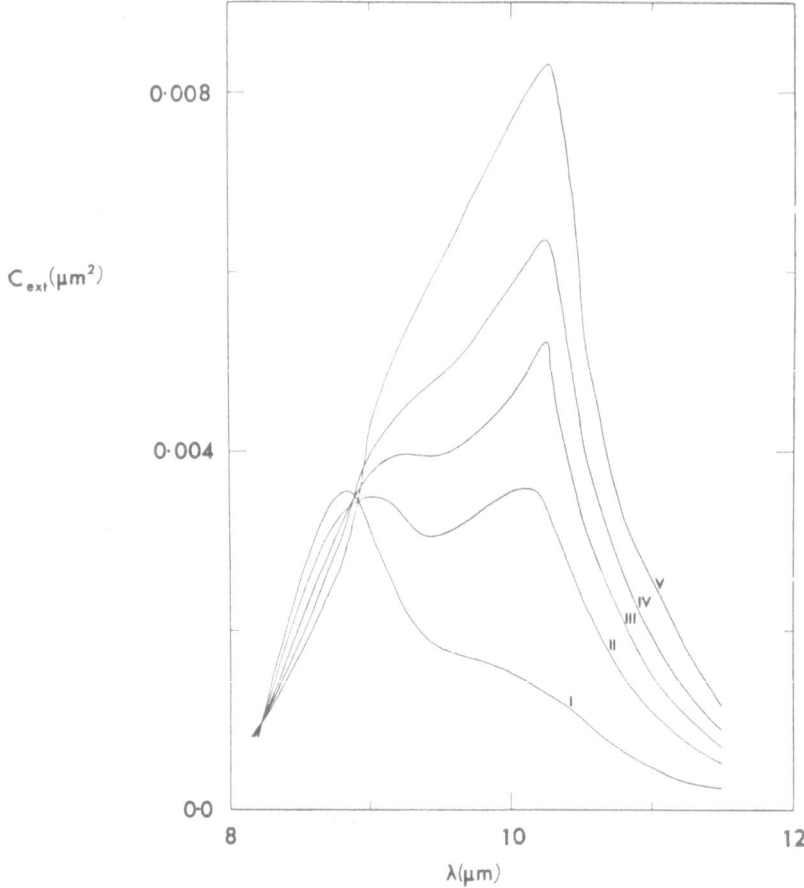

Fig. 5. The extinction cross-sections of the oblate andesite spheroids in several degrees of alignment. Curves I to V are for values cf n of 10.0, 2.0, 1.0, 0.5 and 0.1 respectively; and $b=0.1$ μm and $x=0.1$.

and at both for $0.4 < n < 2.0$. For oblates it is at 10.2 μm for $0.0 < n < 1.0$; at 9.0 μm and 10.2 μm for $n \sim 2.0$; and at 8.77 μm for $n \sim 10.0$. It is values of n between 0.0 and 1.0, in particular those near 0.5, that are the most likely to occur in astronomical situations. Thus we conclude that alignment is unlikely to affect the infrared band for oblate spheroids but will affect its shape for prolate spheroids. This conclusion was also reached for obsidian particles by Morgan and Nandy (1974b) though the twin peak results were less apparent. They also concluded that shape is more important for oblate than for prolate spheroids, and this is borne out here in Figure 4. However, the differences in the shapes of the bands for different silicates are much greater for spheroids than for spheres.

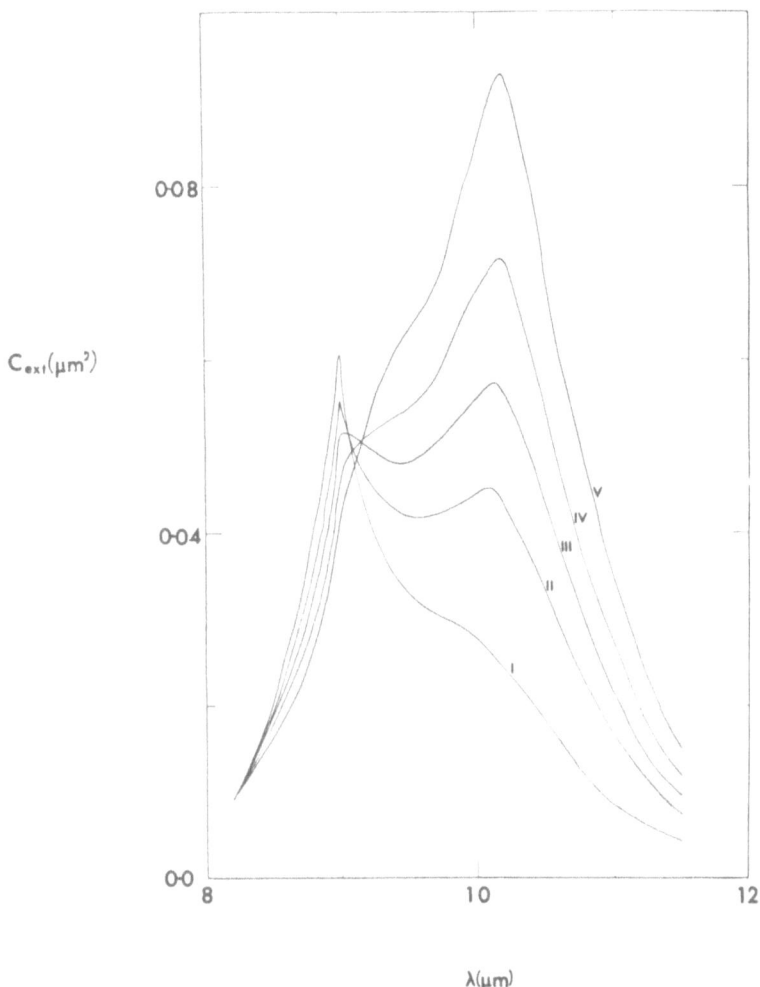

Fig. 6. As Figure 5 for prolate spheroids with $x = 10.0$. Curves I to V are for values of n of 0.1, 0.5, 1.0, 2.0 and 10.0, respectively.

References

Gilra, D. P.: 1971, *The Scientific Results of the OAO*-2, NASA SP-310, p. 295.
Greenaway, D. L., Harbeke, G., Bassani, F., and Tosatti, E.: 1969, *Phys. Rev.* **178**, 1340.
Hackwell, J. A., Gehrz, R. D., and Woolf, N. J.: 1970, *Nature* **227**, 822.
Huffmann, D. R. and Staap, J. L.: 1971, *Nature Phys. Sci.* **229**, 45.
Knacke, R. F., Gaustad, J. E., Gillett, F. C., and Stein, W. A.: 1969, *Astrophys. J.* **155**, L189.
Morgan, D. H. and Nandy, K.: 1974a, *Astrophys. Space Sci.* **29**, 285.
Morgan, D. H. and Nandy, K.: 1974b, *Astrophys. Space Sci.* **31**, 411.
Pollack, J. B., Toon, O. B., and Khare, B. N.: 1973, *Icarus* **19**, 372.
Taft, E. A. and Philipp, H. R.: 1965, *Phys. Rev.* **138**, A197.
Tosatti, E. and Bassani, F.: 1970, *Nuovo Cimento* **65**, 161.
Wickramasinghe, N. C.: 1967, *Interstellar Grains*, Chapman and Hall, London.
Woolf, N. J. and Ney, E. P.: 1969, *Astrophys. J.* **155**, L181.

FAR-ULTRAVIOLET EXTINCTION IN σ SCORPII

THEODORE P. SNOW, JR. and DONALD G. YORK

Princeton University Observatory, Princeton, N.J., U.S.A.

Abstract. It was found earlier from OAO-2 data (Bless and Savage, 1972) that considerable variability with direction in space is present in both the shape and level (relative to $B-V$ color excess) of the interstellar extinction curve in the far ultraviolet. The star σ Sco was shown to be a case of extremely low UV extinction, but there was some question of whether this could be due to scattered nebular light entering the large entrance slit of the Wisconsin spectrometer aboard OAO-2. We have obtained UV data on σ Sco using *Copernicus* (OAO-3), which has an entrance slit on the order of 10^3 times smaller in projected area than that of OAO-2, so that the contribution to the signal from scattered nebular light would be correspondingly smaller. We find very good agreement with the extinction curve of Bless and Savage, confirming the low UV extinction in the line of sight to σ Sco. The curve is extended down to 1000 Å, showing a continued rise towards short wavelengths.

1. Introduction

Earlier studies of ultraviolet interstellar extinction curves (Bless and Savage, 1970, 1972; Stecher, 1965, 1969) have shown that the nature of the extinction below 3000 Å is quite different from that at visible wavelengths in that the level of extinction with reference to $E(B-V)$ is highly variable, and the curve in most stars is dominated by a strong peak of extinction, centered at about 2175 Å and usually referred to as the '2200 Å bump'. Bless and Savage also found that in general the extinction increases rapidly at wavelengths shortward of $7\,\mu^{-1}$; York *et al.* (1973) have reported that the extinction curve continues to rise as far shortward as $10\,\mu^{-1}$.

In a few cases of extremely low UV extinction found with the Wisconsin spectrometer aboard the Orbiting Astronomical Observatory 2 (OAO-2), Bless and Savage (1972) point out the possibility that scattered nebular light entering the large entrance slit of the spectrometer could be the cause of the low extinction which they found. The same argument can be applied to the low UV extinction curve found earlier for $\theta^1 + \theta^2$ Orionis by Carruthers (1969), since the entrance aperture on his rocket-borne instrument was comparable in size to that of the OAO-2 spectrometer. The Princeton telescope-spectrometer aboard *Copernicus* (OAO-3) has an entrance slit with projected dimensions 0″.3 by 39″, which gives it an effective collecting area on the order of 10^3 smaller than those used by Bless and Savage and by Carruthers. Therefore we expect that any contribution to the continuum flux which is due to scattered nebular light would be very much smaller in *Copernicus*, and that a comparison of *Copernicus* extinction curves with those of the Wisconsin group would

allow an assessment to be made of the extent to which such scattered light has affected their derived extinction curves. Besides the overall level of extinction, the shape
and central strength of the 2200 Å bump are also of interest, since this feature has
never before been observed with as small a slit as that of *Copernicus*. Witt and Lillie
(1973) have shown that the bump is predominantly an absorption feature, based on
the fact that there is a pronounced minimum in the diffuse galactic light at 2200 Å,
so we expect that scattered light has probably not affected earlier observations of the
bump.

In the present paper we report on the *Copernicus* ultraviolet extinction curve for
σ Scorpii, one of the stars associated with nebulosity for which Bless and Savage
found an anomalously low level of extinction. Our curve extends over the interval
2850 to 1000 Å (3.5 to 10^{-1}), with a gap between 1900 and 1450 Å, where the sensitivity of *Copernicus* is low.

2. Observations and Data Reduction

The *Copernicus* telescope-spectrometer has been described by Rogerson, Spitzer
et al. (1973). Spectral scans were made of σ Scorpii (B1 III, $E(B-V)=0.40$) and the
comparison star β Canis Majoris (B1 II–III, $E(B-V)=0.01$) including the regions
1000 to 1450 Å at 0.2 Å resolution (phototube U2) and 1885 to 2850 Å at 0.4 Å
resolution (tube V2).

The stray light contribution to the U2 spectra was removed by use of the algorithm
of York *et al.* (1973) as modified by Bohlin (1974). Since the background noise in U2
is generally < 30 cts/14 s integration time and the signal rates > 2000 cts/14 s throughout most of the U2 region for β CMa and σ Sco, no correction was made to the U2
spectra for background due to charged particles.

A correction was necessary, however, for the very high (> 3500 cts/14 s) particle
background counts in the V2 spectra. The V2 background contributions were removed
by use of averaged background data acquired over the first 5000 orbits of the lifetime
of *Copernicus*. Since the overall background level has been found to vary on a time
scale of months, it was necessary to take into account the date of the observation
when deriving the background contribution.

After corrections were made for stray light and particle background, the spectra
were degraded in resolution by averaging them into 5 Å bins. This was done to help
reduce scatter in the extinction curve due to imprecise matching of the many photospheric lines in the UV spectra of hot stars. This procedure did not completely elimin-
ate such scatter, however; the derived extinction curve is noisiest in the regions where
the greatest number of stellar lines occurs, chiefly shortward of 2200 Å. Another
contribution to the scatter comes from small drifts in the pointing of the spacecraft,
which can cause fluctuations of 10% in the signal level on time scales of an hour or
less. Finally, further imprecision was introduced by errors in the subtraction of the
particle background. Very bad points clearly due to a faulty background correction

have been eliminated. In all such cases, the spacecraft was near the South Atlantic Anomaly when the observation was made.

The 5 Å resolution spectra were compared point-by-point to derive the extinction curve according to

$$\frac{E(\lambda - V)}{E(B - V)} = -\frac{2.5}{E(B - V)} \left[\log \left(\frac{S_R}{S_U} \right)_\lambda - C \right], \tag{1}$$

where $(S_R/S_U)_\lambda$ is the ratio of the flux of the reddened star to that of the unreddened star at wavelength λ, and C is a normalization constant given by

$$C = \frac{1}{2.5} (V_U - V_R), \tag{2}$$

where V_U and V_R are the V-magnitudes of the unreddened and reddened stars, respectively. This constant term C normalizes the extinction curve to unity at $\lambda = \lambda_B$, the effective wavelength of the B filter in the UBV system.

3. Results and Discussion

In Figure 1 is depicted the *Copernicus* UV extinction curve for σ Sco, with β CMa as the comparison star. The dashed line indicates the σ Sco extinction curve of Bless and Savage (1972), and the solid line shows for contrast the curve for ζ Oph derived by Bless and Savage and extended with *Copernicus* data to $10\,\mu^{-1}$ by York *et al.* (1973).

The agreement between our curve for σ Sco and that of Bless and Savage is quite good, both in the overall level of the extinction and in the shape of the curve. As in the case of ζ Oph, *Copernicus* data show that for σ Sco the extinction continues to rise steeply towards short wavelengths, down to 1000 Å.

The excellent agreement between the extinction curves for σ Sco derived from OAO-2 and from *Copernicus* data conclusively rules out scattered nebular light as the reason for the Wisconsin observations of anomalously low UV extinction in this star. There is apparently no significant modification of the nature of the observed 2200 Å bump due to scattered light.

Our results therefore tend to support the conclusion of Bless and Savage (1972) and Savage (1973) based on visual observations that the cause of the low extinction in σ Sco and other similar cases is a modification of the grains near the star rather than scattered nebular light. Whether the modification is in the size distribution or the chemical composition, or in some other property of the grains is not clear. If the sharp rise at very short wavelengths is due to a distinct population of very small grains, as has been suggested (Witt, 1973) then it is apparent that this grain population is not entirely absent in the line of sight to σ Sco.

Further *Copernicus* studies are being carried out to determine the nature of grains in and out of dense clouds, to seek evidence for detailed structure in the UV extinction curve, and to determine the nature of the extinction at wavelengths shortward of 1000 Å.

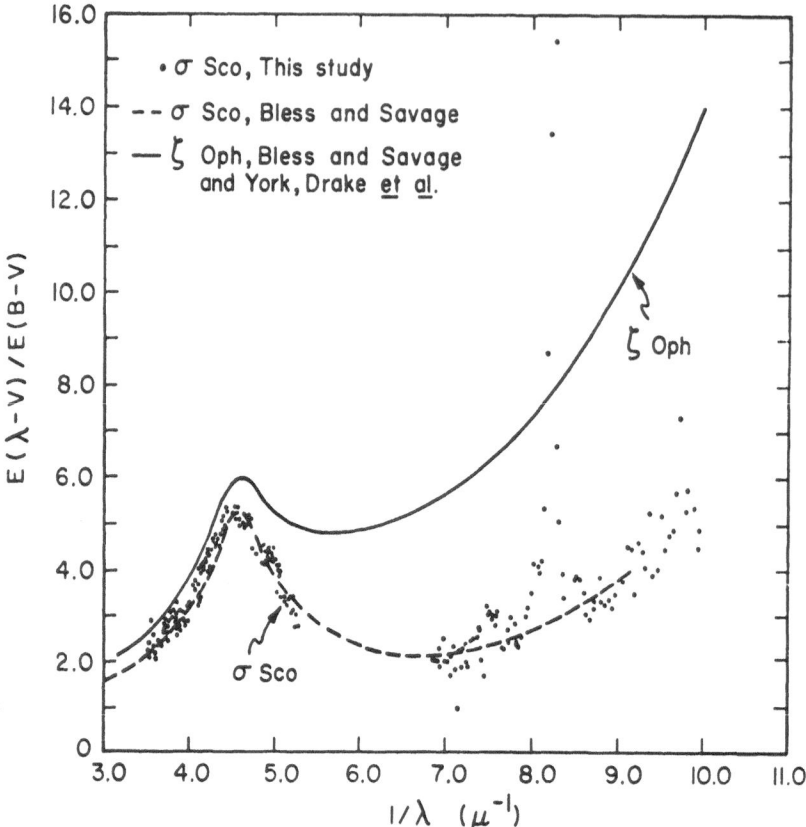

Fig. 1.　The points show the ultraviolet extinction in σ Scorpii, derived in the present study. β CMa is the comparison star. The dashed line indicates the curve for σ Sco derived by Bless and Savage. For contrast, the curve for ζ Oph is also shown, indicated by the solid line. Note that, even though the overall level of extinction in σ Sco is much lower than in ζ Oph, both curves show a steep rise at the shortest wavelengths. The very high-lying points around 8.23 μ^{-1} are due to interstellar Lα absorption in the line of sight to σ Sco.

References

Bless, R. C. and Savage, B. D.: 1970, in L. Hourniaux and H. E. Butler (eds.), 'Ultraviolet Stellar Spectra and Related Ground-Based Observations', *IAU Symp.* **36**, 28.

Bless, R. C. and Savage, B. D.: 1972, *Astrophys. J.* **171**, 293.

Bohlin, R. C.: 1974, unpublished.

Carruthers, G. R.: 1969, *Astrophys. Space Sci.* **5**, 387.

Rogerson, J. B., Spitzer, L., Drake, J. F., Dressler, K., Jenkins, E. B., Morton, D. C., and York, D. G.: 1973, *Astrophys. J. Letters* **181**, L97.

Savage, B. D.: 1973, in J. M. Greenberg and H. C. van de Hulst (eds.), 'Interstellar Dust and Related Topics', *IAU Symp.* **52**, 21.

Stecher, T. P.: 1965, *Astrophys. J.* **142**, 1681.

Stecher, T. P.: 1969, *Astrophys. J. Letters* **157**, L125.

Witt, A. N.: 1973, in J. M. Greenberg and H. C. van de Hulst (eds.), 'Interstellar Dust and Related Topics', *IAU Symp.* **52**, 53.

Witt, A. N. and Lillie, C. F.: 1973, *Astron. Astrophys.* **25**, 397.

York, D. G., Drake, J. F., Jenkins, E. B., Morton, D. C., Rogerson, J. B., and Spitzer, L.: 1973, *Astrophys. J. Letters* **182**, L1.

EXISTENCE AND AMOUNT OF INTERGALACTIC DUST

K.-H. SCHMIDT

Zentralinstitut für Astrophysik der Akademie der Wissenschaften der Deutschen Demokratischen Republik, Sternwarte Babelsberg, G.D.R.

Abstract. The existence of intergalactic dust has been proved by the following observational facts: the decrease of the numbers of distant galaxies and clusters of galaxies behind the central regions of near clusters of galaxies; the different distributions of RR Lyrae stars and galaxies near ι Microscopii (Hoffmeister's cloud); the dependence of colour excesses of galaxies on supergalactic coordinates as well as on the surface density of bright galaxies; the colour index vs redshift correlation of quasi-stellar objects. The densities of intergalactic dust are estimated to be between 5×10^{-30} g cm^{-3} (near the centers of clusters of galaxies) and 2×10^{-34} g cm^{-3} (in general intergalactic space). The grains may be formed either in the early phases of the Universe ($25 < z < 50$) or may be expelled from galaxies by the radiation pressure. The most effective destruction process seems to be the evaporation by soft cosmic rays.

Dust in interstellar space of our own Galaxy has been a familiar concept for several decades. On the other hand, the existence of grains outside stellar systems is still controversial. The aim of this paper has been to compile the observational facts on intergalactic dust and to estimate the density of this matter.

The existence of luminous intergalactic matter of very low surface brightness in the Coma cluster was first suggested by Zwicky (1951, 1952). From the fluctuations of the numbers of galaxies near the galactic north pole the same author concluded the existence of absorbing intergalactic matter. By several authors – especially by Polish astronomers – the statistics of galaxies and clusters of galaxies behind near clusters of galaxies was used to estimate the amount of the extinction by intracluster dust grains. The most extensive investigation in this respect was done by Karachentsev and Lipovetski (1968) who determined the optical depth in the blue and in the red spectral ranges for several near clusters of galaxies. By this method the surface density of galaxies or clusters of galaxies behind a near cluster is determined and compared with the same quantity in the neighbourhood of the near cluster.

The results obtained by this method are influenced by various sources of error. One of these effects can be described as follows: The boundaries of near clusters often have complicated structures, which often are considered as distant clusters. Thus, a higher surface density of distant clusters is simulated.

Another effect concerns the fact that, as a result of a nearby cluster in front of distant clusters the surface density of weak galaxies is increased. Therefore, distant objects are more difficult to recognize. At a sufficiently high surface density of a nearby cluster the distant object will be submerged in the fluctuations. In this way we have to expect a lower surface density of distant clusters in the central regions of near clusters of galaxies. This is exactly what has been observed. But according to Karachentsev and Lipovetski each of the two effects amounts only to about 10 to 20% of the observed values. Therefore, we may conclude that the observed deficiency of distant objects behind nearby clusters is a real effect which may be caused by the obscuration of intra-cluster dust particles.

From the data given by Karachentsev and Lipovetski the mean extinction in the direction of the center of a cluster is equal to $A_B = 0.34 \pm 0.08$ mag. in the blue and $A_R = 0.20 \pm 0.11$ mag. in the red spectral range of the Palomar Observatory Sky Survey. These quantities correspond to a visual intra-cluster extinction per unit length of about 1 mag. Mpc^{-1} and an intra-cluster dust density of about 5×10^{-30} g cm^{-3} in the central region assuming a characteristic particle radius $a = 10^{-5}$ cm and a material density of the grains of about $s = 2.5$ g cm^{-3}. Furthermore, the different values of A_B and A_R refer to a selective extinction. By this fact the assumption of a grain radius of the order of the wavelength of the visible light seems to be justified.

Evidence of an extended intergalactic dust cloud in the direction north of ι Microscopii was given by Hoffmeister (1961) from counts of distant galaxies, colour-excess measurements, and investigations concerning the shapes of galaxies. In order to establish the extragalactic nature of the cloud Hoffmeister compared the distributions of the galaxies and the RR Lyrae variables in the field in question. From the nearly uniform distribution of the RR Lyrae stars we may exclude an obscuring cloud inside our own Galaxy responsible for the 'hole' in the distribution of galaxies. The amount of extinction was estimated to be between 0.5 and 1.2 mag. Because of the large extent this cloud was expected to be a member of the local group of galaxies (Schmidt, 1963). It should be mentioned, however, that the search for 21 cm radiation of neutral hydrogen in the direction of Hoffmeister's cloud was without any success (Varsavsky, 1967).

The existence of intra-cluster dust was supported by Holmberg (1958), who measured colour excesses up to 0.2 mag. in the Virgo cluster. Recently Takase (1972) and De Vaucouleurs et al. (1972) discussed the UBV colours of several hundred members of the local supergalaxy. There seems to be a dependence of the colours on the supergalactic longitude as well as on the supergalactic latitude. Since the effect is not much larger than the mean errors of the photometry the result is not quite certain. In addition to the indicated dependence, there seems to be a weak correlation of the colour excesses with the number of Shapley-Ames galaxies per unit area. Although the effects are small the existence of dust grains in the local supergalaxy is very probable. As with the galaxies the intergalactic dust in the local supergalaxy seems to be distributed in a flattened disk. By colour-excesses the selective character of the intergalactic extinction is demonstrated whereby the radii of the grains are confined.

From the mean redshift and the mean colour excess of the galaxies under consideration the extinction per unit length in the local metagalaxy is about $(2–3) \times 10^{-3}$

mag. Mpc^{-1} assuming the ratio between the extinction in the visual range and the colour excess E_{B-V} as usual on interstellar conditions $A_V/E_{B-V}=3$ and the Hubble constant $H_0=55$ km s^{-1} Mpc^{-1}. This value corresponds to a density of about 10^{-32} g cm^{-3}.

The dependence of the colour excess on metagalactic longitude and latitude as well as the correlation of the colour excess with the surface density of bright galaxies points to a lower dust density in the general intergalactic space than inside the metagalaxy or

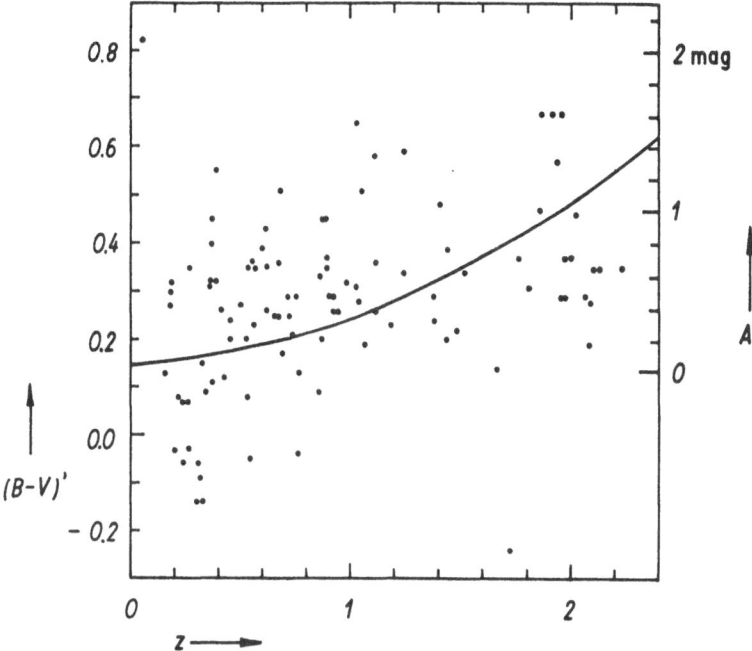

Fig. 1. Colour index $(B-V)'$ of 102 QSO's, corrected for emission lines and galactic interstellar extinction by Evans (1972), plotted vs redshift z (according to Schmidt, 1974). Extinction $A_V(z)$ in the visual range vs redshift z in an Einstein-De Sitter universe is compared with the observations (solid line).

in the clusters. Nevertheless, one should expect any influence of the intergalactic dust grains on the radiation from quasistellar objects. Thus, according to Chiao and Wickramasinghe (1972) the apparent cutoff of QSOs at the redshift $z=2.5$ may be caused by intergalactic extinction.

Another influence of the grains may be the reddening of quasars. Recently Evans (1972) corrected the UBV colours of 102 quasistellar objects for galactic extinction and emission lines. Since the energy distribution in the spectra of the quasars under consideration can be approximated by $F(v) \sim v^{-n}$ with $n=1.02$, on the average, the influence of the K-effect may be neglected. In Figure 1 the corrected colours $(B-V)'$ of the 102 quasars are plotted vs the redshift z. A correlation between both the quantities is indicated corresponding to a general intergalactic extinction of about 10^{-4} mag. Mpc^{-1} in the visual spectral range (Schmidt, 1974).

Otherwise, the correlation in the diagram may be interpreted as an evolutionary effect in the sense that the quasars at large redshifts are redder than those at small redshifts. But there is no way until now to prove this possibility.

Further the objects at large redshifts in the diagram could be more luminous, on the average, than the near quasars. Possibly there is a colour-luminosity relation which is responsible for the correlation in the diagram. In order to test whether the observed correlation is caused by such a selection effect, we looked for a relation between the absolute magnitude M_V and the corrected colour index $(B-V)'$ in the sample of quasi-stellar objects of Evans having redshifts smaller than $z=0.4$. No correlation could be found between these two quantities. Therefore, the best alternative seems to be to interpret the diagram as a result of selective extinction caused by intergalactic grains.

The large dispersion of the relation in the diagram may be attributed mainly to the procedure of correction and to a comparatively wide spread of the normal colour indices of the QSOs. On the other hand we can deduce from this scatter that the inter-galactic dusty matter may be distributed irregularly. In this connection it should be mentioned that in the $(U-B)'$, z-diagram the dispersion is so large that no correlation can be seen.

Now we wish to compare the observed correlation between colour excess and red-shift with the theoretically expected dependence of extinction on redshift. The visual extinction for zero-pressure world models with the cosmological constant $\Lambda=0$ is given by (Oleak and Schmidt, 1975)

$$A_V(z) = 1.086\pi a^2 Q n_0 \frac{c}{H_0} \frac{1}{15q_0^2} \{[6q_0^2(1+z)^2 - 4q_0(1-2q_0)(1+z) +$$
$$+ 4(1-2q_0)^2](1+2q_0 z)^{1/2} - 30q_0^2 + 20q_0 - 4\}. \qquad (1)$$

In this equation it was assumed that the Universe is homogeneously filled with grains. Further, a λ^{-1} law of extinction as a reasonable approximation was assumed. Consideration of the peak in the interstellar extinction curve at λ 2200 Å does not alter the curve significantly. In this connection it should be mentioned that the case of Thomson scattering of electrons (neutral extinction) was considered by Bahcall and Salpeter (1965) and by Bahcall and May (1968).

In Equation (1) a is the grain radius, Q the efficiency factor of the particles for extinction, n_0 the number density of the grains at $z=0$, c the velocity of light, H_0 the present Hubble constant, and q_0 the deceleration parameter. For an Einstein-De Sitter universe ($q_0=\frac{1}{2}$) from Equation (1) it follows that

$$A_V = 1.086\pi a^2 Q n_0 \frac{c}{H_0} \frac{2}{5} [(1+z)^{5/2} - 1]. \qquad (1a)$$

In fitting the observations by Equation (1a) it was assumed that

$$A_V = 3[(B-V)' - (B-V)'_0], \qquad (2)$$

where $(B-V)'_0 = 0.14$ has the meaning of the mean colour index $(B-V)'$ at zero red-shift. Equation (1a) fits the observations best if the condition

$$a^2 Q n_0 = 3.2 \times 10^{-30} \text{ cm}^{-1} \tag{3}$$

is fulfilled (solid line in Figure 1). If we assume the particle radius again $a = 10^{-5}$ cm and the efficiency factor $Q = 2$ as reasonable quantities, the number density of grains at redshifts near $z = 0$ is $n_0 = 1.6 \times 10^{-20}$ cm^{-3}, corresponding to a mass density $\varrho_D \approx 1.7 \times 10^{-34}$ g cm^{-3} if the material density of the particles is $s = 2.5$ g cm^{-3}.

Table I compiles the obtained densities of intergalactic dust. A wide spread of the values in different regions is evident. Obviously the dust density is large in regions of high density of galaxies. It should be mentioned that the estimates on the upper limits of the intergalactic extinction obtained by discussion of the magnitude-redshift relation of bright galaxies in clusters are not in contradiction with the results above, as we shall see later.

TABLE I

Densities of intergalactic dust in different regions

Region	Density (g cm^{-3})
Clusters of galaxies	5×10^{-30}
Local supercluster	10^{-32}
Intergalactic space	2×10^{-34}

Now we shall make some remarks on the origin and the life-time of intergalactic grains. As stated above, Equation (1a), which fits the observed colour excess-redshift relation of the quasistellar objects, is valid for an Einstein-De Sitter universe homogeneously filled with grains. This statement is tantamount to a requirement that the bulk of the dust was formed earlier than $z = 2.5$, because the observations concordant with Equation (1a) attain approximately this redshift. There are two hypotheses concerning the origin of intergalactic grains. Layzer and Hively (1972) supposed that most of the matter in the early universe condensed into stars in the mass-range of 5 to 10 solar masses. These stars would have released most of their energy at redshifts between 25 and 50. Subsequently, during the explosive phase of their evolution, these stars would have ejected the heavy elements from which the grains could have been formed.

Another group of processes concerns the supply of grains from stellar systems. Spitzer (1949) proposed that in supernova outbursts dust particles are blown out of the Galaxy. Further, Schmidt and Van den Bergh (1969) and later Chiao and Wickramasinghe (1972) and Pecker (1972) discussed the ejection of grains from a galaxy by the radiation pressure of stars in the spiral arms of it. It is shown that grains with radii of about 10^{-5} cm have the best chance to escape into the intergalactic space. The grain size of maximum chance of escape somewhat depends on the properties of the particles. We may expect particles of this critical size around 10^{-5} cm to be most abundant in intergalactic space. Since according to Partridge and Peebles (1967) the galaxies in their early phases were by a factor 700 more luminous than now, the ejection of particles by radiation pressure was considerably more effective in the past. Therefore, most of the dust grains in the intergalactic space have been expelled by this process just after the origin of the galaxies. This statement presupposes that the bulk of the interstellar grains was formed in the early phases of a galaxy. This seems justified because at that time the rate of star formation was high. In this connection it should be mentioned that the estimates of Schmidt and Van den Bergh (1969) as well as of Chiao and Wickramasinghe (1972) yield the same order of dust density as follows from the discussion of the colour index – redshift plot of the quasistellar objects.

Obviously, both the hypotheses require the presence of quasi-primordial intergalactic dust. However, in order to preserve the validity of Equation (1a) the life-time of the intergalactic grains has to be longer than, or at least as long as, the Hubble time. Otherwise severe restrictions are imposed on the ejection of grains into the intergalactic space. The life-time of the particles follows from a discussion of the rate of growth and the rate of destruction of these grains. Since the gas density in the intergalactic space is very small, the dust grains do not grow appreciably. On the contrary, a destruction by various processes has to be expected. Obviously, the most effective process which destroys the particles seems to be the evaporation by supra-thermal cosmic rays. Considering interstellar conditions this process was studied in detail by Kimura (1962) and later by Watson and Salpeter (1972). As mentioned above, the intergalactic grains probably have radii of about 10^{-5} cm. In this case the heating of the whole grain by a single cosmic ray particle is not sufficient to evaporate it, as has been shown by Watson and Salpeter. The energy deposited by a supra-thermal particle in a grain first appears as electron ionization and excitation energy within a few lattice spacings of the cosmic-ray path. In a non-metallic grain this electron energy is transferred to the lattice in a cylindrical heat tube. The initial radius of this tube is small enough that the effective temperature within the tube is high. Thus, at least the grain molecules at the ends of the heated tube will evaporate. From the graph given by Watson and Salpeter we estimate the efficiency of this spot-heating mechanism assuming the cosmic ray flux in the intergalactic space to be smaller by one order of magnitude than in the interstellar space. For a grain of radius $a = 10^{-5}$ cm it follows a mean life-time of $\tau_{CR}(0) = 8.3 \times 10^{10}$ years. The investigation of Kimura (1962) yields a similar value if we take into consideration that only the molecules at the ends of the heated tube have a chance of evaporation. Because of the expansion of the universe the mean life-time of an intergalactic grain depends on z as

$$\tau_{CR}(z) = \tau_{CR}(0)/(1 + z)^3. \qquad (4)$$

Probably most of the cosmic rays in the intergalactic space originated in quasi-stellar objects. In this case the assumption of the cosmic ray flux proportional to the number density of quasistellar objects seems to be a reasonable approximation. According to M. Schmidt (1970) the number density of quasars varies as $(1+z)^6$ up to the redshift $z=2.5$. Therefore, in this case the life-time of a grain having a radius $a=10^{-5}$ cm is given by

$$\tau'_{CR}(z) = \tau_{CR}(0)/(1 + z)^6. \tag{5}$$

For a redshift $z=2.5$ in the first case (Equation (4)) the life-time of an intergalactic grain is about 2×10^9 yrs, in the second case (Equation (5)) only about 4×10^7 yrs. Because of the uncertainties it is difficult to decide whether the life-time of an intergalactic dust particle is longer than the Hubble time or not.

Other destruction processes of intergalactic grains have lower efficiencies than the considered spot-heating mechanism by suprathermal particles. So the sputtering by impinging protons of the hot intergalactic gas ($T \approx 10^8$ K) would be comparable with the spot-heating process if the gas density is as large as about 10^{-28} g cm^{-3}. Also the accretion of grains by galaxies is less efficient than the destruction by cosmic ray particles except in dense clusters of galaxies.

If the life-time proportional to $(1+z)^{-6}$ is the correct one at redshifts $z=2$ or 3, then during some 10^7 years the total amount of dust must be renewed in order to satisfy the colour index vs redshift diagram of the quasistellar objects.

At the conclusion some references to implications of intergalactic dust should be made. Naturally, the existence of grains in intergalactic space has various influences. Especially, the relation between apparent magnitudes m of the brightest galaxies in clusters and their redshifts z will be affected in the sense that the deceleration parameter q_0 will be reduced. From a discussion of this m, z-relation an upper limit of the intergalactic extinction at first was estimated by Eigenson (1949) and later by Nickerson and Partridge (1971) and by Sandage (1972). These estimates are not in disagreement with the above results on the density of intergalactic dust as can be seen from Figure 2. In this figure the observed m, z-relation for the brightest members in clusters according to Sandage (1972) is shown (by dots). Further, the theoretical relation for the deceleration parameter $q_0 = \frac{1}{2}$ without intergalactic obscuration is drawn (solid line). For the same deceleration parameter the relation is included considering the extinction which follows from the colour index vs redshift plot of the quasars (dashed line). Obviously, it is impossible to decide from this relation for or against the existence of an intergalactic extinction.

Another implication concerns the opacity of the Universe in the optical range. From the colour index vs redshift diagram of the quasistellar objects we cannot conclude on the dependence of the extinction on the redshift beyond $z=2.5$. This is a matter of the formation and the life-time of the grains. Moreover, in this respect one has to consider the wavelength-dependence of the extinction at large redshifts, which presumably does not vary as λ^{-1}. Rather one should expect only a weak dependence of the extinction on the wavelength because then the size of the grains is large against the wavelength.

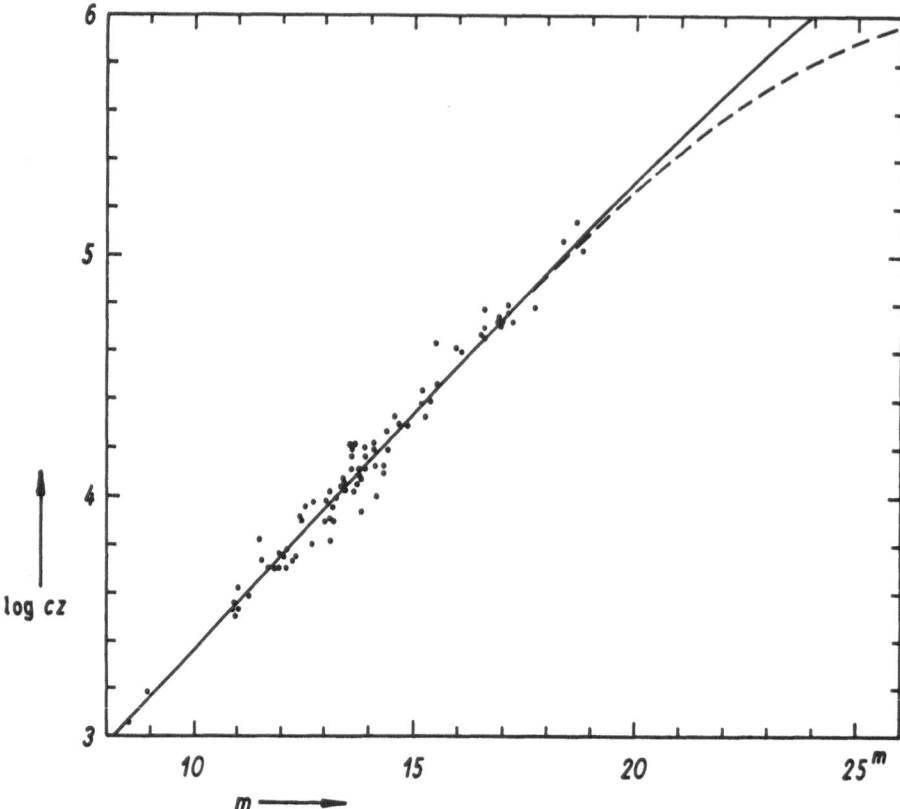

Fig. 2.   Magnitude-redshift relation for brightest members in clusters according to Sandage (1972) (dots) and for an Einstein-De Sitter universe ($q_0=\frac{1}{2}$) without intergalactic extinction (solid line), and with intergalactic extinction according to Figure 1 (dashed line).

But if we assume the intergalactic extinction as derived from the colour index vs redshift diagram of the QSO's it is difficult to explain the cutoff of the quasistellar objects at $z=2.5$ as proposed by Chiao and Wickramasinghe (1972). In order to realize the cutoff – if it is real at all – a steeper ascent of the extinction with the redshift beyond $z=2.5$ is necessary. In this connection, an ineffective search for young galaxies (Davis and Wilkinson, 1973) points to a large intergalactic extinction at greater redshifts. Because of the large distance of these young galaxies having redshifts of about $z=7$ their light may be almost completely absorbed by intergalactic grains.

References

Bahcall, J. N. and May, R. M.: 1968, *Astrophys. J.* **152**, 37.
Bahcall, J. N. and Salpeter, E. E.: 1965, *Astrophys. J.* **142**, 1677.
Chiao, R. Y. and Wickramasinghe, N. C.: 1972, *Monthly Notices Roy. Astron. Soc.* **159**, 361.
Davis, M. and Wilkinson, D.: 1973, *Bull. Amer. Astron. Soc.* **5**, 320.
De Vaucouleurs, G., De Vaucouleurs, A., and Corwin, H. G.: 1972, *Astron. J.* **77**, 285.
Eigenson, M. S.: 1949, *Astron. Zh.* **26**, 278.
Evans, A.: 1972, *Monthly Notices Roy. Astron. Soc.* **160**, 407.

Holmberg, E.: 1958, *Medd. Astron. Obs. Lund Ser.* II, No. 136.
Hoffmeister, C.: 1961, *Z. Astrophys.* **55**, 46.
Karachentsev, I. D. and Lipovetski, V. A.: 1968, *Astron. Zh.* **45**, 1148.
ʹKimura, H.: 1962, *Contr. Dep. Astron. Univ. Tokyo* No. 62.
Layzer, D. and Hively, R.: 1973, *Astrophys. J.* **179**, 361.
Nickerson, B. G. and Partridge, R. B.: 1971, *Astrophys. J.* **169**, 203.
Oleak, H. and Schmidt, K.-H.: 1975, *Astron. Nachr.*, in press.
Partridge, R. B. and Peebles, P. J. E.: 1967, *Astrophys. J.* **147**, 868.
Pecker, J.-C.: 1972, *Astron. Astrophys.* **18**, 253.
Sandage, A.: 1972, *Astrophys. J.* **178**, 1.
Schmidt, K.-H.: 1963, *Astron. Nachr.* **287**, 33.
Schmidt, K.-H.: 1974, *Astron. Nachr.* **295**, 163.
Schmidt, K.-H. and van den Bergh, S.: 1969, *Astron. Nachr.* **291**, 115.
Schmidt, M.: 1970, *Astrophys. J.* **162**, 371.
Spitzer, L.: 1949, *Phys. Rev.* **76**, 583.
Takase, B.: 1972, *Publ. Astron. Soc. Japan* **24**, 295.
Varsavsky, C. M.: 1967, *Trans. IAU* **XIII B**, 189.
Watson, W. D. and Salpeter, E. E.: 1972, *Astrophys. J.* **174**, 321.
Zwicky, F.: 1951, *Publ. Astron. Soc. Pacific* **63**, 61.
Zwicky, F.: 1952, *Publ. Astron. Soc. Pacific* **64**, 242.

DIFFUSE INTERSTELLAR BAND FORMATION
IN DENSE CLOUDS

THEODORE P. SNOW, JR.*‡

Princeton University Observatory, Princeton, N.J., U.S.A.

and

JUDITH G. COHEN

Kitt Peak National Observatory†, Tucson, Ariz., U.S.A.

Abstract. Measurements of the strengths of the diffuse interstellar bands at 4430, 5780 and 5797 Å show that the bands tend to be weak with respect to extinction in dense interstellar clouds. Data on 10 stars in the ϱ Ophiuchi cloud complex show further that the diffuse band-producing efficiency of the grains decreases systematically with increasing grain size. It is concluded that the diffuse bands are not formed in the mantles which accrete on the grains in interstellar clouds, but that they could be produced in the cores of grains or in some molecular species.

1. Introduction

Most speculation concerning the origin of the unidentified diffuse interstellar bands has centered on models in which the bands are produced in solid grains. This emphasis arises primarily because of the generally good correlation between band strengths and interstellar reddening, and because of various difficulties with other types of sources.

It is a general feature of absorption bands produced in small solid particles that several properties of the bands, such as the profiles or the central wavelengths, are subject to significant variations as a function of grain size and density of absorbers (see Aannestad and Purcell, 1973, and references cited therein). The purpose of the present study is to determine the dependence of the band strengths on mean particle size, in interstellar regions where it is believed that significant variations in grain size exist.

2. The Data

Photographic spectra were obtained of 13 stars embedded in dark clouds (i.e., stars in prominent cloud complexes having $E_{BV} \geqslant 0^m\!\!.50$), 14 stars lying behind less dense

---

* Visiting Astronomer, Kitt Peak National Observatory, which is operated by the Associated Universities for Research in Astronomy, Inc., under contract with the National Science Foundation.

‡ Part of this study was carried out by the author at the University of Washington.

† Operated by the Associated Universities for Research in Astronomy, Inc., under contract with the National Science Foundation.

portions of dark cloud complexes, and over 50 distant supergiants whose color excesses are due to their long lines of sight through several intervening clouds. The dark cloud stars in this study lie primarily in the ϱ Ophiuchi and Perseus II clouds. Spectra of 5 stars in IC 348, a very dense region in the Per II complex, were provided by Dr S. E. Strom.

Equivalent widths were measured for the relatively narrow diffuse features at 5780 and 5797 Å, and the central depth of the very broad 4430 Å band was determined for each star. Standard errors are estimated to be 15% of the true equivalent width of $\lambda\lambda$ 5780 and 5797 (or 10 mÅ for features weaker than 40 mÅ), and 1.5% of the continuum level for λ 4430.

Our data for distant supergiants and stars lying behind optically thin ($E_{BV} <$ 0^{m}50) portions of dark clouds were augmented by diffuse band measurements of other authors, care being taken in the case of λ 4430 to reduce the central depths to a common system by statistical techniques such as those described by Deeming and Walker (1967) and by Snow (1973).

3. Results

3.1. COMPARISON OF CLOUD STARS WITH DISTANT SUPERGIANTS

Plots of band strength against E_{BV} show that $\lambda\lambda$ 4430 and 5780 are systematically weakened in the dark cloud stars, as indicated in Figure 1, where it is seen that most dark cloud stars lie well below the mean correlation based on distant supergiants. This is consistent with Wampler's (1966) study of λ 4430, but not with the recent finding of Bromage and Nandy (1973) that the band strengths in the VI Cygni OB association (Cygnus OB2) correlate normally with E_{BV}. It should be pointed out, however, that the VI Cygni association is in the galactic plane and lies nearly 2 kpc from the Sun (Walborn, 1973), whereas the ϱ Oph and Per II complexes are considerably nearer and are well out of the plane. Hence the large color excesses in VI Cygni are probably due to the long line of sight through foreground material, and not to extinction in dense clouds like the ϱ Oph cloud or Per II. Polarization data (Coyne *et al.*, 1974) show nothing unusual for the grains in the direction of VI Cygni, whereas for the ϱ Oph region (Carrasco *et al.*, 1973) and IC 348 (Strom, 1974) it is evident that grain sizes increase with increasing cloud density. Hence the conclusions of Bromage and Nandy concern an interstellar region unlike those considered in the present study.

Plots of λ 5780 vs λ 4430 and λ 5797 vs λ 5780 show no systematic deficiency of one band with respect to the others in dark clouds, as compared with the distant supergiants.

3.2. THE ϱ OPHIUCHI COMPLEX

We have diffuse band data for 10 of the stars in the ϱ Oph dark cloud complex for which Carrasco *et al.* (1973) give photoelectric and polarimetric data pertaining to

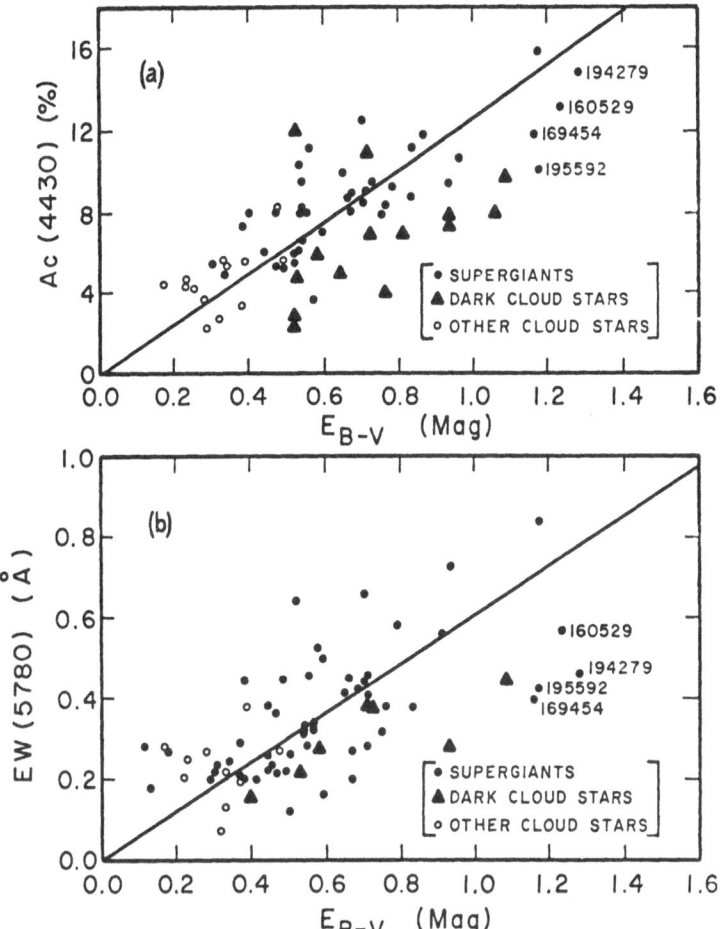

Fig. 1. (a) The correlation of λ 4430 central depth with color excess. The solid line indicates the mean correlation based on the distant supergiants alone, forced through the origin. Most of the stars in dense clouds (triangles) lie well below this line, indicating that λ 4430 in dense clouds is weak with respect to color excess. (b) The corresponding plot, showing the equivalent width of λ 5780 vs color excess. The same trend is evident as for λ 4430; the correlation between these two bands appears the same, both in and out of dark clouds (see text).

the mean interstellar grain size. It is found that two indicators of grain size, the ratio E_{VK}/E_{BV} (roughly the ratio of total to selective extinction) and the wavelength of maximum polarization, both increase with increasing optical length (as indicated by E_{VK}), showing that the mean grain size increases with increasing cloud density. This is consistent with the suggestion of Routly and Spitzer (1952) and the more recent conclusions of Cohen (1973) and Wallerstein and Goldsmith (1974) that grains in interstellar clouds can grow by accreting atoms onto their surfaces, since the greatest grain growth would be expected in the densest portions of clouds.

Our data show that the diffuse band-producing efficiency of the grains in the ϱ Oph complex is greatest in the outer portions of the cloud, and that the band strength per grain (as indicated by the ratio S/E_{VK}, where S is the band strength in appropriate

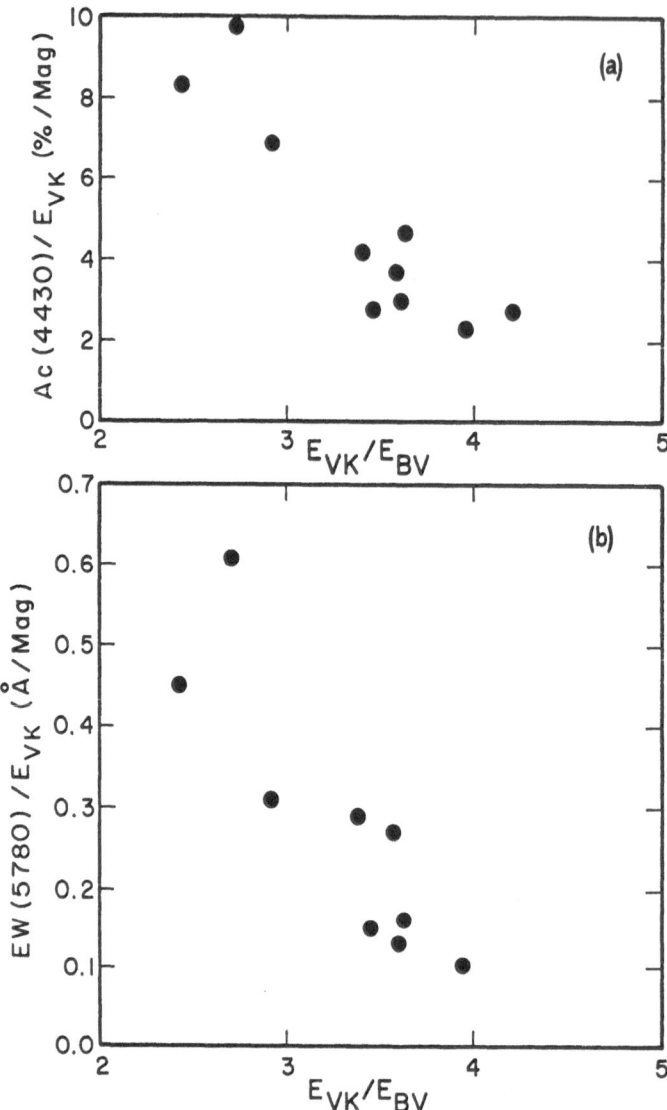

Fig. 2.   (a) The strength of λ 4430 per grain plotted against the ratio of total to selective extinction for stars in the ϱ Oph cloud complex. It is seen that in the cloud interior, where mantles have apparently accreted onto the grains, the diffuse band-producing efficiency of the grains is lowest. (b) A similar plot, showing the systematic decrease of λ 5780 strength per grain with increasing ratio of total to selective extinction. The trend for both bands is similar if other indicators of grain size, such as the wavelength of maximum polarization, are plotted in the horizontal axis.

units) decreases systematically with increasing grain size. In Figure 2 is shown the dependence of the strengths per grain of $\lambda\lambda$ 4430 and 5780 on the ratio of total to selective absorption. The trend is the same if λ_{max}, the wavelength of maximum polarization, is plotted in place of E_{VK}/E_{BV}. Data from Merrill *et al.* (1937) on the strength of the diffuse feature at 6284 Å show that its strength per grain also decreases with increasing grain size for the 5 ϱ Oph cloud stars included in that study.

Our data are of insufficient quality to allow detailed inferences to be made concerning possible variations in the profiles of the diffuse bands as a function of grain size. No gross effects are seen, however; no shift can be seen in the central wavelength of $\lambda\,4430$ to an upper limit of roughly 5 Å (considerably less for the more narrow bands), and no prominent emission features appear. It would be of very great interest to make a high quality study of the profiles of the diffuse bands in regions containing peculiar dust grains.

4. Discussion

The deficiency of diffuse band strengths in dark clouds and the systematic decrease of band-producing efficiency with increasing particle size strongly imply that the bands are not produced in the mantles of interstellar grains. They could originate in the cores of grains, assuming that the optical properties of the cores could be sufficiently smothered by the accretion of mantles in dense cloud interiors, or they could be produced by some agent which is most abundant in the outer portions of interstellar clouds. This agent could be a distinct population of grains, or some molecular species.

The possibility that some or all of the diffuse bands are produced by molecules has been generally disregarded. It is interesting to note, however, that the findings in the present study would be consistent with a band origin in a species of molecule which either requires a substantial ultraviolet flux for formation or contains hydrogen, since molecules of these kinds would not be formed efficiently in the interiors of dense clouds. For example, the data of Cohen (1973) show that both CH and CH^+ are most abundant with respect to E_{VK} in the outer portions of the ϱ Oph complex. Our observed lack of gross variations in band profiles with grain size would also be consistent with a molecular origin. The suggestion of Herzberg (1967) that predissociation transitions could produce the unstructured breadth of the diffuse bands may be unlikely, however, as shown by arguments such as those of Wilson (1964) that any process which destroys the band carrier requires an unreasonably high formation rate. On the other hand, there are kinds of molecular transitions which could take place in simple molecules consisting of cosmically abundant elements, which would produce broad absorption bands without detectable fine structure, and which would not destroy the molecules (Smith, 1974).

Acknowledgements

We acknowledge with pleasure the kind assistance of Dr S. E. Strom, who provided diffuse band data on 5 stars, and Dr G. Wallerstein, who provided plates of some of the ϱ Oph cloud stars. One of us (T.P.S.) carried out a portion of this study while he was a Predoctoral Research Associate at the University of Washington, where this work was supported in part by the Graduate School Research Fund of the University of Washington, and in part by National Science Foundation grant GP28881. The work done at Princeton was supported by National Aeronautics and Space Administration contract NAS5-1810.

References

Aannestad, P. A. and Purcell, E. M.: 1973, *Ann. Rev. Astron. Astrophys.* **11**, 309.

Bromage, G. E. and Nandy, K.: 1973, *Astron. Astrophys.* **26**, 17.

Carrasco, L., Strom, S. E., and Strom, K. M.: 1973, *Astrophys. J.* **182**, 95.

Cohen, J. G.: 1973, *Astrophys. J.* **184**, 149.

Coyne, G. V., Gehrels, T., and Serkowski, K.: 1974, *Astron. J.* **79**, 581.

Deeming, T. J. and Walker, G. A. H.: 1967, *Z. Astrophys.* **66**, 175.

Herzberg, G.: 1967, in H. van Woerden (ed.), 'Radio Astronomy and the Galactic System', *IAU Symp.* **31**, 91.

Merrill, P. W., Sanford, R. F., Wilson, O. C., and Burwell, C. G.: 1937, *Astrophys. J.* **86**, 274.

Routly, P. M. and Spitzer, L.: 1952, *Astrophys. J.* **115**, 227.

Smith, W. H.: 1974, private communication.

Snow, T. P.: 1973, *Astron. J.* **78**, 913.

Strom, S. E.: 1974, private communication.

Walborn, N. R.: 1973, *Astrophys. J. Letters* **180**, L35.

Wallerstein, G. and Goldsmith, D.: 1974, *Astrophys. J.* **187**, 237.

Wampler, E. J.: 1966, *Astrophys. J.* **144**, 921.

Wilson, R.: 1964, *Publ. Roy. Obs. Edinburgh* **4**, 67.

INTERSTELLAR EXTINCTION AND DIFFUSE
ABSORPTION FEATURES

J. DORSCHNER

Universitäts-Sternwarte Jena, G.D.R.

Abstract. The equivalent width of the λ 2175 Å band, W_{2175}, well known as the big bump in the interstellar extinction curves, has been found to be closely correlated with the colour excess E_{B-V} as well as with the extinction differences E_{8-6} and E_{9-7} defined to characterize quantitatively the steep slopes of the extinction curves in the far ultraviolet.

The equivalent widths of the $\lambda\lambda$ 5780 and 5797 Å diffuse lines show good correlation with E_{B-V}. The correlations of W_{5780} and W_{5797} with E_{8-6} resp. E_{9-7} are, however, rather weak. Correlations between W_{2175} and W_{5780} and between W_{2175} and W_{5797} are indicated.

The results have been qualitatively interpreted in favour of the dust model consisting of a mixture of small silicate grains and larger silicate grains coated by molecular mantles.

1. Introduction

Rocket and satellite observations have revealed that the course of the interstellar extinction curves in the ultraviolet is characterized by two surprising features: the big bump centred about $\lambda^{-1} = 4.6 \, \mu m^{-1}$ and the steep slope in the far ultraviolet beyond $\lambda^{-1} = 7 \, \mu m^{-1}$ (cf. the review article by Wickramasinghe and Nandy, 1972).

Following Greenberg's (1973) suggestion the observed extinction curves can be schematically separated into three main components:

(i) the visual portion continuing into the near UV;

(ii) the absorption band at λ 2175 Å superimposed to the first component and, probably, to the spurs of the third one, too;

(iii) the steeply rising far ultraviolet portion.

The diffuse absorption features modifying the extinction curves at certain narrowly limited wavelength intervals can be considered as a fourth contribution.

In this paper, which is an extension of a former study (Dorschner, 1974), our aim has been to examine the extent to which these four components, represented by appropriate quantities, are correlated with each other. A search for such correlations and estimates of their quality should be expected to give information on whether the interstellar dust constitutes a rather homogeneous medium by its chemical composition and by the origin of the grains, or a mixture of several ingredients of different origin and chemistry.

2. Observational Data

For the correlation analysis carried out in the next section only those stars were selected for which the course of the extinction curve in the ultraviolet was completely measured. The extinction curves were taken from Bless and Savage (1972) and from the review article by Wickramasinghe and Nandy (1972) containing the data of interest published before the OAO results.

In an earlier paper (Dorschner, 1973) the equivalent width of the λ 2175 Å band was published. This quantity is used here to characterize appropriately the strength of this band. As quantities measuring the slopes of the extinction curves in the far ultraviolet the extinction differences between the wave numbers $\lambda^{-1}=8\ \mu m^{-1}$ and $\lambda^{-1}=6\ \mu m^{-1}$, E_{8-6}, as well as between $\lambda^{-1}=9\ \mu m^{-1}$ and $\lambda^{-1}=7\ \mu m^{-1}$, E_{9-7}, have been determined. The visual extinction has been taken into account by the colour excess E_{B-V}. Finally, the equivalent widths of the $\lambda\lambda$ 5780 and 5797 Å diffuse lines have been included. The corresponding data have been taken from the catalogue published by Wu (1972). Table I contains all data used in the following analysis.

TABLE I

Observational data

Star	W_{2175} (Å)	W_{5780} (Å)	W_{5797} (Å)	E_{B-V} (mag.)	E_{8-6} (mag.)	E_{9-7} (mag.)
φ Per	130	0.00	0.00	0.14	0.12	0.24
ζ Per	320	0.20	0.11	0.29	0.87	1.57
ε Per	90	0.07	0.00	0.09	0.24	0.56
ξ Per	360	0.22	0.10	0.33	0.46	1.22
α Cam	390	0.21		0.30	0.63	
$\theta^{1,2}$ Ori	180			0.28	0.08	0.70
σ Ori	60	0.00	0.00	0.06	0.19	
139 Tau	190	0.15		0.12	0.40	
χ^2 Ori	>400	0.41	0.18	0.46		
HD 47 129	>370			0.36		
HD 48 099	>300			0.28		
π Sco	90	0.12		0.07	0.10	
δ Sco	180	0.28	0.13	0.18	0.25	0.59
β^1 Sco	300	0.25	0.11	0.20	0.42	1.00
σ Sco	460	0.38	0.12	0.40	0.28	0.64
ϱ Oph	460			0.46	0.64	
τ Sco	40	0.16	0.10	0.04	0.02	0.07
ζ Oph	270	0.25		0.28	0.70	1.43
κ Aql		0.13	0.00	0.28	0.48	0.90
19 Cep	>310			0.34		
λ Cep	>470	0.47	0.17	0.55		

3. Correlations

Four groups of diagrams have been studied:
   (i) Far ultraviolet extinction (E_{8-6}, E_{9-7}) vs visual extinction (E_{B-V});
   (ii) λ 2175 Å band (W_{2175}) vs extinction (E_{B-V}, E_{8-6}, E_{9-7});
   (iii) $\lambda\lambda$ 5780, 5797 Å diffuse lines (W_{5780}, W_{5797}) vs extinction;
   (iv) λ 2175 Å band (W_{2175}) vs $\lambda\lambda$ 5780, 5797 Å lines (W_{5780}, W_{5797}).

Because there was some suspicion that the quantity E_{8-6} could be influenced by the minimum following the bump towards shorter wavelengths, the quantity E_{9-7}, which is expected to be free of this influence, was used for the sake of control. In the diagrams under investigation, however, no systematic deviations could be found when E_{8-6} was used instead of E_{9-7}. Therefore, the quantity E_{8-6}, which could be determined for a greater number of stars than was the case for E_{9-7}, has been preferentially used in the following figures.

Figure 1 shows the correlation present between E_{8-6} and E_{B-V}. Because of their flat course in the far ultraviolet region, the stars σ Sco and θ Ori have been regarded as outsiders as already pointed out by Bless and Savage (1972) and confirmed in the case of σ Sco by Snow and York (1975). These stars were, therefore, marked by open symbols in all diagrams.

For the representation of stars of the Scorpius–Ophiuchus region triangles were used. As Wu (1972) pointed out, these stars show exceptionally strong diffuse lines.

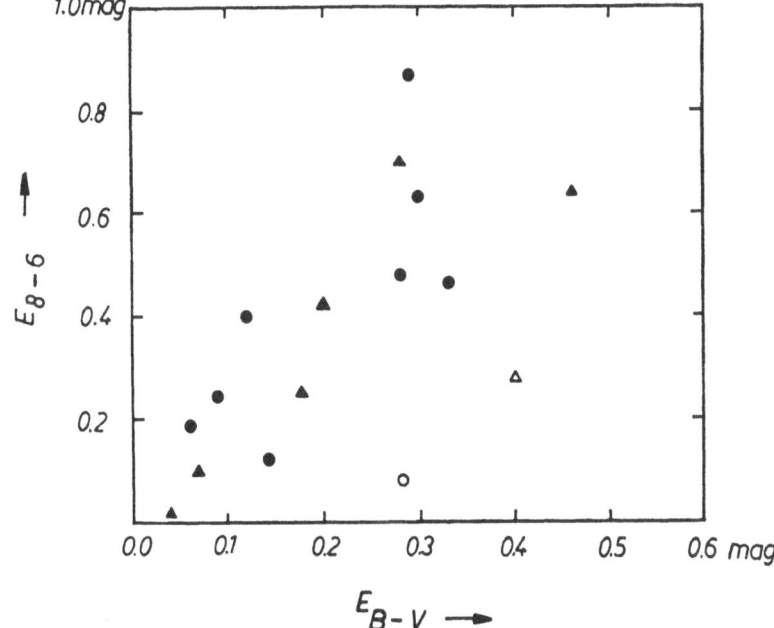

Fig. 1.   Extinction difference E_{8-6} (for definition see in the text) vs colour excess E_{B-V}. Triangles: Scorpius-Ophiuchus stars; closed circles: other stars; open triangle: σ Sco; open circle: θ Ori.

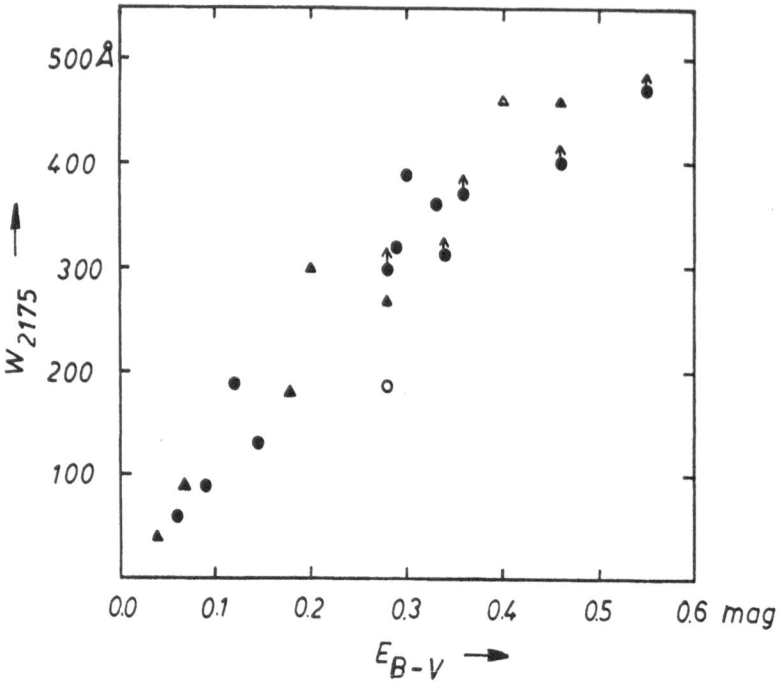

Fig. 2.   Equivalent width W_{2175} of the λ 2175 Å band vs E_{B-V}. Closed circles with an arrow are lower limits of W_{2175}. The other dots have the same meaning as in Figure 1.

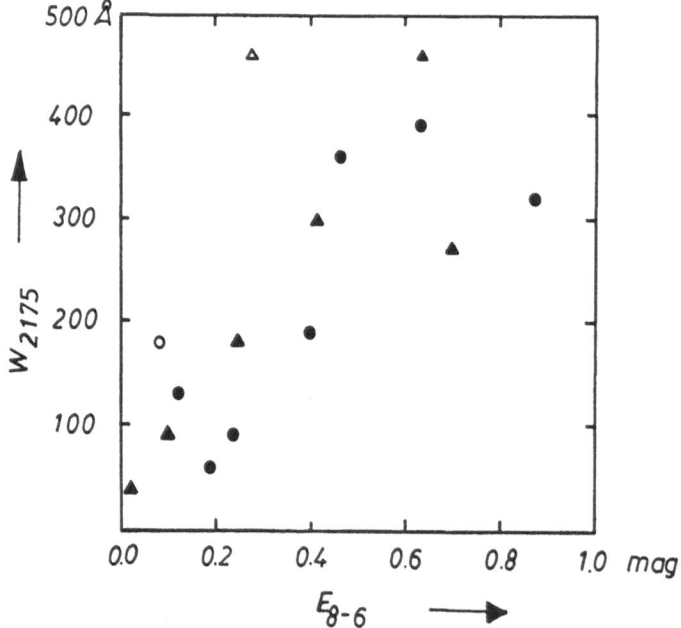

Fig. 3.   Equivalent width W_{2175} vs extinction difference E_{8-6}.

This property can easily be noted on the Figures 4, 5 and 6. This finding is not necessarily in contradiction of the results presented by Snow and Cohen (1975) because these authors selected stars with $E_{B-V} \geqslant 0.5$ mag. which lie more deeply within or even behind the Scorpius–Ophiuchus dust cloud, whereas the stars used in the present investigation apparently lie near the front side of the cloud.

In Figures 2 and 3 plots of W_{2175} vs E_{B-V} resp. E_{8-6} are shown. Obviously, a close correlation of the strength of the λ 2175 Å band with the colour excess is present. The corresponding correlation with E_{8-6} is not so close. This fact cannot be attributed only to photometric errors in the far ultraviolet (Code, 1974). Probably, a significant difference in the quality of the correlations shown in Figures 2 and 3 is present.

In Figure 4 the equivalent width of the λ 5780 Å diffuse line, W_{5780}, is plotted against the colour excess E_{B-V}. In this diagram, the outlying star σ Sco behaves quite normally, where other members of the Scorpius-Ophiuchus group are shifted in a systematic manner to the direction of increasing W_{5780}. Altogether, a reliable correlation is indicated. Contrary to this, W_{5780} correlates very poorly with E_{8-6} resp. E_{9-7}. As suggested by Figure 5 this correlation can be expected to become closer if the Scorpius-Ophiuchus stars could be separated. Considering the small number of stars altogether available in this investigation, a further subdivision of the data is not advisable. The possibility that regional variations are present, however, deserves further attention.

Similar results have also been found for the λ 5797 Å line. However, this was to be

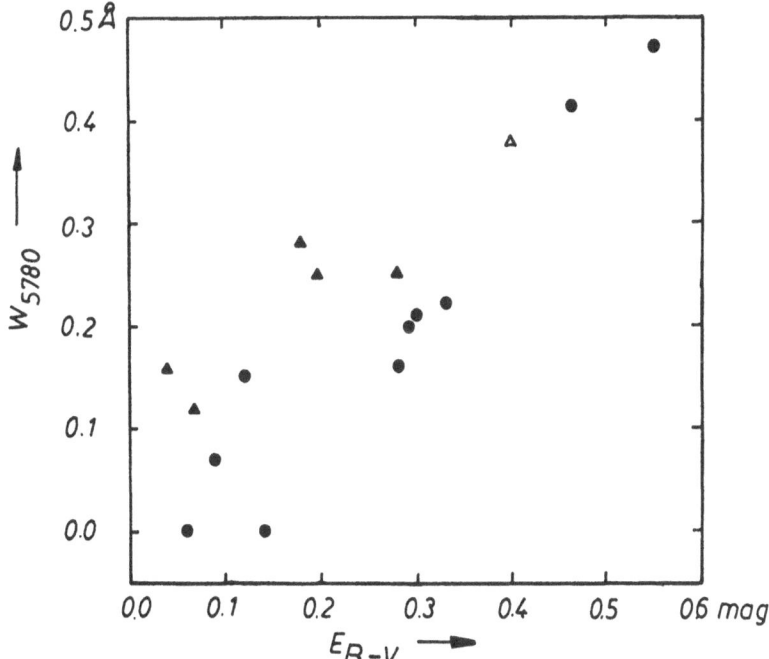

Fig. 4.   Equivalent width W_{5780} of the λ 5780 Å diffuse interstellar line vs E_{B-V}.

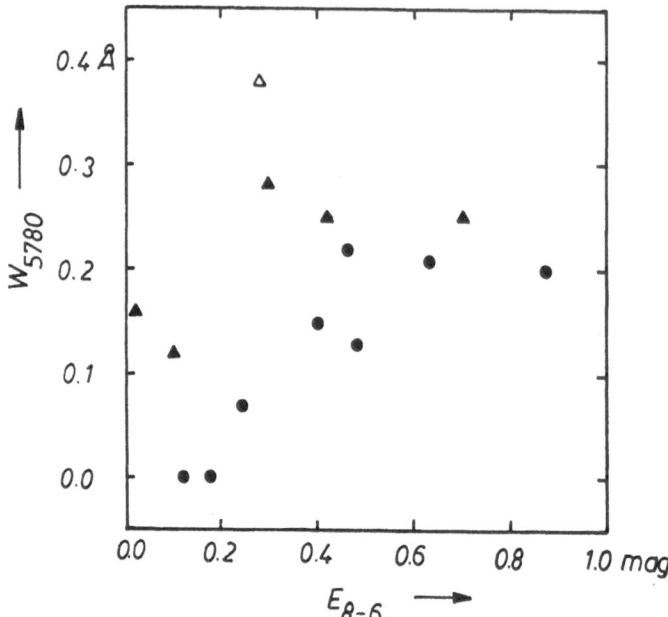

Fig. 5. Equivalent width W_{5780} vs extinction difference E_{8-6}. The Scorpius-Ophiuchus stars are conspicuously shifted toward rising W_{5780} and, thus, obliterate the correlation probably present in the remaining stars.

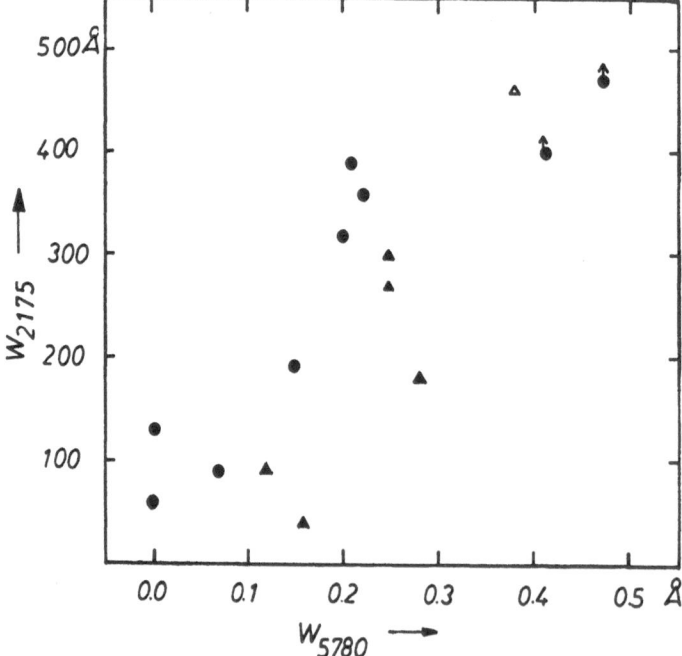

Fig. 6. Equivalent width W_{2175} vs equivalent width W_{5780}.

TABLE II

Correlation coefficients

The correlation coefficient r between the variables x and y is given
by the formula $r(x, y) = \text{cov}\,(x, y)[\text{var}\,(x)\,\text{var}\,(y)]^{-1/2}$.

y vs x	$r(x, y)$		
	all points in the diagram used except those marked by an arrow	additionally σ Sco and θ Ori excluded	all points in the diagram used including those marked by an arrow
E_{8-6} vs E_{B-V}	0.62	0.82	
E_{9-7} vs E_{B-V}	0.63	0.88	
W_{2175} vs E_{B-V}	0.94	0.96	0.91
W_{2175} vs E_{8-6}	0.69	0.86	
W_{2175} vs E_{9-7}	0.64	0.87	
W_{5780} vs E_{B-V}	0.84	0.84	
W_{5780} vs E_{8-6}	0.37	0.53	
W_{5780} vs E_{9-7}	0.35	0.54	
W_{2175} vs W_{5780}	0.76	0.66	0.83

expected because the equivalent width of the λ 5797 Å line is tightly correlated with that of the λ 5780 Å line (Wu, 1972).

Figure 6 shows the plot of W_{2175} vs W_{5780} indicating the existence of a correlation. Even if we exclude σ Sco and the circles with an arrow, which are lower limits of W_{2175} only, the correlation does not vanish, but its quality becomes worse, because of the small number of stars in the diagram.

For each diagram the correlation coefficient has been determined in order to obtain a statistical quantity measuring the degree of dependence between the two stochastic variables plotted vs each other. The values of this quantity are listed in Table II. In interpreting the values of the correlation coefficients, r, one has to remember that the observational errors present in both variables have not been taken into consideration. This implies that the 'true' correlation coefficient is a little larger than the quantity r listed in Table II (cf. Murdin, 1972).

4. Discussion

Because of the small number of stars for which the data used in this investigation were available, the findings presented in this paper need further confirmation by more extensive observational material before reliable conclusions can be drawn. Here, only a few qualitative conclusions shall be drawn which may stimulate discussion about the nature of the dust grains.

The correlations between the components of the extinction curves as mentioned in

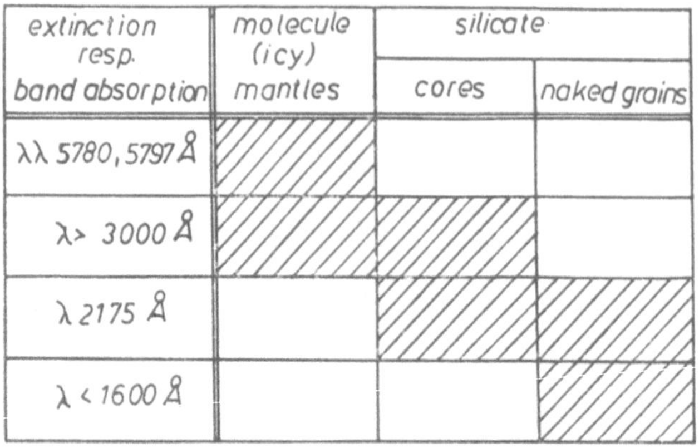

extinction resp. band absorption	molecule (icy) mantles	silicate	
		cores	naked grains
$\lambda\lambda$ 5780, 5797 Å	▨		
$\lambda >$ 3000 Å	▨	▨	
λ 2175 Å		▨	▨
$\lambda <$ 1600 Å			▨

Fig. 7. Schematic representation of the qualitative conclusions. Hatched areas show by which component the extinction resp. line or band absorption noted on the left-hand side is assumed to be mainly caused.

the introduction seem to point to a homogeneous dust model rather than to a mixture of several sorts of dust which originated independently of each other. The results presented can be qualitatively understood if one takes as a basis the dust model emphasized by Greenberg for some years (cf. Greenberg, 1973, 1974). The basic ideas can be outlined as follows: A spectrum of small silicate grains is injected into the interstellar space by any 'smoking' or otherwise dust-producing cosmic objects. The small-size tail of the distribution of the particle radii remains uncoated – except in dense dust clouds, where the grains are protected against the stellar UV radiation. The larger grains become coated by molecule mantles of different thickness – forming, in a sense, 'classical' or 'dirty ice' grains. The core-mantle grains are mainly responsible for the visual portion of the extinction curve whereas the far ultraviolet extinction is mainly caused by the tiny, uncoated silicate grains. Probably, the λ 2175 Å band is produced by the cores as well as by the uncoated grains whereas the $\lambda\lambda$ 5780 and 5797 Å lines seem to originate mainly in the mantles. Figure 7 shows a schematic representation of these tentative conclusions, which must, of course be, scrutinized by corresponding model calculations.

References

Bless, R. C. and Savage, B. D.: 1972, *Astrophys. J.* **171**, 293.
Code, A. D.: 1974, personal communication.
Dorschner, J.: 1973, *Astrophys. Space Sci.* **25**, 405.
Dorschner, J.: 1974, *Astron. Nachr.* **295**, 147.
Greenberg, J. M.: 1973, in J. M. Greenberg and H. C. van de Hulst (eds.), 'Interstellar Dust and Related Topics', *IAU Symp.* **52**, 3.
Greenberg, J. M.: 1974, *Astrophys. J. Letters* **189**, L81.

Murdin, P.: 1972, *Monthly Notices Roy. Astron. Soc.* **157**, 461.
Snow, T. P. and Cohen, J. G.: 1975, *Astrophys. Space Sci.* **34**, 33.
Snow, T. P. and York, D. G.: 1975, *Astrophys. Space Sci.* **34**, 19.
Wickramasinghe, N. C. and Nandy, K.: 1972, *Rep. Prog. Phys.* **35**, 157.
Wu, Ch.-Ch.: 1972, *Astrophys. J.* **178**, 681.

DIFFUSE BAND EXTINCTION AND POLARIZATION
IN CORE-MANTLE GRAINS*

J. MAYO GREENBERG** and SEUNG-SOO HONG**

State University of New York at Albany and Dudley Observatory, Albany, New York

Abstract. Diffuse band shapes in both extinction and polarization are calculated for interstellar core-mantle particles for varying size distributions of mantle thickness. It is shown that no matter whether the source of the bands is in the silicate cores or the accreted icy mantles the polarization shapes are highly asymmetric for all mantle thicknesses. The extinction band shapes are significantly less asymmetric although the effect is clearly present. The only apparent possibility for producing symmetric band shapes in the dust grains is in the very small bare particles in interstellar space which, if they are aligned and produce the $\lambda\,2200$ band, must exhibit a strong polarization effect in this region.

1. Introduction

Ever since the unidentified diffuse interstellar bands were first discovered (Merrill, 1934) they have inspired a wide range of speculation and observations with, as yet, no definitive answer as to their cause. At the end of this paper a selection of references on the diffuse bands is given, grouped according to observations, theory, etc.

The situation, as reviewed by Herbig in 1966 (Herbig, 1967) was that there were 26 recognized bands between 4400 Å and 6700 Å, the strongest and most studied being at $\lambda\,4430$. These bands are generally 5 Å to 25 Å wide with the $\lambda\,4430$ line extending perhaps 50 Å or more. Subsequent investigation has revealed a number of new structures in the extinction curve with widths in the 20–200 Å range, and we now recognize structures as broad as 1000 Å in the visual part of the extinction curve. There thus seems to be a hierarchy of structures extending from 6700 Å to 2200 Å, and whether all of these are due to the same agent or different agents in the interstellar medium is yet to be answered.

The general correlation of the diffuse bands with the extinction is not adequate proof that they are produced by some material in the grains. One method of examining this question is to observe the polarization within the bands because it is unlikely that some agents (like molecules of O^-) outside the grain will be aligned to produce the same degree of polarization. Another method is to search for some structure within the line profiles which appears not only in extinction but also in polarization. If one matches this with the theoretical prediction of such structure by a model for the interstellar grains the case for the diffuse band agent being an intrinsic component of the grains is strong indeed. If, in addition, some feature of the band structure correlates theoretically and observationally with variations in the broad extinction characteristics, such as those indicating different grain sizes, the evidence should be conclusive. Note that this

* Work supported in part by NASA Grant NGR-33-011-043.
** Present address: Huygens Laboratory, University of Leiden, The Netherlands.

does not yet determine the exact cause of the bands but only whether or not they are a manifestation of the grains themselves. Hopefully, such evidence, if positive, would provide information on the chemical and physical properties of the grains.

All previous calculations of predicted band shapes have been made on homogeneous models of grains. In this paper we shall compare the band shapes as produced either in the cores or the mantles of grains whose cores are some basic silicate and whose mantles are optically like those of a generalized material made up of oxygen, carbon, and nitrogen in various combinations with hydrogen. The specific calculations are for a λ 4430 band but many of the results are obviously applicable to other bands so long as the broad optical properties of the core and mantle material in the neighborhood of the band are similar to those around λ 4430. For homogeneous particles, in going from 4000 Å to 6000 Å, the ratios of the diameter to the wavelength changes by a factor of $\frac{2}{3}$ so that the particles are effectively smaller at the longer wavelengths. For core-mantle particles we have to consider in addition the effect of varying the ratio of mantle to core diameters and this is all we have done so far (except, see later, the 'effective' size to wavelength change introduced by comparing perfect full and perfect spinning alignment).

2. The Model

In a series of papers, Greenberg and Hong (1974a, 1974b) have examined the observational consequences of a bimodal model of interstellar grains which consists of a population of silicate core-modified ice mantle particles in the 0.1 μm size range and a population of very small refractory particles in the 0.005 μm size range. It is unlikely that the very small particles are the diffuse band carriers. If they are, it can be shown that there would be no obvious reason to expect internal structure in the bands such as has been observed. (See note at end.) We therefore limit ourselves to calculations on the core-mantle component.

In order simultaneously to predict extinction and polarization structure within the bands, the calculations are performed for concentric circular cylindrical particles. This is the only representation of non-spherical core-mantle particles which can currently be readily calculated. The core size is taken to be a constant $a_c = 0.06$ μm. The mantle sizes are either taken as constant or distributed in size according to

$$n(a_m) = \exp\left[-5\left(\frac{a_m - a_c}{a_t}\right)^3\right].$$

Both perfect picket-fence alignment (pf) and perfect spinning Davis-Greenstein alignment (dg) are considered.

The wavelength dependence of the index of refraction of the core or mantle material is obtained by application of the Clausius-Mosotti relation for dilute imbedded absorbers. Both slightly absorbing ($m'' = 0.05$) and clear ($m'' = 0$) basic representation of the core and mantle materials were considered but the line shapes were insignificantly different even though the magnitudes were changed. Since we are primarily

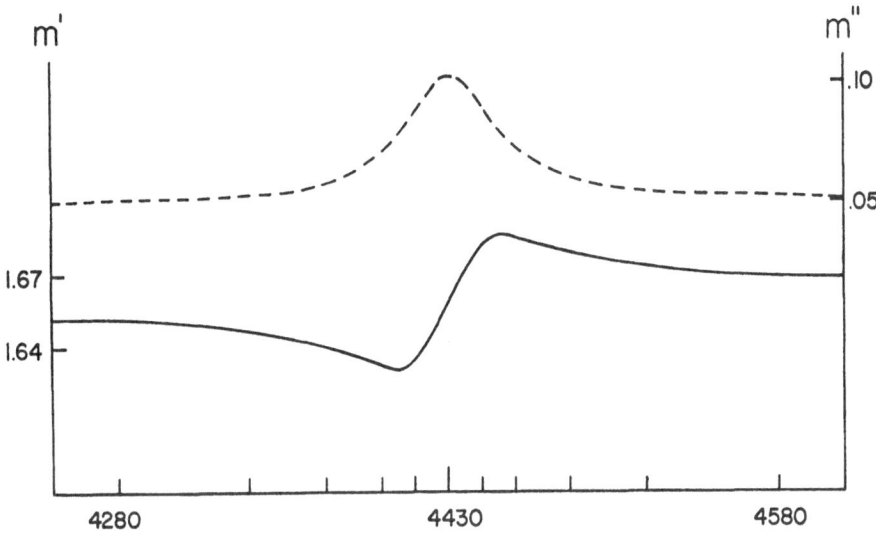

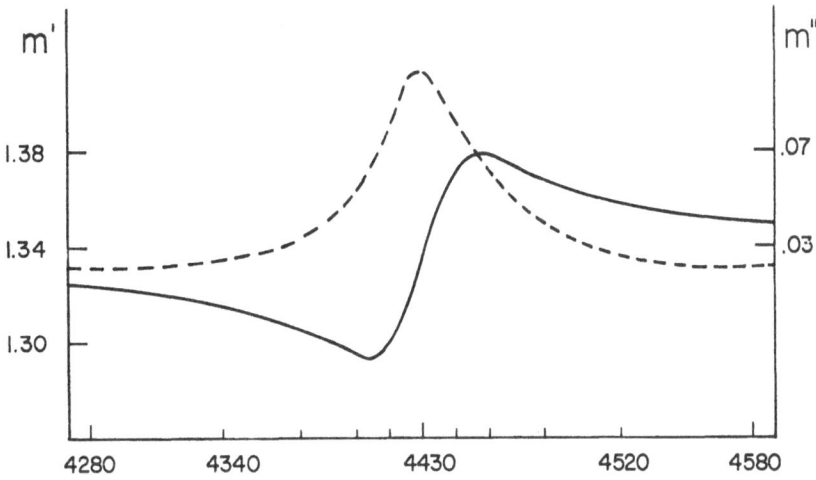

Fig. 1.   The variations of real (m', solid line) and imaginary (m'', broken line) parts of the refractive index around 4430 Å as used in this work. The upper diagram represents the case when the absorbers are imbedded in the core, and the lower one in the mantle.

interested in line shapes we present only our results for dirty core and mantle materials. The respective indices of refraction are shown in Figure 1.

3. Results

The four size distributions chosen were those for $a_i = 0.10\ \mu m$, $0.14\ \mu m$, $0.18\ \mu m$, $0.20\ \mu m$. For $a_i = 0.10\ \mu m$ the average mantle thickness is only about $0.03\ \mu m$. How-

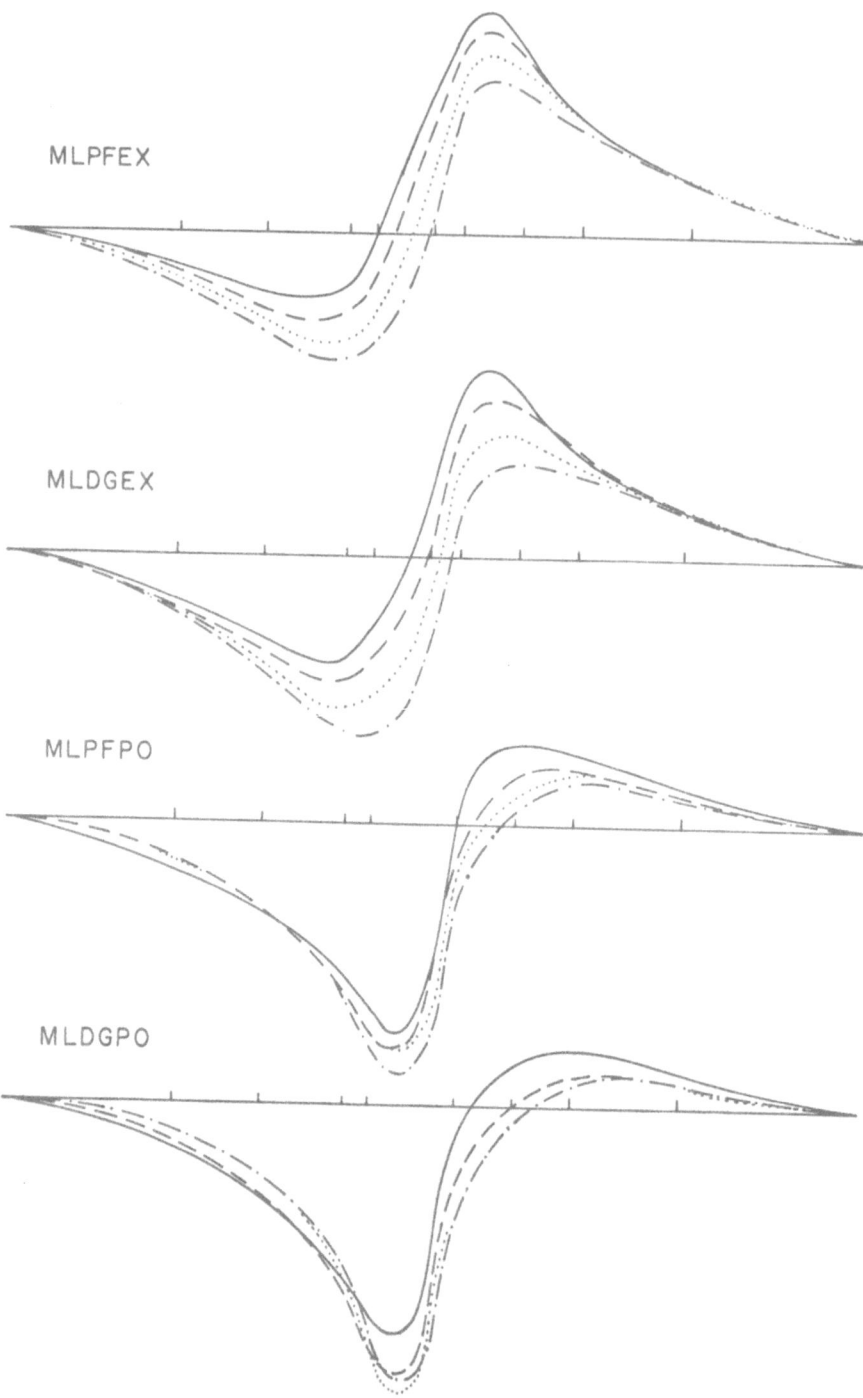

Fig. 2.   The trends of extinction (EX) and polarization (PO) variations, when the absorbers are in the mantle (ML), with varying degrees of alignment (PK, picket-fence; DG, Davis-Greenstein) and with varying mantle thickness. Solid line, dashes, dash-dots and dots are for $a_i = 0.10\,\mu m$, $0.14\,\mu m$, $0.18\,\mu m$, and $0.20\,\mu m$ respectively.

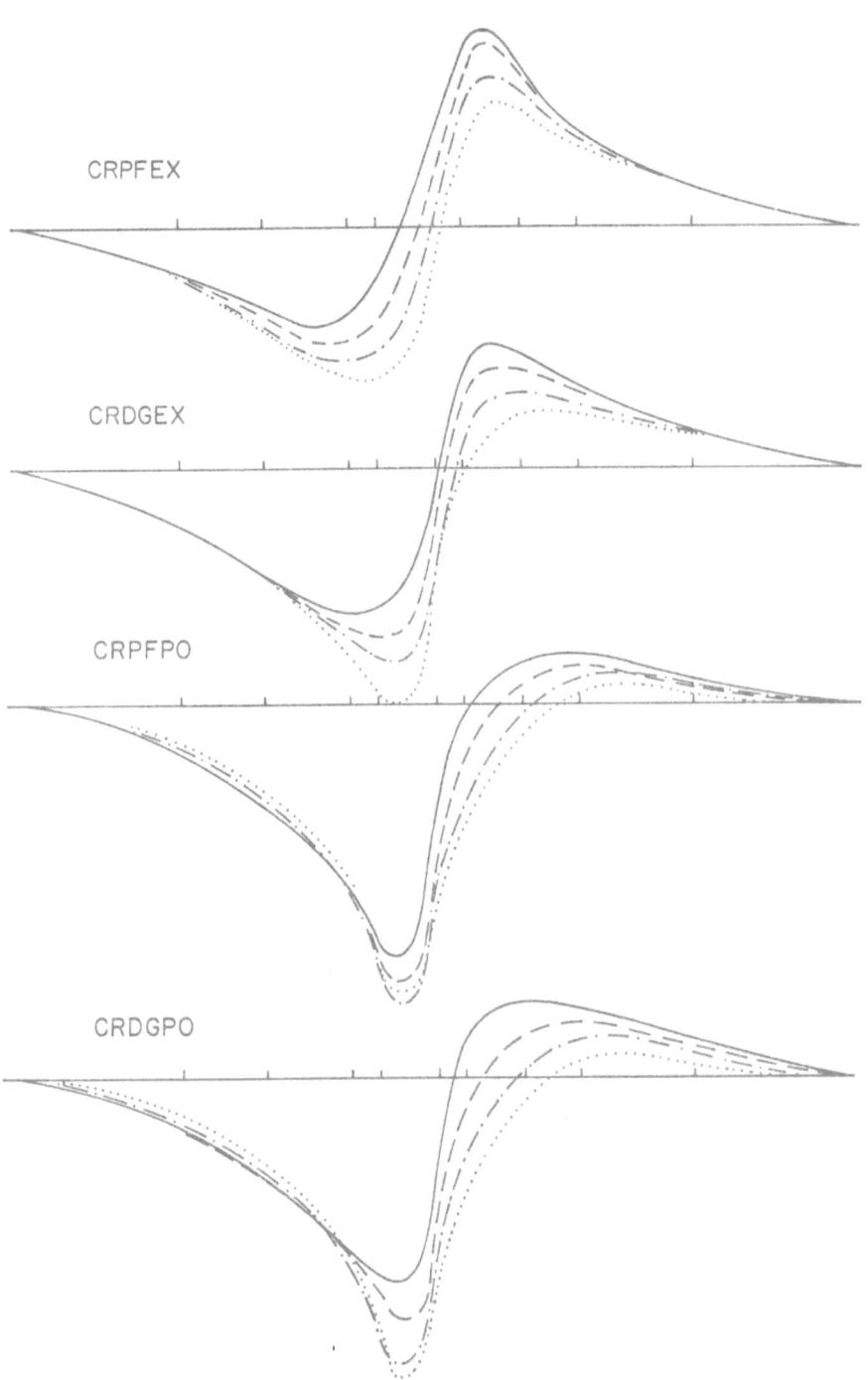

Fig. 3. Same as Figure 2, except that the absorbers are in the core (CR).

ever, even this represents of the order of twice the volume of the core. Figures 2 and 3 show the trends of the shapes with varying mantle thickness when the absorber is in the mantle or the core respectively. We see that an asymmetry in the shape of the extinction band is already quite apparent even for the smallest a_t considered and that the amount of 'emission' relative to additional extinction increases with increasing a_t. The general difference between the case of pf and dg alignment is that the latter, for obvious reasons, acts as if the particles are effectively somewhat larger or, equivalently, the wavelength is shorter. Thus in establishing a trend for fixed mantle thickness and increasing band wavelength we may roughly use the type of change introduced by going from dg to pf alignment.

As in an earlier paper (Greenberg and Stoeckly, 1971) we define the total swing σ as the sum of the absolute values of the increments above and below divided by the reference value. An asymmetry factor, α, is defined as the larger increment (positive or negative) divided by the absolute value of the difference between the larger and smaller increments.

The polarization curves show that no matter whether the absorber is in the core or mantle there is a greater depression at the shorter wavelengths than an excess at the longer wavelengths. This means that, for polarization, $\alpha < 0$ for all the particle shapes we have investigated including single particle sizes ranging from a mantle radius of $0.08 \ \mu m$ to $0.30 \ \mu m$ (not shown).

In Figures 4 and 5 we see the trends in σ and α^{-1} for various particle configurations. We have chosen α^{-1} rather than α because it is a more convenient way of demonstrating

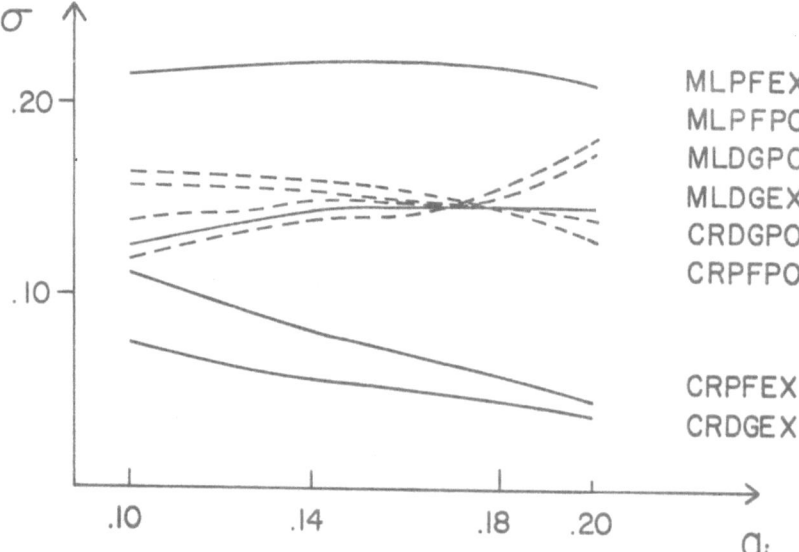

Fig. 4. The trends of the total swing (σ) with varying mantle thickness and for various configurations as indicated by the same abbreviations as in Figures 3 and 4. Solid lines are for extinction and broken lines for polarization. Size parameters a_t in units of μm.

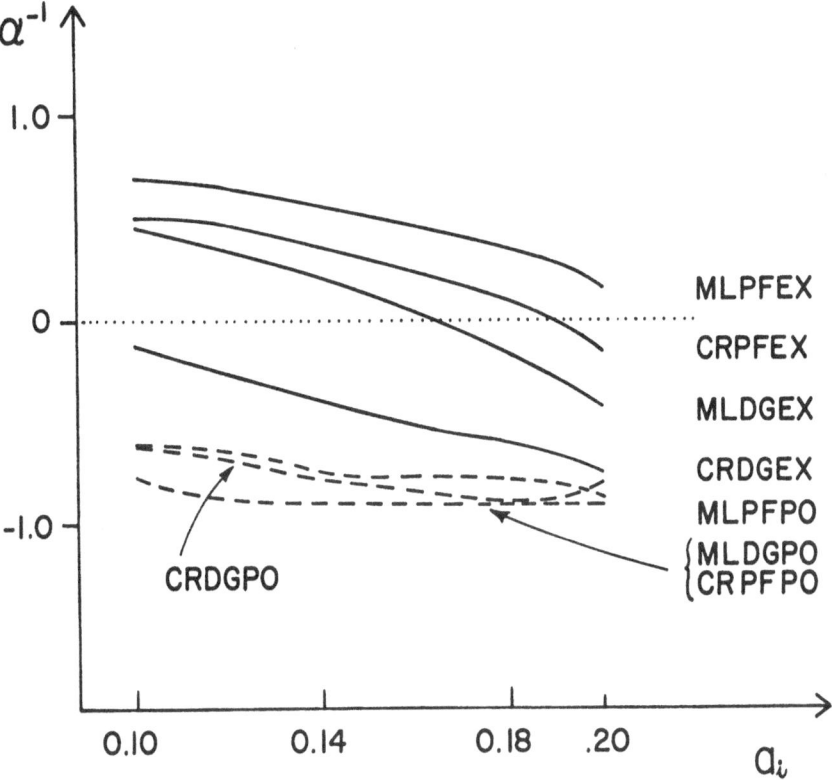

Fig. 5. Same as Figure 4 except for the inverse asymmetry factor (α^{-1}). Note that MLDGPO and CRPFPO differ so little from each other that they cannot be shown separately.

the crossover in the sign of the asymmetry factor if it occurs. The total swing in extinction is significantly larger when the band is produced in the mantle than when it is produced in the core. However, the total swing in polarization is hardly affected by whether the band is of core or mantle origin.

The inverse asymmetry factor appears to be a better theoretical discriminant than the total swing in distinguishing between core and mantle source.

4. Conclusions

If the source of a diffuse band originates by absorption in the interstellar grains, whether in the cores or mantles, there is significant structure in both the extinction and polarization curves. For given sizes and conditions the structure is more asymmetric for polarization than for extinction. For the cases computed, the degree of asymmetry in the polarization is roughly independent of all the size parameters and conditions. This would seem to imply that the polarization asymmetry should appear equally at the longest wavelength diffuse band as at $\lambda\,4430$. The extinction asymmetry is quite model dependent and it is to be expected that at 6700 Å the band would be almost symmetric if the band were produced in a relatively thin mantle but not if it is produced in the core.

Note added in proof. Herbig (1975) finds that all the bands are quite symmetric. If this is absolutely true, one may then say that the diffuse bands were not produced in the classical dust grains. They may, however, be produced in the very small bare particles (Greenberg and Hong, 1974a) or yet another source. If the bands are produced in the bare particles and if any polarization effects are observed in the bands, this implies that the bare particles are aligned, and the overall wavelength dependence of polarization should be very large in the far ultraviolet.

References

Greenberg, J. M. and Hong, S. S.: 1974a, in F. S. Kerr and S. C. Simonson (eds.), 'Galactic Radio Astronomy', *IAU Symp.* **60**, 155.
Greenberg, J. M. and Hong, S. S.: 1974b, in A. F. M. Moorewood (ed.), 'H II Regions and the Galactic Centre', *Proc. of the 8th ESLAB Symp.*, p. 153.
Greenberg, J. M. and Stoeckly, R.: 1971, *Nature* **230**, 15.
Herbig, G. M.: 1967, in H. van Woerden (ed.), 'Radio Astronomy and the Galactic System', *IAU Symp.* **31**, 85.
Herbig, G. M.: 1975, *Astrophys. J.* **196**, 129.
Merrill, P. W.: 1934, *Publ. Astron. Soc. Pacific* **46**, 206.

The following papers and reviews may serve as a useful bibliography. We divide the bibliography into three categories: Observations, Sources, and Models and Reviews.

Observations

A'Hearn, M. F.: 1971, *Astron. J.* **76**, 264.
A'Hearn, M. F.: 1972, *Astron. J.* **77**, 302.
Bromage, G. E.: 1971, *Nature* **230**, 172.
Bromage, G. E. and Nandy, K.: 1973, *Astron. Astrophys.* **26**, 17.
Bromage, G. E., Brück, M. T., and Nandy, K.: 1971, *Astron. Nachr.* **293**, 39.
Brück, M. T. and Nandy, K. N.: 1968, *Nature* **220**, 46.
Brück, M. T., Nandy, K., and Seddon, H.: 1969, *Physica* **41**, 128.
Buscombe, W. and Kennedy, P. M.: 1968, *Monthly Notices Roy. Astron. Soc.* **139**, 417.
Chaldu, R. and Honeycutt, R. K.: 1973, *Publ. Astron. Soc. Pacific* **85**, 87.
Deming, T. J. and Walker, G. A.: 1967, *Z. Astrophys.* **66**, 175.
Gammelgaard, P. and Rudkjobing, M.: 1973, *Astron. Astrophys.* **27**, 261.
Hayes, D. S., Mavko, G. E., Radick, R. R., Rex, K. M., and Greenberg, J. M.: 1973, in J. M. Greenberg and H. C. van de Hulst (eds.), 'Interstellar Dust and Related Topics', *IAU Symp.* **52**, 83.
Herbig, G. M.: 1966, *Z. Astrophys.* **64**, 512.
Honeycutt, R. K.: 1972, *Astron. J.* **77**, 24.
Kellman, S. A.: 1970, *Publ. Astron. Soc. Pacific* **82**, 1368.
Kristenson, H. and Rudkjobing, M.: 1965, *Publications de l'observatoire de Haute-Provence* **48**, 107.
Martin, P. G. and Angel, J. R. P.: 1974, *Astrophys. J.* **188**, 517.
Mavko, G. E., Hayes, D. S., Greenberg, J. M., and Hiltner, W. A.: 1974, *Astrophys. J.* **187**, L117.
Merrill, P. W.: 1934, *Publ. Astron. Soc. Pacific* **46**, 206.
Murdin, P.: 1972, *Monthly Notices Roy. Astron. Soc.* **157**, 461.
Nandy, K. and Seddon, H.: 1970, *Nature* **227**, 264.

Rudkjobing, M.: 1970, *Astrophys. Space Sci.* **6**, 157.
Seddon, H.: 1967, *Nature* **214**, 258.
Seddon, H.: 1968, *Nature* **217**, 932.
Seddon, H.: 1969, *Nature* **222**, 757.
Snow, Jr., T. P.: 1973, *Astrophys. J.* **78**, 913.
Snow, Jr., T. P. and Cohen, J.: 1974, *Astrophys. J.* **194**, 313.
Snow, Jr., T. P. and Wallerstein, G.: 1972, *Publ. Astron. Soc. Pacific* **84**, 492.
Stoeckly, R. and Dressler, K.: 1964, *Astrophys. J.* **139**, 240.
Walker, G. A. H.: 1963, *Monthly Notices Roy. Astron. Soc.* **125**, 141.
Walker, G. A. H.: 1966, *Observatory* **86**, 117.
Walker, G. A. H., Hutchings, J. B., and Younger, P. F.: 1969, *Astron. J.* **74**, 1061.
Walker, G. A. H., Hutchings, J. B., and Younger, P. F.: 1970, in L. Houziaux and H. E. Butler (eds.), 'Ultraviolet Stellar Spectra and Ground-Based Observations', *IAU Symp.* **36**, 52.
Wampler, E. J.: 1966, *Astrophys. J.* **144**, 921.
Wilson, R.: 1958, *Astrophys. J.* **128**, 57.
Wolstencroft, R. D. and Kemp, J. C.: 1974, preprint.
Wu, C. C.: 1972, *Astrophys. J.* **178**, 681.
York, D. G.: 1971, *Astrophys. J.* **166**, 65.

Sources

Dorschner, J.: 1970, *Astron. Nachr.* **292**, 107.
Dorschner, J.: 1971, *Nature* **231**, 124.
Duley, W. W.: 1963, *Physica* **41**, 134.
Duley, W. W. and Graham, W. R. M.: 1971, *Astron. Nachr.* **293**, 33.
Graham, W. R. M. and Duley, W. W.: 1970, *Publ. Astron. Soc. Pacific* **82**, 1030.
Herzberg, G.: 1967, in H. van Woerden (ed.), 'Radio Astronomy and the Galactic System', *IAU Symp.* **31**, 91.
Huffman, D. R.: 1970, *Astrophys. J.* **161**, 1157.
Manning, P. G.: 1970, *Nature* **226**, 829.
Manning, P. G.: 1970, *Nature* **227**, 1123.
Manning, P. G.: 1971, *Nature* **230**, 131.
Manning, P. G.: 1972, *Nature* **239**, 87.
Manning, P. G.: 1973, *Nature* **245**, 73.
Martin, P. G.: 1970, *Nature* **228**, 844.
McIntyre, H. A. J. and Williams, D. A.: 1970, *Monthly Notices Roy. Astron. Soc.* **148**, 53.
Rudkjobing, M.: 1969a, *Astrophys. Space Sci.* **3**, 102.
Rudkjobing, M.: 1969b, *Astrophys. Space Sci.* **5**, 68.
Runciman, W. A.: 1970, *Nature* **228**, 843.
Wickramasinghe, N. C., Ireland, J. G., Nandy, K., Seddon, H., and Wolstencroft, R. D.: 1968, *Nature* **217**, 412.
Wolstencroft, R. D., Ireland, J. G., Nandy, K., and Seddon, H.: 1963, *Monthly Notices Roy. Astron. Soc.* **144**, 245.

Models and Reviews

Bromage, G. E.: 1972, *Astrophys. Space Sci.* **18**, 449.
Bromage, G. E.: 1973, *Astrophys. Space Sci.* **15**, 420.
Greenberg, J. M. and Hong, S. S.: 1974, in T. Gehrels (ed.), 'Planets, Stars, and Nebulae Studies with Photopolarimetry', *IAU Colloq.* **23**, 916.
Greenberg, J. M. and Stoeckly, R.: 1971, *Nature* **230**, 15.
Hayes, D. S., Movko, G. E., Radick, R. R., Rex, K. H., and Greenberg, J. M.: 1973, in J. M. Greenberg and H. C. van de Hulst (eds.), 'Interstellar Dust and Related Topics' *IAU Symp.* **52**, 83.
Herbig, G. H.: 1967, in H. van Woerden (ed.), 'Radio Astronomy and the Galactic System', *IAU Symp.* **31**, 85.

Kelly, A.: 1971, *Astrophys. Space Sci.* **13**, 211.
Krishna Swamy, K. S.: 1972, *Astrophys. Space Sci.* **16**, 75.
Merrill, P. W.: 1934, *Publ. Astron. Soc. Pacific* **46**, 206.
Nandy, K. and Seddon, H.: 1971, *Astron. Nachr.* **293**, 37.
Wickramasinghe, N. C. and Nandy, K.: 1971, *Nature* **229**, 36.

A REVIEW OF RECENT OBSERVATIONS OF
DUST IN H II REGIONS

LINDSEY F. SMITH

Max-Planck-Institut für Radioastronomie, Bonn, Germany

Abstract. The upper limit for the absorption cross section σ_H^{ext}, of dust in H II regions in the wavelength range 912–504 Å derived by Mezger *et al.* (1974), is compatible with that expected for large dust grains, and a gas-to-dust ratio equal to that in the general interstellar medium. The albedo of the small grains must be high for $\lambda > 504$ Å. This restriction is lifted if the visual extinction cross section of the grains in H II regions is less than that for grains in the general interstellar medium. New observations of the Orion Nebula indicate that the visual extinction cross section is within a factor 2 of the value in the general interstellar medium.

1. Introduction

Dust in H II regions may be directly observed by emitted IR radiation and by scattered stellar radiation. It may be indirectly observed through extinction of stellar and nebular radiation and through its effect on the ionization structure of the nebula. From such data, one may derive absorption cross sections for both the far ultraviolet (UV) and for the visual wavelengths.

The dust in H II regions may be modified by stellar radiation, high kinetic temperature of the gas, and/or by recent star formation. A comparison of the properties of dust inside and outside of dense H II regions offers an observational test of whether significant modification occurs. I refer to regions outside of dense H II regions as the 'general' interstellar (IS) medium.

We distinguish between absorption, scattering and extinction cross sections: $\sigma^{ext} = \sigma^{abs} + \sigma^{sca}$, where σ is a cross section of a single grain, in cm^2, and all three cross sections are functions of wavelength. If x is the number of grains per H-atom, the total cross section per H-atom of the medium is $\sum_i x_i \sigma_i$, summed over all sizes and types of grains. $\sum_i x_i \sigma_i / m_H$ is the total cross section per gram of the IS medium. The efficiency, Q, is the ratio of σ to the geometric cross section of the grain.

Section 2 is a review of the derivation of absorption cross sections. The very low value for the region λ 912–504 Å may be compatible with the theory for large dust grains, but it places stringent limits on the properties of small grains. The limits are relaxed if the *visual* extinction cross section of grains is smaller in H II regions than in the general IS medium. It should be possible to detect such a difference with visual and radio observations (Section 3). The observations suggest that the visual extinction cross section of dust in the Orion Nebula is the same as that in the general IS medium.

2. Absorption Cross Sections

The relative extinction cross sections for the general IS medium have been determined down to 1000 Å from OAO-A2 and Copernicus results (York *et al.*, 1973; Bless and Savage, 1972). The absolute scale of the extinction cross sections may be determined from the observed ratio of the column density of hydrogen to the visual extinction. For example, Jenkins and Savage (1974) derive $N_H/E_{B-V}=7.5\times10^{21}$ atoms cm^{-2} mag^{-1}. Using the relation $A_V=1.086\,\tau_v$, and assuming a ratio of total-to-selective absorption of 3.0, one derives a total extinction cross section in the visual, $x_1\sigma_v^{ext}=3.7\times10^{-22}$ cm^2 H-atom^{-1}. This value is an average over the lines of sight to many stars observed by OAO and Copernicus. It applies to the general IS medium in the vicinity of the Sun. In setting $\sum_i x_i\sigma_i=x_1\sigma_v$, we have assumed that visual extinction is due to only one type of grain, with a number density x_1 relative to H. The smaller grains which are responsible for extinction in the UV do not contribute significantly to the extinction at visual wavelengths. The inverse, however, is not true; the large grains are expected to have $Q^{abs}=1$ at short wavelengths. Q_v^{ext} lies between 1 and 2, depending on the type of grain. For the core-mantle grains proposed by Greenberg (1974), $Q_v^{ext}\approx1.5$. Thus, at short wavelengths, one expects a contribution to $\sum_i x_i\sigma_i^{abs}=x_1\sigma_v^{ext}/(1.5\pm0.5)$ by the large grains. We now drop the symbol '$\sum$'; short wavelength cross sections are understood to be sums over several types of particles.

For wavelengths >912 Å, the absorption cross sections may be determined from the extinction cross sections, and the albedo, $a=\sigma^{sca}/\sigma^{ext}$. a may be derived from observations of the diffuse galactic light or of reflection nebulae. At visual wavelengths (van de Hulst and de Jong, 1969; Mathis, 1973; Mattila, 1970; Witt and Lillie, 1973) the derived values of the albedo converge to $a(B)=0.6\pm0.1$; $a(U)>a(B)>a(V)$, with $a(U)$ about 10 percent greater than $a(V)$.

In the UV, Witt and Lillie (1973), from observations of the diffuse galactic light, derived the albedoes and hence the absorption cross sections for $\lambda>1500$ Å. These values apply to the general IS medium. The value of $x\sigma^{abs}$ at 1500 Å is $\sim0.4x_1\sigma_v^{ext}$, which is less than the minimum contribution, $\sim0.5x_1\sigma_v^{ext}$, expected from the large grains.

Mezger, Smith, and Churchwell (1974, hereafter referred to as MSC) used the degree of ionization and the amount of IR radiation observed from H II regions to derive the absorption cross section below 912 Å. The results are given in terms of mean effective cross section $x\sigma_u^{abs}$, $x\sigma_H^{abs}$, $x\sigma_{He}^{abs}$ for the wavelength regions $\lambda>912$, $912>\lambda>504$, $504>\lambda$, respectively. They find that the absorption cross section $x\sigma_{He}^{abs}$ ($\lambda<504$ Å) is high, greater than $x_1\sigma_v^{ext}$; the ratio $x\sigma_{He}^{abs}/x\sigma_H^{abs}$ is greater than 4; and hence, $x\sigma_H^{abs}$ is small, less than $x_1\sigma_v^{ext}$. $x\sigma_u^{abs}$ is uncertain, mainly because of the importance, in this wavelength range, of multiple scattering.

Three points are important:

(1) The derivation of low $x\sigma_H^{abs}$ is independent of Witt and Lillie's observations and strongly supports their deduction of a low absorption cross section from 1800 to 1500 Å.

(2) The rise to high absorption must be fast and occur close to 504 Å. One can imagine a second absorption peak similar in strength (see below) and steepness of rise to the absorption peak at $\lambda\,2200$.

(3) MSC normalized all cross sections to $x_1\sigma_v^{ext}$ using the Savage and Jenkins (1972) value of N_H/E_{B-V} which yields $x_1\sigma_v^{ext}=5.5\times10^{-22}$ cm^2 H-atom^{-1}, 1.5 times the improved estimate from Jenkins and Savage (1974). Since the derivation of the UV absorption cross sections does not depend on the value of $x_1\sigma_v^{ext}$, the ratio of the cross sections is increased by 1.5. I now give a review of the derivation of the cross sections and their upper limits.

Churchwell *et al.* (1974) observed that the number ratio of He$^+$/H$^+$ in H II regions varies and, in particular, is <0.02 for H II regions in the immediate vicinity of the galactic center. They also noticed a correlation between low observed ratio of He$^+$/H$^+$ and high IR flux.

One assumes that all Lα photons are absorbed by the dust and the energy reradiated in the IR. However, the IR energies observed are 4–15 times the amount available in the Lα photons. Harper and Low (1971) suggested that other stellar photons are also being absorbed by the dust. If He-ionizing photons are preferentially absorbed by the dust, the observed correlation could easily be explained.

To demonstrate that selective absorption by dust is the most reasonable explanation, one argues as follows: The observed ratio He$^+$/H$^+$ = Ry, where y is the cosmic He/H ratio by number, and R is the ratio of the volumes of the He$^+$ and H$^+$ regions weighted by the square of the proton densities. Variation of the observed He$^+$/H$^+$ ratio could be due to: (1) variations of y from one H II region to another; (2) ionization by low temperature stars yielding $R<1$; or (3) preferential absorption of the Helium-ionizing photons by dust, also yielding $R<1$.

Variation of y offers no good explanation of the correlation of He$^+$/H$^+$ with the IR excess. Further, recent observations (Pauls *et al.*, 1974) show that the He$^+$/H$^+$ ratio in Sgr A West is probably near normal (i.e. ~0.1); thus it is unlikely that small y is responsible for the low observed He$^+$/H$^+$ ratio in the other H II regions near the galactic center.

Ionization by low temperature stars could, in principle, explain a correlation between small values of R and high IR excess. However, it is unlikely that the giant H II regions are ionized by very large numbers of cooler stars. Further, we know of one specific case, viz. the Orion Nebula, where $R\lesssim0.9$ and the exciting star is an O6 star.

Thus, modification of the ionizing radiation by dust is the most reasonable explanation for *all* the observations. Selective absorption of helium-ionizing photons by dust has also been suggested by Leibowitz (1973) and by Jura and Wright (1974).

MSC derive the ratio of the absorption cross sections from the observed correlation between R and IR excess (the excess energy over that expected from Lα heating alone, expressed in units of the Lα energy available). All H II regions are assumed to be ionized by a stellar flux like that produced by a cluster of stars following the Salpeter (1955) original luminosity function and having Auer and Mihalas (1972) model atmospheres. This assumption needs only to apply statistically, since no results depend on individual H II regions.

The predicted relationship between R and the IR excess depends only on the cross section ratios, $x\sigma_{He}^{abs}/x\sigma_{H}^{abs}$ and $(1+s)x\sigma_{u}^{abs}/x\sigma_{H}^{abs}$, where $(1+s)$ is a factor allowing for multiple scattering. There is a range of values of the two ratios which fit the (very low accuracy) observations. However, *no* fit is obtained, i.e. the correlation cannot be explained, unless $x\sigma_{He}^{abs}/x\sigma_{H}^{abs} > 4$.

Absolute values for the cross sections may then be derived in the following way: From an observation of a He$^+$/H$^+$ ratio that is less than 0.1, a value for R and hence for the optical depth, τ_{He}, of the dust to the He-ionizing radiation may be derived. (There is a dependence of τ_{He} on the ratio $x\sigma_{He}^{abs}/x\sigma_{H}^{abs}$, which is small if the ratio is large.) The optical depth $\tau_{He} = x\sigma_{He}^{abs}N_H$, where N_H is the column density of hydrogen from the star to the boundary of the H II region. N_H can be estimated from radio observations with reasonable statistical accuracy. Thus, $x\sigma_{He}^{abs}$ may be derived, and MSC obtain

$$x\sigma_{He}^{abs} \approx (8 \pm 1) \times 10^{-22}(x\sigma_{He}^{abs}/x\sigma_{H}^{abs})/(x\sigma_{He}^{abs}/x\sigma_{H}^{abs} - 1) \leqslant$$
$$\leqslant 12 \times 10^{-22} \text{ cm}^2 \text{ H-atom}^{-1}.$$

This is Equation (25) of MSC modified in the sense of not using the visual cross section as a zero point. The upper limit follows from $x\sigma_{He}^{abs}/x\sigma_{H}^{abs} > 4$. It is a little greater than the value ($\sim 9 \times 10^{-22}$ cm^{-2} H-atom^{-1}) at λ 2200 Å.

The maximum value for $x\sigma_{H}^{abs}$ also corresponds to a ratio of $x\sigma_{He}^{abs}/x\sigma_{H}^{abs} = 4$ such that

$$x\sigma_{H}^{abs} \leqslant 3 \times 10^{-22} \text{ cm}^2 \text{ H-atom}^{-1}$$

This is to be compared with

$$x_1\sigma_v^{ext} = 3.66 \times 10^{-22} \text{ cm}^2 \text{ H-atom}^{-1}$$

for the general IS medium (derived at the beginning of this section).

The upper limit for $x\sigma_{H}^{abs}$ is compatible with the expected value, $\sim 0.7x_1\sigma_v^{ext}$ for the contribution from large grains. However, this requires that the small grains, responsible for most of the extinction at short wavelengths, contribute little to the absorption cross section; i.e. the small grains must have a relatively high albedo (see Section 4).

If $x\sigma_{He}^{abs}/x\sigma_{H}^{abs} = 7$, as suggested by MSC, then the absorption cross sections are about a factor of 2 lower and are incompatible with the value of $x_1\sigma_v^{ext}$. However, this value of $x_1\sigma_v^{ext}$ is derived from Jenkins and Savage's (1974) observations and applies to the general IS medium in the solar neighbourhood. If the number or size of the large grains were smaller in H II regions, then $x_1\sigma_v^{ext}$ would also be smaller and the incompatibility would be resolved.

3. Dust in H II Regions vs Dust in the General IS Medium

I now review the observations relating to the value of $x_1\sigma_v^{ext}$ in H II regions.

Observations (Savage, 1973) indicate that stars associated with nebulosity may have weaker than average UV extinction. θ Ori is the most extreme example, showing a weak λ 2200 bump as well as low extinction at still shorter wavelengths. The visual part of the extinction curve is attributed (e.g. Greenberg, 1973) to grains of radius $\sim 0.1\ \mu$. The short wavelength extinction rise is attributed to a large number of grains with radius $\sim 0.01\ \mu$. The number ratio of small grains to large grains is about 10^3. The observed weak UV extinction may be explained (Witt, 1973) by a decrease of the number ratio of small grains to large grains by a factor of about 2.

Johnson (reviewed in 1968) has claimed that $A_V/E_{B-V} > 3$ in many regions. A number of these observations have been re-interpreted (see e.g. review by Elsässer, 1969). However, for some objects, such as θ Ori, which lie in bright nebulosity, the high values of A_V/E_{B-V} remain (Johnson, 1967). Witt (1973) did not allow for $A_V/E_{B-V} > 3$ in his model fits. Its effect is to lower the UV extinction relative to the visual extinction, implying an even greater reduction in the relative numbers of small grains than suggested by Witt.

O'Dell and Hubbard (1965) derived the dust distribution in the Orion Nebula from Hα, Hβ and visual continuum observations. Comparison with Menon's (1961) gas density distribution implies that the gas-to-dust ratio varies by a factor 25, the dust being 5 times under-abundant near the Trapezium and 5 times over-abundant at a distance of $15'$ from the stars. The results are, however, subject to the uncertainties in the assumed distribution of the gas which was derived from observations with $6'$ resolution.

I obtain rather different results from a comparison of Hα, Hβ and 3.7 cm isophotes of the Orion Nebula. (The Hα and Hβ photographs were obtained in collaboration with the group of Courtès with the 193-m telescope of Haute Provence Observatory. The 3.7 cm isophotes were obtained by J. Wink with the NRAO interferometer.) Single dish observations by Schraml and Mezger (1969) at 2 cm were extrapolated with a λ^{-2} wavelength dependence to 3.7 cm, for continuum features too large to be seen with the interferometer. The resolution is $3''$ for Hα, Hβ and 3.7 cm and $2'$ for the 2 cm observations. The results confirm the conclusion of Münch and Persson (1971), based on Hα and Hβ photometry, that the dust and gas must be well mixed.

I find that the ratio $N(H^+)/\tau_\alpha$ is everywhere the same, from 0.5 to $4'$ (0.07 to 0.6 pc) from the Trapezium and is equal to the value for the general IS medium implied from Jenkins and Savage's (1974) value of N_H/E_{B-V}. The reduction is preliminary and errors could amount to a factor of 2. However, I believe that a difference of a factor 5 is excluded. This implies that $x_1\sigma_v^{ext}$, the extinction cross section due to the large particles, is constant within the nebula and equal (within a factor 2) to the value in the general IS medium.

It should also be mentioned that several groups of authors (e.g. Wright, 1973; Soifer *et al.*, 1972; Harper and Low, 1971) have derived the gas-to-dust ratio in H II regions from a comparison of IR and radio observations. Determination of the mass of dust from IR observations depends on the (σ^{abs}, λ) relation and also on the grain temperature to the 5th power. The grain temperatures are *not* known. In H II regions, there is certainly a range of temperatures that contributes significantly to the emitted radiation. The colour temperature is, therefore, physically meaningless; the effective temperature that should be used in a mass determination is not clear. An extreme example: The very high value of 10^5 for the gas-to-dust ratio derived for M17 by Wright (1973) results from the assumption that all the dust inside the H II region has a temperature equal to the 5–10 μ colour temperature of 240 K, whilst most of the radiation at 20–100 μ is supposed to come from dust at a temperature of 75 K, located outside the H II region. However, if one assumes that all the dust is inside the H II region, one derives a gas-to-dust ratio of 100.

4. Conclusions

There are two possible ways of fitting the current theory to the observations:

(1) The total cross section of large grains is the same in H II regions as in the general IS medium; their absorption efficiency is unity at short wavelengths. The extinction cross section of small grains is probably less in H II regions than in the general IS medium by a factor $\geqslant 2$; the albedo of the small grains for $\lambda > 504$ Å must be relatively high, ~ 0.7 at 1150 Å to give the extinction level observed for θ Ori. If the low UV extinction (coupled with large A_V/E_{B-V}), such as observed for θ Ori, are not characteristic features of the dense H II regions, the required albedo is much higher. (Note that the far UV absorption cross sections apply to H II regions that are mostly larger and denser than the Orion Nebula.)

(2) The extinction cross section of large grains is reduced in H II regions by a factor $\leqslant 2$ (i.e. the large grains are reduced in number or size). The extinction cross section of the small grains is reduced by a correspondingly larger factor.

References

Auer, L. H. and Mihalas, D.: 1972, *Astrophys. J. Suppl.* **24**, 193.
Bless, R. C. and Savage, B. D.: 1972, *Astrophys. J.* **171**, 293.
Churchwell, E., Mezger, P. G., and Huchtmeier, W.: 1974, *Astron. Astrophys.* **32**, 283.
Elsässer, H.: 1969, in L. V. Mavridis (ed.), *Structure and Evolution of the Galaxy*, D. Reidel, Holland, p. 70.
Greenberg, J. M.: 1973, in J. M. Greenberg and H. C. van de Hulst (eds.), 'Interstellar Dust and Related Topics', *IAU Symp.* **52**, 3.
Greenberg, J. M.: 1974, *Astrophys. J.* **189**, L81.

Harper, D. A. and Low, F. J.: 1971, *Astrophys. J.* **165**, L9.

Jenkins, E. B. and Savage, B. D.: 1974, *Astrophys. J.* **187**, 243.

Johnson, H. L.: 1967, *Astrophys. J.* **150**, L39.

Johnson, H. L.: 1968, *Stars and Stellar Systems* 7, 167.

Jura, M. and Wright, E. L.: 1974, *Astrophys. J.* **187**, 473.

Leibowitz, E. M.: 1973, *Astrophys. J.* **181**, 369.

Mathis, J. S.: 1973, *Astrophys. J.* **186**, 815.

Mattila, K.: 1970, *Astron. Astrophys.* **9**, 53.

Menon, T. K.: 1961, *Publ. N.R.A.O.* **1**, 1.

Mezger, P. G., Smith, L. F., and Churchwell, E.: 1974, *Astron. Astrophys.* **32**, 269.

Münch, G. and Persson, S. E.: 1971, *Astrophys. J.* **165**, 241.

Pauls, T., Mezger, P. G., and Churchwell, E.: 1974, *Astron. Astrophys.* **34**, 327.

Salpeter, E. E.: 1955, *Astrophys. J.* **121**, 161.

Savage, B. D.: 1973, in J. M. Greenberg and H. C. van de Hulst (eds.), 'Interstellar Dust and Related Topics', *IAU Symp.* **52**, 21.

Savage, B. D. and Jenkins, E. B.: 1972, *Astrophys. J.* **172**, 491.

Schraml, J. and Mezger, P. G.: 1969, *Astrophys. J.* **156**, 269.

Soifer, B. T., Pipher, J. L., and Houck, J. R.: 1972, *Astrophys. J.* **177**, 315.

van de Hulst, H. C. and de Jong, T.: 1969, *Physica* **41**, 151.

Witt, A. N.: 1973, in J. M. Greenberg and H. C. van de Hulst (eds.), 'Interstellar Dust and Related Topics', *IAU Symp.* **52**, 53.

Witt, A. N. and Lillie, C. F.: 1973, *Astron. Astrophys.* **25**, 397.

Wright, E. L.: 1973, *Astrophys. J.* **185**, 569.

York, D. G., Drake, J. F., Jenkins, E. B., Morton, D. C., Rogerson, J. B., and Spitzer, L.: 1973, *Astrophys. J.* **182**, L1.

NUCLEATION AND GROWTH OF DUST GRAINS*

E. E. SALPETER

Physics Dept. and Center for Radiophysics and Space Research, Cornell University

Abstract. Classical nucleation theory must be modified for the condensation of small, solid dust grains. Ion nucleation is unimportant and exchange reactions with gas molecules are required. For carbon-rich stellar atmospheres, hydrocarbons act as intermediaries for converting acetylene gas into carbon particles. For oxygen-rich atmospheres, magnesium silicate grains form by surface nucleation. Sputtering and coalescence of grains is discussed.

* Published 1974 in *Astrophys. J.* **193**, 579.

FORMATION AND FLOW OF DUST GRAINS IN COOL
STELLAR ATMOSPHERES*

E. E. SALPETER

Physics Dept. and Center for Radiophysics and Space Research, Cornell University

Abstract. Stars are considered which would have no mass loss in the absence of dust grains, but are sufficiently cool for grains to condense in the atmosphere. Stars of 'high' luminosity lose gas and dust together at a copious rate; stars of 'medium' luminosity lose mainly grains without gas, and the atmosphere may be denuded of grain-forming material. Formulae are given for mass flow rates and for final velocities. A servomechanism is suggested which may lead to fairly large ($\sim 0.1\,\mu$), scattering silicate grains. An instability is suggested for cool M giants and supergiants, which leads to a patchy distribution of surface temperature (grains in cool patches, high flux in warm patches).

* Published 1974 in *Astrophys. J.* **193**, 585.

PHYSICAL ADSORPTION OF HYDROGEN ON
INTERSTELLAR GRAPHITE GRAIN SURFACES

R. F. WILLIS and B. FITTON

Surface Physics Group, Astronomy Division, European Space Research Organisation,
ESTEC, Noordwijk, Holland

Abstract. We review existing *single-particle* theories concerning parameters of importance which determine the kinetics of hydrogen molecule formation and ejection from cold ($T_{gr} \lesssim 20$ K) graphite grain surfaces. The nature of the *single-particle* quantum states of low mass gas atoms and molecules in a periodic surface lattice potential is considered. Contributions to the physical adsorption potential due to dynamic polarizability effects arising from the *long-range* collective valence-electron charge-density oscillations (plasmons) of the substrate are discussed. *Short-range* electron correlation effects at the surface may lead to the formation of a 'quasimolecular state' of adsorbed H_2 with a bond length ~ 3.5 Å and a reduced bond energy ~ 0.075 eV. It is proposed, that one consequence of this dynamical screening of the adsorbed molecules is that they are ejected normal to the grain surface with velocities $\lesssim 20$ km s^{-1} and not necessarily in a high vibrational state. Similar dynamical effects could be important in determining activation processes and long-range ordering in monolayer films of adsorbed H_2. The astrophysical consequences of these *many-body* effects are discussed in the light of recent experimental and observational results.

1. Introduction

In interstellar space the formation of diatomic molecules, such as H_2, is thought to proceed mainly by adsorption and subsequent recombination of atoms on the surfaces of solid dust particles (Van de Hulst, 1949). Three-body recombinations in the gas phase (Dalgarno and McCray, 1972) are too slow at the low densities of interstellar gas ($n_H \sim 0.1$ to 10^3 H atoms cm^{-3}) to account for the observed density of interstellar H_2 (Carruthers, 1970; Spitzer *et al.*, 1973). Most of the interstellar matter is in the form of atomic hydrogen gas at temperatures of the order of 100 K and about 1% of the interstellar mass is in the form of solid dust grains (diameters of the order of 10^{-4} cm), with much lower temperatures, $T_g \sim 6$ to 30 K. A major constituent of interstellar grains is thought to be graphite (Hoyle and Wickramasinghe, 1962) so that physisorption of H atoms on cold graphite particle surfaces is likely to be a dominant process in the formation of interstellar H_2.

Several authors (Gould and Salpeter, 1963; Knaap *et al.*, 1967; Williams, 1968; Augason, 1970; Hollenbach and Salpeter, 1971) have discussed processes by which hydrogen atoms become physically adsorbed on graphite grain surfaces, diffuse across the surface, combine and are desorbed. Several difficulties arise, the most serious of which are the uncertainties about the nature of the adsorption energies for atomic and molecular hydrogen and the ejection mechanism for the H_2 molecules.

In most previous theoretical treatments, the only contribution to the physical adsorption potential which has been considered is that due to van der Waals long-range dispersive forces, the two-body potential between a single crystal-atom and the gas atom being summed over all lattice atoms in the surface (Knaap *et al.*, 1967; Hollenbach and Salpeter, 1970). Augason (1970) pointed out that additional *electrostatic* fields could produce higher adsorptive potentials. The origin of such forces in graphite grains was not discussed but the fact that ortho ($J=1$)- and para ($J=0$)- hydrogen are separated on graphite led King and Benson (1966) to suggest that electrostatic interactions partly determine the adsorption energy. In order for ortho-para separation to occur, the adsorbed molecules must be subject to hindered rotation since they differ only in their ground state rotational energy. Hindered rotation requires a surface force-field more directional than van der Waals dispersive forces and an adsorptive potential, $U \gtrsim 400$ K (0.8 kcal mole^{-1}).

In the present paper, we begin (Section 2) by briefly reviewing the existing theory concerning the parameters of importance which determine the kinetics of molecule formation on grain surfaces. In Section 3, we consider the nature of *single-particle* quantum states of low mass gas atoms and molecules in a periodic surface-lattice potential with reference to numeric results obtained for helium (Hagen *et al.*, 1971). We show that the origin of a *dynamic* electrostatic contribution to the adsorption potential for graphite surfaces is in the *long-range collective* oscillations of the valence-electron gas, i.e., the 'plasmon' modes at the surface. We carry the discussion further by speculating how other effects due to the *short-range* dynamical nature of the substrate electrons can greatly lower the bond energy of a hydrogen molecule oriented parallel to the surface.

An important consequence of this short range dynamical screening, Section 4, is that H_2 molecules can be ejected essentially normal to the surface with a translational kinetic energy of several eV. In Section 5, we discuss the possibility that similar effects could be important in determining activation processes and perhaps in long-range ordering of the adsorbed species. Collective phenomena associated with the adsorbed hydrogen layer could lead to a rich complexity of phase transitions with increasing coverage at low grain temperatures, $T_{gr} \lesssim 20$ °C. In Section 6, we speculate that the role of many-body collective effects, which have hitherto not been considered, could have important astrophysical consequences. In particular, they could explain the recent observations from the *Copernicus* satellite (Spitzer *et al.*, 1973) of the rotational lines of newly formed H_2 molecules. The latter appear to be ejected from the grains with a translational kinetic energy of about 3 eV in their *ground* electronic and vibrational state with considerable rotational energy, $J=4$, 5 and 6 (Spitzer and Cochran, 1973).

2. Kinetics of H_2 Formation on Grain Surfaces

We briefly review the case of a finite-size grain having of the order of 10^6 adsorption sites and binding energies U_A and U_M for atomic and molecular hydrogen respectively (Gould and Salpeter, 1963; Knaap et al., 1966). Parameters which determine the adsorption and recombination of H atoms on the surface are the residence time of an atom on a grain τ_R, and the time between successive captures, τ_c, given by

$$\tau_R = \frac{1}{v_0} \exp\left(\frac{U_A}{RT_{gr}}\right) \tag{1}$$

and

$$\tau_c = (\alpha n_H \langle v \rangle \sigma)^{-1}, \tag{2}$$

where v_0 is the characteristic vibration frequency, 10^{12}–10^{13} s^{-1}; R the gas constant; T_{gr} the grain temperature ($\sim$6 to 30 K); U_A the atomic H adsorption potential, which is equal to the surface adsorption potential, V_{min}, minus the zero point energy; α is the sticking coefficient between 0.1 and 1; n_H the number density of H atoms, $n_H \sim 10^3$ atoms cm^{-3} in an interstellar dust cloud; $\langle v \rangle$ the average velocity of H atoms at 100 K equal to 1.45×10^5 cm s^{-1}; and σ the cross-sectional area of the grain, 10^{-8} cm^2. In order for a H$_2$ molecule to form, the 'residence time' must be greater than the 'capture time', $\tau_R > \tau_c$, and expressions (1) and (2) can be equated to determine the maximum grain temperature $(T_{gr})_{max}$ below which molecules will form

$$(T_{gr})_{max} = \frac{U_A}{R}\left(\ln \frac{v_0}{\alpha n_H \langle v \rangle \sigma}\right)^{-1}. \tag{3}$$

The probability of finding two atoms on the same grain at any instant is given by

$$\gamma = \frac{\alpha \langle v \rangle n_H \sigma}{v_0} \exp\left(\frac{U_A}{RT_{gr}}\right). \tag{4}$$

It can be seen that U_A is the critical parameter, as both $(T_{gr})_{max}$ and γ are dependent upon the nature of the adsorption potential. However, the adsorption potential of atomic hydrogen has not been accurately measured on any surface to date.

Augason (1970) makes the assumption that the ratio of the atomic and molecular adsorption potentials, U_A/U_M, is the same as the ratio of their *static* polarizabilities, 0.84. Using a value of $U_M = 0.96$ kcal mole^{-1} measured for H$_2$ on carbon black at 2700 °C (Ross and Oliver, 1964), he derives $U_A = 0.81$ kcal mole^{-1} (400 K)* for the adsorption energy of atomic hydrogen on a perfect graphite surface. This yields a value of $(T_{gr})_{max} = 12$ K, using a value of $v_0 = 10^{13}$ s^{-1} for the characteristic vibration frequency for H on graphite. Day (1972) has derived a value of the sticking coefficient, $\alpha = 0.3$, from experimental measurements of the accommodation coefficient for H$_2$ on graphite surfaces in the temperature range 77 to 273 K. Using this value and Augason's parameters, Equation (3) gives a value for $(T_{gr})_{max} = 11.53$ K. Measurements by Lee (1971, 1972, 1975) of the adsorption energies for H$_2$ on ice, solid CO and 'dirty' graphite surfaces in the temperature range 4 K to 40 K, how-

---

* Adsorption energies are given in kcal mole^{-1} and °K since *both* are commonly used in the current literature; electron volts (eV) and ergs units are also included for reference purposes in Table I.

ever, indicate that U_A for atomic hydrogen on clean graphite surfaces is probably greater than the above assumed value, $U_A = 0.8$ kcal mole^{-1}.

3. Quantum States and the Mechanism of Physical Adsorption of H· Atoms on a Perfect Graphite Surface

In view of the above uncertainties concerning the value of U_A, attempts have been made to calculate the adsorption potential by determining the potential energy of an H atom as a function of position on a grain lattice by summing-up the van der Waals pair interactions. Knaap *et al.* (1966) obtained a range of U_A between 0.3 and 0.5 kcal mole^{-1} (150 to 250 K) for atomic hydrogen adsorption on the surface of a 'perfect' ice crystal. Hollenbach and Salpeter (1970) performed a more detailed calculation and arrived at values of 0.89 kcal mole^{-1} (450 K) and 1.09 kcal mole^{-1} (550 K) for the adsorption ground state energies of atomic and molecular hydrogen, again for an ice crystal surface. They concluded that similar values should apply for a graphite grain, and represent *lower* limits on the binding energies. These calculations of the physical adsorption energy are essentially a *single-particle* approach to the problem. In the next few paragraphs we consider in some detail this and other factors which determine the values of U_A and U_M, limiting our discussion primarily to the specific case of a graphite lattice surface although many of the conclusions will be applicable to solid surfaces in general.

3.1. SINGLE-PARTICLE STATES

The interaction of a *single* H· atom with a graphite substrate can be written as a sum over all carbon atoms of a two-body gas-carbon potential. The substrate is assumed to be a semi-infinite, graphite basal plane surface, defects and thermal vibrations of the lattice being ignored. The graphite solid consists of a parallel arrangement of hexagonal basal plane lattices, staggered as shown in Figure 1a. The potential energy of a physically adsorbed H· atom is expected to be virtually identical for both an *ABAB* (Figure 1a) or for the other conventional *ABC ABC*-type stacking structure.

Assuming that the two-body potential is of the Leonard-Jones (12–6) type, the potential energy of an H· atom with a position vector **r** is given by

$$V(\mathbf{r}) = \sum_i 4\varepsilon_0 \left[\left(\frac{\varrho_0}{\varrho_i} \right)^{12} - \left(\frac{\varrho_0}{\varrho_i} \right)^6 \right], \tag{5}$$

where ϱ_i is the distance between the H atom and the ith carbon atom and where the sum extends over all carbon atoms. The van der Waals constants ε_0 and ϱ_0 for H/graphite and H$_2$/graphite are not known with any certainty. An estimate of ε_0 can be obtained from the electron polarizabilities of the two atoms involved. Hollenbach and Salpeter (1970) estimate values of the order of $\varepsilon_0 = 60$ to 100 K and $\sigma \approx 3$ Å. The behaviour of $V(\mathbf{r})$ as a function of height Z above the crystal surface is of the form shown in Figure 2a. The results are those for an accurate numeric calculation

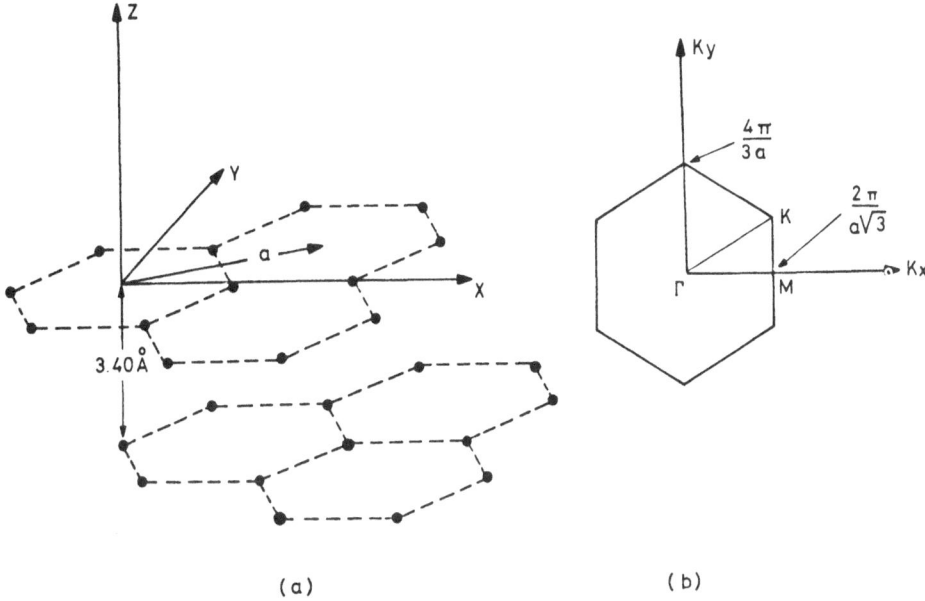

Fig. 1.   (a) *AB*-stacking structure of hexagonal basal (0001) planes of graphite lattice. (b) Brillouin zone corresponding to the lattice of adsorption sites.

of the potential energy of a *helium* atom on graphite for the three specific locations in the basal plane indicated (Hagen *et al.*, 1971). In the case of He^3 and He^4, the potential varies little around the perimeter of the hexagonal adsorption cell, the energy minimum being located at the centre. The potential well in the direction perpendicular to the surface is very deep and narrow so that excitations in the Z-direction are energetically well separated, whilst periodic variations in $V(\mathbf{r})$ across the surface are small, causing excitations across the surface to occur in wide *bands*.

The adatom single particle quantum states are found by solving the Schrödinger equation

$$\left[-\frac{\hbar^2}{2m} \nabla^2 + V(\mathbf{r}) \right] \psi(\mathbf{r}) = E\psi(\mathbf{r}). \tag{6}$$

If the actual pair-interaction potential $V(\mathbf{r})$ is replaced by a harmonic-oscillator potential and a fully localized oscillator wavefunction is used, an upper limit to the zero point energy is obtained (Knaap *et al.*, 1966). In this approximation the ground state energy of an adsorbed low-mass atom is

$$E(0) = V_{min} + \hbar\omega_{xy} + \tfrac{1}{2}\hbar\omega_z. \tag{7}$$

However, as pointed out by Hollenbach and Salpeter (1970), the real potential is periodic and the real wave functions for the adsorbed states are not perfectly localized, i.e. the adatom ground-state wave-function is of the '*band*' type. The result is that the zero point energy – $(\hbar\omega_{xy}+\tfrac{1}{2}\hbar\omega_z)$ is reduced from a value of about 600 K (Knaap *et al.*, 1966) to a value of 300 K (Hollenbach and Salpeter, 1970), with a

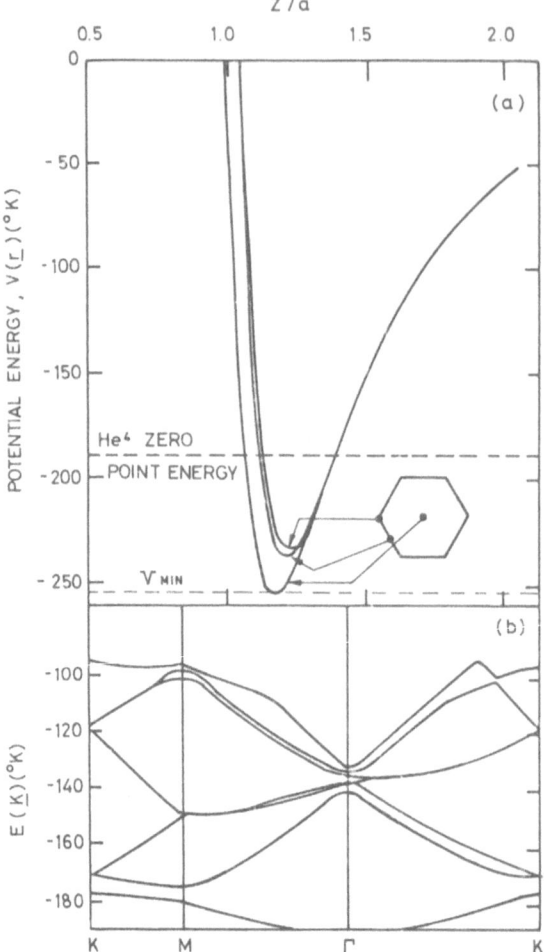

Fig. 2. (a) Potential energy curves of a physisorbed helium atom showing $(V(\mathbf{r}))$ as a function of height of the atom z above a graphite crystal surface for three specific locations in the $x-y$ plane. (b) Energy bands for an adsorbed helium atom along the $KM\Gamma K$ contour representing the principal symmetry directions of the graphite lattice Brillouin zone showing the lowest seven bands. The results are those of Hagen *et al.* (1971) and represent an accurate numeric calculation for the case of an He⁴ atom on the graphite basal plane. Results of a similar form are expected for H and H_2 on graphite with $V_{min} \sim 800$ K; $V_{zero} \sim 300$ K and $U \sim 500$ K. The band width of the lowest energy band, $\Delta E_b^{(0)} \sim 5\text{--}20$ K (see text).

consequent substantial increase in the ground state adsorption energy, $E(0)$. These values are considerably larger than those determined for He⁴ shown in Figure 2a.

A *complete* description of the adatom quantum states requires that Equation (6) be solved for values of $\mathbf{k}$ throughout the reduced two-dimensional hexagonal Brillouin zone by the use of band theory methods (Milford and Novoco, 1971) while the z-dependence suggests the expansion of $\psi(\mathbf{r})$ as a linear combination of orthonormal func-

tions having a *discrete* spectrum.* The energy eigenvalues are most conveniently
displayed by plotting $E(\mathbf{k})$ vs $\mathbf{k}$ along the high symmetry directions $KM\Gamma K$, the points
of high symmetry, Γ, K and M being those shown in Figure 1b. Again it is instructive to consider the results which have been determined for He^4, Figure 2b (Hagen
et al., 1971). Since we are concerned with low temperature behaviour, Figure 2b
shows only the lowest seven bands up to some 80 K above the ground state band.
The energy band structure is characterized by very wide band widths and very narrow
band gaps, indicative of He^4 adatoms being highly mobile across the surface.

The surface mobility of a low mass adatom, such as H or He, is determined by
quantum mechanical tunnelling through the lattice potential barrier which is temperature independent for $T_{gr} \lesssim 50$ K. The diffusion time τ_D required for an atom to
move to an adjacent lattice site is finite even at zero temperature and of the order of

$$\tau_D \approx \frac{4\hbar}{\Delta E_b^{(0)}}, \tag{8}$$

where $\Delta E_b^{(0)}$ is the energy width of the lowest energy band. $\Delta E_b^{(0)}$ for He^4 on graphite
is of the order of 10 K (Figure 2b). Hollenbach and Salpeter (1970) obtained values
of similar magnitude for H and H$_2$ on ice, i.e., 20 K for H/H$_2$O and 5 K for H$_2$/H$_2$O.
As a consequence, they conclude that recombination can occur almost spontaneously
if a second atom is captured while a first atom is adsorbed on the surface.

3.2. Dynamic Electrostatic Polarizability Effects

Atoms on the surface of a conductor are affected by very strong electric fields, which,
on graphite, is of the order of 10^6 esu cm^{-2} (Williams, 1968). Such fields have been
observed to induce strong dipoles (of the order of a debye) in adsorbed atoms and
molecules oriented perpendicular to and with positive ends away from the surface.
For gases with no permanent dipole moment, the adsorption energy is increased by
an amount

$$\Phi = -\tfrac{1}{2}\mu E_z^2, \tag{9}$$

where μ is the principal moment of polarizability of the adsorbed molecule and E_z
is the electric field intensity normal to the surface (Augason, 1970). There is a *repulsive*
force between two adsorbed atomic dipoles which raises a potential barrier of about
0.5 eV at a distance of 1 Å. Fluctuations in the surface fields produce fluctuations in
these dipoles, which give rise to an additional *attractive* force, as indicated in the
potential energy diagram, Figure 3.

The origin of these fields at the surface of a conductor is in the behaviour of the
valence-electron charge density. Charge density oscillations occur due to the *long-
range* electrostatic or Coulomb interactions between the valence electrons of the solid
(Pines, 1956). A schematic of a cut through a solid conductor surface showing the

---

* The potential $V(\mathbf{r})$, Equation (5), cannot be written as separate functions of z plus x and y. The
result is some *slight* mixing of motion parallel to the plane with motion perpendicular to the plane.

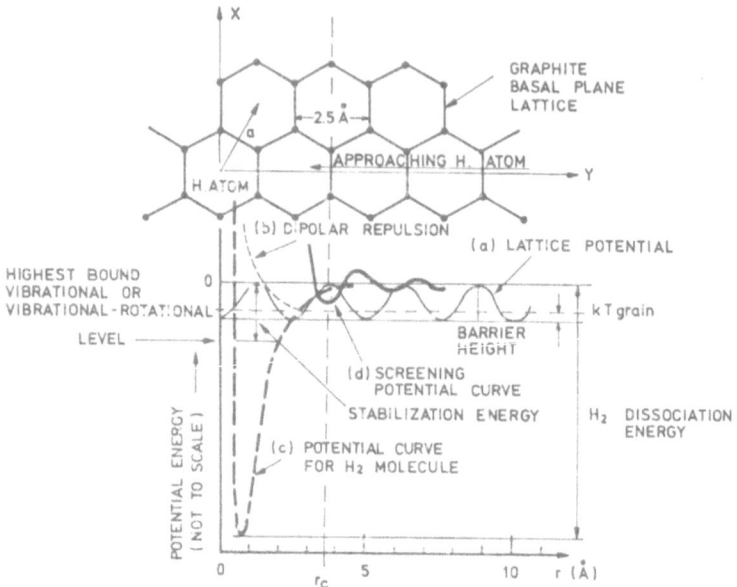

Fig. 3.   Schematic diagram of the various potential fields experienced by two approaching H atoms
physisorbed on a graphite surface.

charge density fluctuations and the electric field lines associated with such a 'plasmon'
oscillation of energy $\hbar\omega_p$ and wavelength λ_p is shown in Figure 4a. Similar longi-
tudinal oscillations of the valence electron gas, analogous to sound waves, are observed in graphite. The frequency of the plasmon oscillations is given approximately
by

$$\omega_p^2 = \frac{4\pi n e^2}{m} \approx \frac{3e^2}{mr_s^3}, \tag{10}$$

where n is the number density of valence electrons per unit volume, e is the electronic
charge, m the free electron mass and r_s is a parameter defined as the radius of a
sphere whose volume is the mean volume per electron. For graphite, plasmon oscillations occur at energies $\hbar\omega_p \approx 7$ eV and 25 eV, associated with oscillations of the
π valence electrons alone, and the $\pi + \sigma$ valence electrons in total, respectively (Willis
et al., 1973). Equation (10) yields values of $n_\pi \approx 3.6 \times 10^{22}$ electrons cm^{-3}, $r_s^\pi \approx 1.87$ Å
and $\omega_p^\pi \approx 1.1 \times 10^{16}$ rad s^{-1} for the π valence band electrons; $n_{\pi+\sigma} \approx 4.6 \times 10^{23}$ electrons cm^{-3}, $r_s^{\sigma+\pi} \approx 0.81$ Å, and $\omega_p^{\pi+\sigma} \approx 3.8 \times 10^{16}$ rad s^{-1} for the complete $\pi + \sigma$
valence band electrons.

   The Coulomb interaction of the electrons can be divided into two parts: a *long-
range* part, whose effect is described by the above plasmon oscillations, and a residual
short-range 'screened' interaction, whose range is about 1 Å. At the surface of a
conductor, the valence electron density is believed to vary in a complex manner, as
indicated in Figure 4b, the number density $n(z)$ varying rapidly from an approximately
constant value inside the solid to zero outside *over a distance of* 1 Å *or so* (Lang and

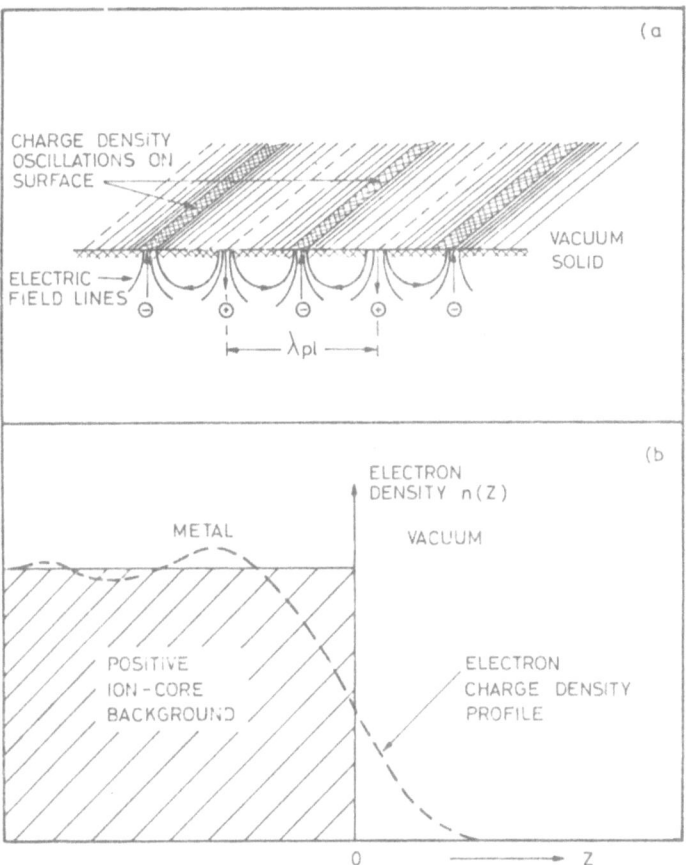

Fig. 4.   (a) Schematic diagram of electron charge density oscillations and electric field lines associated with a single 'plasmon' wave of energy $\hbar\omega_p$ and wavelength λ_p at the surface of a free-electron conductor. (b) Electron charge density profile at the surface of a free-electron solid showing the electron density extending a short distance (1 to 2 Å) into vacuum.

Kohn, 1970). The role of such effects in the determination of the attractive and repulsive contributions to the physisorption interaction energy has only lately been recognized. Recent calculations for He atoms on metals (Kleiman and Landman, 1973) indicate that an accurate determination of the dynamic polarizability is essential in calculations of the physical adsorption energy.

The detailed shape of the electron charge density profile at a surface (Figure 4b) can influence the way in which the covalent bond of a hydrogen molecule is formed in the region where the substrate electrons spill-out from the surface. For an inhomogeneous electron gas near a surface, there is a critical electron density at a critical distance Z_c (~a few ångströms) below which a screened proton will have a bound state. Further out from the surface, $Z > Z_c$, and below this critical electron density, something akin to the start of covalent bond formation will occur. Such considerations have recently led Brown et al. (1974) to conclude that the typical form of the molecular interaction energy, shown in Figure 3, will be replaced by a screened oscillatory interaction for a H_2 'molecule'* oriented parallel to the surface. The important qualitative points to emerge are:

(i) the energy required to separate the H· atoms (protons) to an infinite distance *parallel* to the surface is vastly reduced. For a *metal* with an $r_s \approx 1.4$ Å, corresponding to a *bulk* plasmon energy, $\hbar\omega_p \approx 11.2$ eV, Brown *et al.* (1974) deduce a reduction from approximately 4.7 eV for the 'free' H_2-molecule to 0.075 eV at the surface.

(ii) a principal minimum in the screened oscillatory interaction at 3.5 Å results in the bond length being considerably increased compared with the value of 0.74 Å for the 'free' H_2 molecule.

3.3. POTENTIAL ENERGY DIAGRAM AND SURFACE RECOMBINATION

The system of potential fields for the interaction of two approaching H· atoms on a graphite surface is summarized in Figure 3, following the scheme of Williams (1968). The magnitudes of the relevant energies are given in Table I.

TABLE I

Energy	°K	kcal mole^{-1}	ergs	eV
U_A	~450	0.89	6.2×10^{-14}	0.039
U_M	~550	1.09	7.6×10^{-14}	0.047
H_2 dissociation energy	~5.10^4	100	6.9×10^{-12}	4.478
Graphite lattice potential energy	~300	0.6	4.1×10^{-14}	0.026
kT_{gr}	~30	0.06	4.1×10^{-15}	0.003
Stabilization energy	~1000	2.0	1.4×10^{-13}	0.086
Graphite optical phonon energy	~300	0.6	4.1×10^{-14}	0.026
Screening potential energy	~100	0.2	1.4×10^{-14}	0.009

Neglecting for a moment the dynamic effects described in Section 3.2 the atoms have a binding energy U_A and are thermalized to the surface temperature $kT_{gr} \ll U_A$. Should two atoms approach along the attractive part of the potential curve for molecular H_2 formation, they do not fly apart immediately after collision as in the gas phase, due to presence of the lattice potential barrier. The potential energy released (the H_2 dissociation energy) appears as translational kinetic energy causing the atoms to vibrate $1/D$ times before flying apart once more, D being the quantum mechanical diffusion coefficient for the lateral barrier ($D \sim 10^{-3}$ for $T_{gr} \approx 10$ K). During this excited stage, which lasts for 10^{-12} s or so, the assumption is made that in order to stabilize the H_2 molecule, a small amount of 'stabilization energy' must be dissipated, transferring the molecule into a high vibration or vibration-rotation level of the ground state. Abrupt exothermic removal of the remaining binding energy then

---

* The term 'covalent bond' is no longer an appropriate description of the interaction between two H· atoms in such a potential. In this state they may be regarded as a loosely-bound 'quasi-molecule' at the surface.

results in the prompt and energetic ejection of the molecule from the surface in a high rotational quantum state $(J \sim 8)$ and a vibrational state $(v \sim 12)$ (Hollenbach and Salpeter, 1970).

However, as Williams (1968) pointed out, a difficulty with this model is that since no radiative rotation-vibration transition can take place in a time 10^{-12} s, the stabilizing energy must be removed by interaction with the lattice. There is a low probability of this occurring since the requirement is that the substrate must have a phonon frequency close in energy to the energy it is desired to remove from the excited surface molecule. Also, since this is a *resonance* condition, both stabilizing and unstabilizing collisions can occur with similar probabilities. Williams (1968) concludes that collisions between lattice electrons and the surface 'molecule-imperfection' are likely to be more effective, but again the de-excitation cross-section for a non-polar molecule makes this process of low probability.

If the interaction between the H· atoms is determined largely by the dynamic screened oscillatory potential curve, shown superimposed in Figure 3, the probability of chemical recombination *at the surface* is considerably reduced. The atoms will come together initially and form a weak 'screened' form of covalent bond of length ~ 3.5 Å and energy ~ 0.075 eV.

4. Ejection Mechanism for H_2

There is some observational evidence to support the formation of H_2 molecules on metal surfaces having an increased H–H separation and a much weakened bond energy. Definite vibrational frequencies of H_2 and D_2 in the proximity of a metal surface have been observed in low energy electron scattering experiments on H_2 adsorbed on tungsten metal surfaces by Propst and Piper (1967). The curvature near the first minimum of the screened interaction potential, Figure 3, is such that reasonable agreement is expected with the energy losses observed in the high resolution electron scattering measurements.

This 'quasi-molecule' will bounce about on the surface due to its thermal energy and interaction with lattice phonons. The problem represents a complex three-dimensional calculation, but we envisage an escape process similar to that illustrated in Figure 5. At some instant, the loosely-bound quasimolecule will move out beyond the critical screening distance, Z_c, causing the bond length to contract about its centre-of-mass due to the decay of the short-range charge-density screening. Co-valent bond formation occurs and a considerable amount of the overall H_2 binding energy will appear as centre-of-mass translational kinetic energy, the molecule being ejected normal to the surface.

An estimate of the overall translational kinetic energy is given by

$$E_{kin} = V_0 + 2kT_{gr} - \Delta E_s - E_{rot}, \tag{11}$$

where V_0 is the effective molecular dissociation energy (heat of dissociation)=

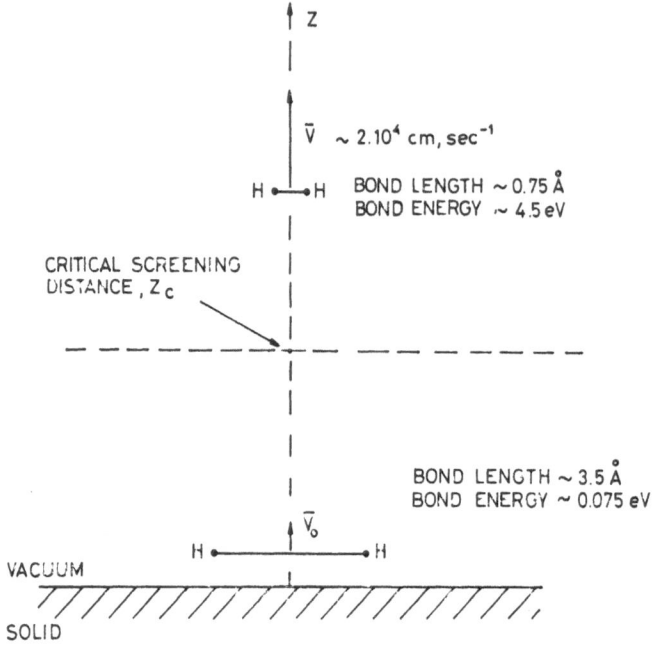

Fig. 5.   Schematic diagram illustrating a possible effect of substrate-electron screening on covalent bond formation and the subsequent desorption of an H_2 molecule.

4.478 eV or $\sim 5 \times 10^4$ K (Herzberg, 1950); $kT_{gr} \lesssim 30$ K; ΔE_s is the energy transferred to the lattice per 'bounce' ≈ 100 K and $E_{rot} \approx 2000$ K, the rotational kinetic energy (Hollenbach and Salpeter, 1970). Since $(2kT_{gr} - \Delta E_s)$ represents a negligible amount in comparison to V_0, $(E_{kin} + E_{rot}) \approx$ several electron volts. Equation (11) ignores any energy dissipated due to the decay of the charge-density screening at the surface. However, this is likely to be small also and of the order of 100 K since the depth of the first maximum of the oscillatory potential, Figure 3, has been estimated to be only of this order of magnitude (~ 0.01 eV, Brown et al., 1974). The essential point to emerge is that the H_2 molecule will leave normal to the surface with a translational kinetic energy of 3 to 4 eV in a high rotational quantum state $J \approx 8$ but not necessarily in a high vibrational or vibration-rotational state.

The observations of van Willigen (1968) and Dabiri et al. (1971) on the spatial distribution of H_2 desorbed from *metal* surfaces following recombination of atomic hydrogen at elevated temperatures (~ 1100 K) provide experimental evidence for such a mechanism. The measured angular distributions appear to be strongly peaked in the direction *normal* to the surface.

5. Long-Range Co-operative Effects in Adsorbed H_2 Monolayers

At low enough temperatures, 6 K $\lesssim T_{gr} \lesssim 10$ K, in conditions typical of dense ($n_H \sim 10^3$ atoms cm^{-3}), dark dust clouds, it is conceivable that the lifetime of the quasi-molecules on the grain surface will be sufficiently long for sub-monolayer coverages to build up. Under such circumstances, quasimolecule–quasimolecule interactions will become important and the co-operative motions within the monolayer will have an effect on molecule formation and ejection processes. It is not inconceivable that the oscillatory screening (Section 3.2) could play an important role in the ordering or in the activation processes taking place in the two-dimensional layer.

Recent heat capacity and specific heat measurements of physisorbed monolayers on graphite surfaces have revealed a rich-complexity of two-dimensional phase relationships and phase transitions (Dash, 1974). It has been suggested (Ginzburg and Sobyanin, 1972) that hydrogen monolayers adsorbed on graphite might well exhibit superfluidity, as observed in helium monolayers. Recent specific heat measurements of para-hydrogen monolayers* on graphite at low coverages (Bretz and Chung, 1974) do indicate that solidification of the layer has been suppressed to below ~ 1 K. Quasimolecule formation of the type discussed above might be expected to give rise to hydrogen films behaving in such a manner.

Theoretical considerations are complicated however by lack of detailed understanding of the dynamical aspects of the charge-density screening at surfaces (Lang and Kohn, 1970), and the fact that condensed molecular H_2 properties cannot be accurately predicted by any phenomenological pair-interaction potential known to date (England *et al.*, 1974).

6. Astrophysical Consequences

From the foregoing analysis, it is clear that the major contribution to the physisorption energy is the van der Waals energy, so that the arguments presented to date on the overall kinetics of H_2 molecule formation remain true. The adsorption energy is likely to be of the order of 500 to 1000 K, the upper limit being due to electrostatic effects (Section 3.2; see also, the result of Lee (1975)), but the recombination efficiency of H atoms on *perfect* crystal surfaces will remain high only for $T_{gr} \lesssim 20$ K. Since it is practically impossible for a grain to cool below 5 to 7 K in the typical interstellar radiation field (Field, 1969; Purcell, 1969), this limits the temperature range for H_2 molecule formation to $5 \lesssim T_{gr} \lesssim 20$ K. Such cool conditions are likely to exist only in the central regions of dense interstellar dust clouds.

Observations on H_2 molecular lines from the *Copernicus* satellite (Spitzer *et al.*, 1973) indicate that in many cases, rotational excitation of interstellar H_2 is apparently due to *newly formed* molecules, which leave the grains with about 3 eV of translational kinetic energy in high J-rotational states ($J \approx 5$) but in the *ground* electronic and vibrational state, producing the relatively wide Lyman lines observed in some stars. The ejection mechanism proposed in Section 4 would give rise to the *fast* ejection of H_2 molecules from grain surfaces with velocities having an upper limit ~ 20 km s^{-1} in dense dust cloud regions. Such a mechanism does not require that the H_2 molecule be stabilized into a high vibration or vibrational-rotational state prior to ejection from the surface (Williams, 1968). Eventual interaction of the

---

* Molecular hydrogen converts to the $J = 0$ state (*para*-hydrogen) once it is physisorbed onto graphite.

newly formed H_2 molecules with ultraviolet photons could then produce spectral lines of the type observed. Also, such a mechanism could provide a means of heating the gas in central dark cloud regions with high molecular to atomic ratios approaching 1. However, it remains for infrared observations of such regions (Gull and Harwit, 1971; Dalgarno and Wright, 1972), to establish the validity of such a proposed mechanism.

7. Concluding Remarks

One of the main points which we wished to introduce in the present paper was that of the role of *collective* effects in the mechanism of the formation and ejection of H_2 molecules from cold grain surfaces. The electron correlation effects associated with the lengthening of the H_2 covalent bond (Section 3) may be regarded as a precursor mechanism to *chemical* adsorption, which occurs at high temperature ~ 1000 K and should be specific to free-electron-like conductors such as metals, semi-metals (graphite) and semiconductors; we expect quite different behaviour for insulators.

However, it should be stressed that the ideas discussed in Sections 3, 4 and 5 are to some extent speculative since they derive from the newly developing and somewhat controversial field of many-body phenomena in solids. The fact that such effects exist is not at all in dispute but quantitative estimates of their magnitude, particularly their role in the physisorption of light gas atoms at surfaces is the subject of current debate. Experiments are currently in progress in our laboratory in an effort to establish the nature and magnitude of such effects. In the meantime, we hope that the treatment which we have given is appropriate to this subject at its present stage of development and will serve to stimulate readers to further consideration of the relative importance of such phenomena.

Acknowledgements

The authors wish to acknowledge the invaluable discussions which they had with Prof. E. E. Salpeter and Drs T. P. Snow and T. J. Lee during the course of this conference, and with our colleagues Prof. A. A. Lucas and Dr B. Feuerbacher on plasmon behaviour in solids.

References

Angason, G. C.: 1970, *Astrophys. J.* **162**, 463.
Bretz, M. and Chung, T. T.: 1974, *J. Low Temp. Phys.* **17**, 479.
Brown, J. S., Brown, R. C., and March, N. H.: 1974, *Phys. Letters* **47A**, 489.
Caruthers, G. R.: 1970, *Astrophys. J. Letters* **161**, L81.
Dabiri, A. E., Lee, T. J., and Stickney, R. E.: 1971, *Surface Sci.* **26**, 522.
Dalgarno, A. and McCray, R. A.: 1972, *Ann. Rev. Astron. Astrophys.* **10**, 375.
Dalgarno, A. and Wright, E. L.: 1972, *Astrophys. J.* **174**, L49.
Dash, J. G.: 1974, *Low Temp. Phys. – LT 13*, **1**, 19.
Day, K. L.: 1972, in J. M. Greenberg and H. C. van de Hulst (eds.), 'Interstellar Dust and Related Topics', *IAU Symp.* **52**, 311.

England, W., Etters, R., Raich, J., and Danilowicz, R.: 1974, *Phys. Rev. Letters* **32**, 758.

Field, G. B.: 1969, *Monthly Notices Roy. Astron. Soc.* **144**, 411.

Ginzburg, V. L. and Sobyanin, A. A.: 1972, *JETP Letters* **15**, 242.

Gould, R. and Salpeter, E. E.: 1963, *Astrophys. J.* **138**, 393.

Gull, T. R. and Harwit, M. O.: 1971, *Astrophys. J.* **168**, 15.

Hagen, D. E., Novaco, A. D., and Milford, F. J.: 1971, *Proc. Symp. on Adsorption-Desorption Phenomena*, Florence, Italy, Academic Press, New York, 1972.

Herzberg, G.: 1950, *Molecular Spectra and Molecular Structure*, Vol. I, Van Nostrand, New York.

Hollenbach, D. and Salpeter, E. E.: 1970, *J. Chem. Phys.* **53**, 79.

Hollenbach, D. and Salpeter, E. E.: 1971, *Astrophys. J.* **163**, 155.

Hoyle, F. and Wickramsinghe, N. C.: 1962, *Monthly Notices Roy. Astron. Soc.* **124**, 417.

King, J. and Benson, S.: 1966, *J. Chem. Phys.* **44**, 1007.

Kleiman, G. G. and Landman, U.: 1973, *Phys. Rev.* **B8**, 5484.

Knaap, H. F. P., Van den Meijenberg, C. N. J., Beenaker, J. J. M., and Van de Hulst, H. C.: 1966, *Bull. Astron. Inst. Neth.* **18**, 256.

Lang, N. D. and Kohn, W.: 1970, *Phys. Rev.* **B1**, 4555.

Lee, T. J., Gowland, L., and Reddish, V. C.: 1971, *Nature Phys. Sci.* **231**, 193.

Lee, T. J.: 1972, *Nature Phys. Sci.* **237**, 99.

Lee, T. J.: 1975, *Astrophys. Space Sci.* **34**, 123.

Milford, F. J. and Novaco, A. D.: 1971, *Phys. Rev.* **A4**, 1136.

Pines, D.: 1956, *Rev. Mod. Phys.* **28**, 184.

Propst, F. M. and Piper, T. C.: 1967, *J. Vac. Sci. Technol.* **4**, 53.

Purcell, E. M.: 1969, *Astrophys. J.* **158**, 433.

Ross, S. and Oliver, J. P.: 1964, *On Physical Adsorption*, Interscience, New York.

Spitzer, L. and Cochran, W. D.: 1973, *Astrophys. J.* **186**, L23.

Spitzer, L., Drake, J. F., Jenkins, E. B., Morton, D. C., Rogerson, J. B., and York, D. G.: 1973, *Astrophys. J.* **181**, L116.

Van de Hulst, H. C.: 1949, *Rech. Astron. Obs. Utrecht* **11**, Part 2.

Van Willigen, W.: 1968, *Phys. Letters* **28A**, 80.

Williams, D. A.: 1968, *Astrophys. J.* **151**, 935.

Willis, R. F., Fitton, B., and Painter, G.: 1974, *Phys. Rev.* **B9**, 1926.

EXTINCTION AND POLARIZATION MODELS

N. C. WICKRAMASINGHE

Dept. of Applied Mathematics and Astronomy, University College, Cardiff, Wales

Abstract. Currently favoured models for interstellar grains are reviewed in relation to observational criteria which bear on their optical properties.

1. Introduction

The search for the composition of interstellar grains has continued for over four decades. Although there is yet no general consensus amongst astronomers as to their precise nature, several definite trends have emerged from recent work (Wickramasinghe and Nandy, 1972; Aannestad and Purcell, 1973). In this summary I shall confine my attention to information which can be gleaned from astronomical observations relating to the optical properties of interstellar dust. To narrow down the range of possible contenders, we first note that the observed mean visual extinction coefficient $\kappa_V \sim 1$ mag/kpc demands that the mass density of grain material responsible for extinction is $\sim 10^{-26}$ g cm^{-3}. This conclusion is only weakly dependent on the precise composition of grain material provided grain dimensions are $\sim 10^{-5}$ cm. With a mean hydrogen density of $\sim 10^{-24}$ g cm^{-3} in the galactic plane this imposes an important constraint on grain models – viz., the elements comprising the grains must make up $\sim 1\%$ by mass of all interstellar matter. Grain models thus far proposed which satisfy this requirement to a greater or lesser extent include iron, graphite, ices, silicates and crystalline organic polymers (Wickramasinghe, 1974, 1975). It is worth noting that both iron and silicates may present some difficulty in this context because the constituent materials are at least marginally under-abundant. We assess the status of these various models in relation to available data, particularly with regard to observations of interstellar extinction and polarization.

2. Summary of Data

The wavelength dependence of interstellar extinction is now available for a large number of stars over the wavelength range $\sim 3 \mu$–1000 Å (Bless and Savage, 1972). These observations are depicted in Figure 1 in the form of a 'mean' extinction curve. The salient features are an apparent 'knee' at $\lambda = 4300$ Å separating two linear segments, a hump at $\lambda = 2200$ Å, followed by a minimum at ~ 1650 Å and a continued rise into the further ultraviolet.

The wavelength dependence of linear polarization, although it is available only over a somewhat narrower wavelength region, provides an important additional criterion. Polarization curves vary significantly from star to star, but an 'average' curve

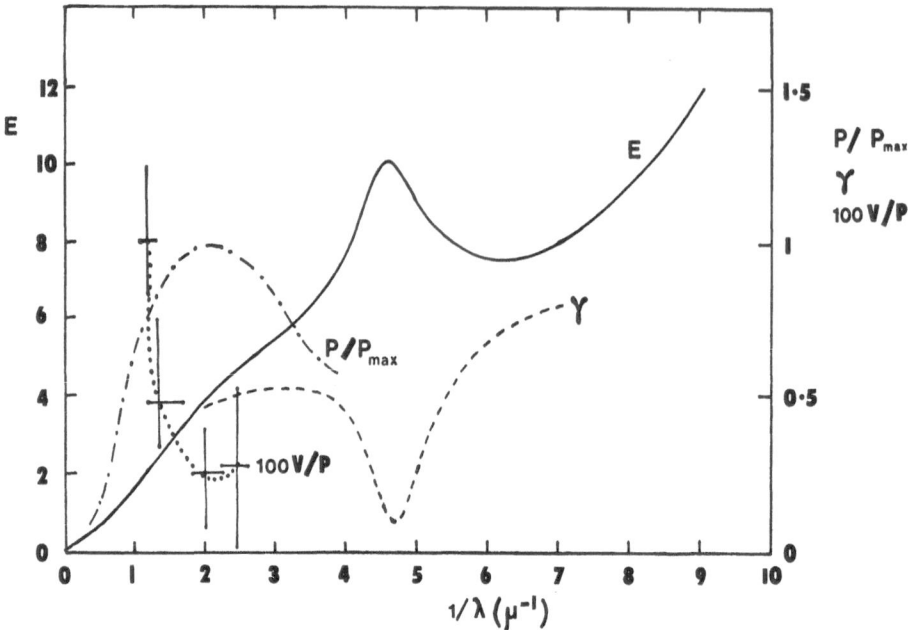

Fig. 1. Summary of mean observational data relating to grains. Interstellar extinction (solid curve,
E), albedo (dashed curve, γ), linear polarization (dot-dash curve, P/P_{max}) and circular polarization
(crosses joined by dotted curve, 100 V/P).

is shown in Figure 1 (Serkowski, 1967, 1973; Coyne and Wickramasinghe, 1969).

A further observation which serves as an important discriminant is the wavelength
dependence of interstellar circular polarization (Martin, 1973; Martin *et al.*, 1973).
This data is also shown in Figure 1, along with data on the albedo of interstellar
grains. The latter information is obtained from studies of reflection nebulae and the
diffuse galactic light (Witt and Lillie, 1973).

3. Model Calculations

For a spherical grain model comprised of an isotropic material of known optical
constants the Mie formulae may be used to calculate optical cross-sections $C_{ext}(a, \lambda)$,
$C_{sca}(a, \lambda)$ and $\gamma = C_{sca}/C_{ext}$ where a is the radius and λ is the wavelength. The measured
extinction $\Delta m(\lambda)$ is then related to $C_{ext}(a, \lambda)$ (or $\bar{C}_{ext}(\lambda)$ for a size-distribution) by

$$\Delta m(\lambda) \propto C_{ext}(\lambda)$$

(see, for example, Wickramasinghe, 1967).

This enables a direct comparison to be made between observation and theory for
spherical grain models.

For grains in the form of 'infinite cylinders' two cross-sections C_E, C_H may be
computed for a given value of a (and prescribed values of optical constants). The sub-

scripts E, H refer to electric and magnetic vectors parallel to the grain axes. The polarization in a simple picket fence-type alignment model is then

$$P(\lambda) \propto C_E(\lambda) - C_H(\lambda).$$

Another quantity calculable for cylindrical grains is a measure of circular polarization for aligned grains defined by

$$\zeta = V/P,$$

where V is the usual Stokes parameter (Martin, 1973). This parameter can also be deduced observationally from studies of sources with strong linear polarization such as the Crab Nebula.

It was hoped for many years that a single grain material – e.g. ice – could explain all the observed optical characteristics of grains. This hope was not realized. Impure ice grains which satisfactorily account for the extinction curve and the wavelength dependence of polarization in the wavelength range $2\,\mu > \lambda > 2600$ Å (Serkowski, 1973) if they are in the form of needles of mean radius $\sim 0.2\,\mu$, fail to account for the

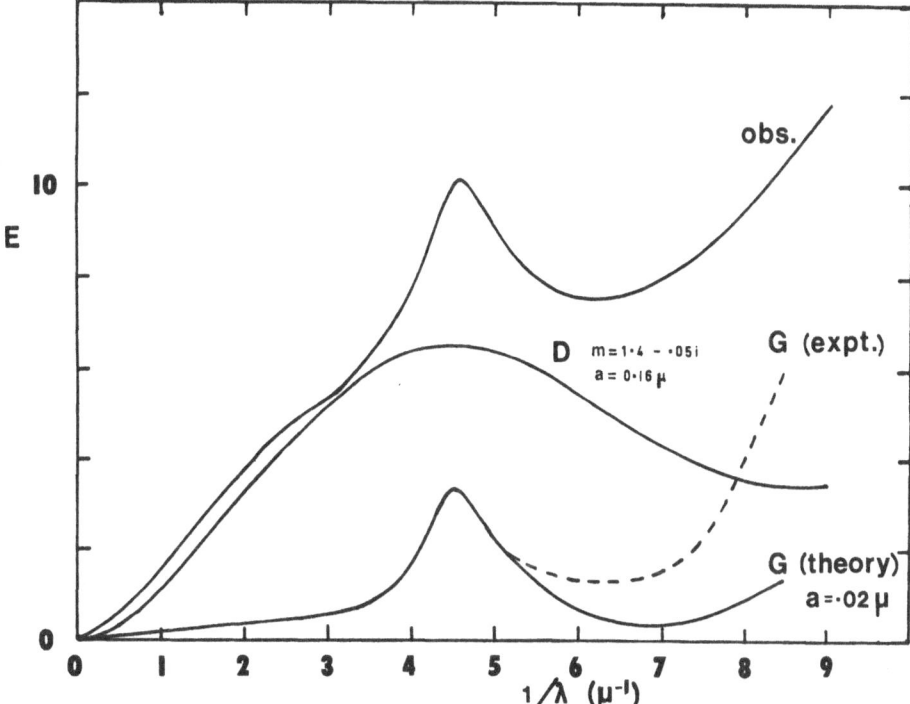

Fig. 2.   Relative contributions to total extinction ('obs') from a 2-component grain model. 'G' stands for graphite spheres of radii $\sim 0.02\,\mu$ 'D' refers to a component consisting of dielectric needles with refractive index 1.4–0.05i. The dashed segment in 'G' included an enhancement factor in the far ultraviolet extinction inferred from measurements of Day and Huffman (1973). This two component model gives good agreement with the observed wavelength dependences of linear and circular polarization as depicted in Figure 1. The mass ratio of graphite to dielectric in this model is $1:7.5$ assuming a unit specific gravity for the dielectric material.

rise in the observed extinction including the 2200 Å hump in the far ultraviolet. Ice grains, or indeed any other types of grain with strongly dielectric properties if they are in the form of slender needles, can also produce the observed wavelength dependence of circular polarization. A similar situation exists for silicate or polyoxymethylene needles of mean radius $\sim 0.15\,\mu$. On the other hand, graphite or iron grains are not able to account simultaneously for the observed extinction and polarization (both linear and circular) in the visible spectral region.

However, strong evidence for the existence of graphite grains is found in the far ultraviolet extinction curve. The observed conspicuous hump in extinction at $\lambda \simeq$ 2200 Å matches closely that predicted on the basis of theory and experiment for small graphite spheres of radii $\lesssim 0.02\,\mu$ (Wickramasinghe and Guillaume, 1965; Gilra, 1972). The steep rise in extinction at $\lambda < 1700$ Å is not reproduced in calculations of extinction by small graphite grains using bulk refractive index data and the Mie formulae. However, the relevance of such a computational procedure may be in doubt (Wickramasinghe et al., 1974). It is interesting to note that experimental results of Day and Huffman (1973) and Huffman (1975) do in fact show both a 2200 Å hump as well as a far-ultraviolet extinction rise (see Figure 2). Thus, there is fairly impressive evidence for the existence of a graphite component of the grains which dominates the extinction in the ultraviolet at wavelengths $\lambda < 2600$ Å.

The mean albedo and phase function of interstellar grains in the optical spectral region tends to rule out a dominant contribution from small metallic (or graphite) grains over this waveband (see Wickramasinghe and Nandy, 1972). Data in the further ultraviolet, including a dip in albedo at 2200 Å, goes in favour of the dominance of a graphite component.

In summary the likely contributors to extinction and other optical characteristics of grains over various wavelength regions is indicated in Table I and shown schematically in Figure 2.

TABLE I

	λ	Main contributor
Extinction		Dielectric
Linear polarization	2–$0.3\,\mu$	needles (D):
Circular polarization		radii $0.16\,\mu$, if $m = 1.4$–$0.05i$
Albedo		
Extinction	0.3–$0.1\,\mu$	Graphite particles:
Albedo		radii $0.02\,\mu$

The curve marked 'D' represents contribution to extinction from a dielectric component of refractive index $m = 1.4$–$0.05i$, in the form of slender needles of radii $0.16\,\mu$. The curves marked 'G' refer to the contribution from graphite spheres of radii $0.02\,\mu$; the solid curve is calculated from bulk optical constants whilst the dashed segment includes an enhancement factor in the far ultraviolet inferred from Day and Huffman (1973).

A composite grain model with these two different types of particle existing separately in the proportion $D:G=7.5:1$ by mass could explain all the available data*. The dielectric component $[D]$ is as yet poorly identified. The existence of a significant H_2O ice component is ruled out by the negative results for the 3.1 μ ice band in the case of several stars which are believed to have strong interstellar extinction (Knacke et al., 1969). Similarly, at least marginal evidence against silicates being the main contributor to extinction is provided by cosmic abundance criteria (Caroff et al., 1973). On the other hand, spectral features at ~ 10 and 18 μ observed for heated dust grains in regions such as the Trapezium nebula may be indicative of a silicate component of dust which could constitute at least part of the component designated 'D' (Woolf and Ney, 1969).

In the view of the author a more likely possibility is that silicate grains ejected from cool stars are suitable sites for the polymerization of 'organic' molecules such as H_2CO (Wickramasinghe, 1974) and that such material could provide a major contribution to 'D'. Formaldehyde polymers which tend to form long whiskers because of their molecular morphology are ideally suited to explaining many of the features of 'D' in the infrared and visual spectral regions (Wickramasinghe, 1975). For example, they are strongly dielectric ($n \simeq 1.4$) in the optical region and possess 10 and 18 μ spectral bands which are in satisfactory agreement with observational data (Cook and Wickramasinghe, 1975; Cooke, 1975). A recent analysis of cometary data indicates that grains producing 10 and 18 μ emission bands in comets have a vaporization temperature of $\sim 500\ K$ – significantly lower than that appropriate to silicates, but in good agreement with that of crystalline organic polymers (Mendis and Wickramasinghe, 1975). If cometary and interstellar dust have a similar origin, a large fraction of interstellar dust may consist of grains of material similar to crystalline organic polymers. My tentative conclusion is that interstellar dust consists in the main of a mixture of graphite grains (nearly spherical) and silicate grains coated with crystalline organic polymers.

* This mass ratio assumes a unit specific gravity for the unidentified dielectric component 'D'.

References

Aannestad, P. A. and Purcell, E. M.: 1973, Ann. Rev. Astron. Astrophys. **11**, 309.

Bless, R. C. and Savage, B. D.: 1972, Astrophys. J. **171**, 293.

Caroff, L. J., Petrosian, V., Salpeter, E. E., Wagoner, R. V., and Werner, M. W.: 1973, Monthly Notices Roy. Astron. Soc. **164**, 295.

Cooke, A.: 1975, Astrophys. Space Sci., in press.

Cooke, A. and Wickramasinghe, N. C.: 1975, Proc. Roy. Astron. Soc. Conference on Far Infrared Astronomy, held at Windsor, England, July 9–11, 1975 (ed. M. Rowan-Robinson), Pergamon, in press.

Coyne, G. V. and Wickramasinghe, N. C.: 1969, Astron. J. **74**, 1179.

Day, K. L. and Huffman, D. R.: 1973, Nature **243**, 50.

Gilra, D. P.: 1972, 'The Scientific Results from the Orbiting Astronomical Observatory (OAO-2)' ed. by A. D. Code, NASA SP-310.

Huffman, D. R.: 1975, *Astrophys. Space Sci.* **34**, 175.

Knacke, R. F., Cudaback, D. D., and Ganstad, J. E.: 1969, *Astrophys. J.* **158**, 151.

Martin, P. G.: 1973, in J. M. Greenberg and H. C. van de Hulst (eds.), *Interstellar Dust and Related Topics*, D. Reidel Publ. Co., Dordrecht-Holland.

Martin, P. G., Illing, R., and Angel, J. R. P.: 1973, *Ibid.*, p. 169.

Mendis, D. A. and Wickramasinghe, N. C.: 1975, *Astrophys. Space Sci.*, in press.

Serkowski, K.: 1966, *Astrophys. J.* **144**, 857.

Serkowski, K.: 1973, in J. M. Greenberg and H. C. van de Hulst (eds.), *Interstellar Dust and Related Topics*, D. Reidel Publ. Co., Dordrecht-Holland.

Wickramasinghe, N. C.: 1974, *Nature* **252**, 462.

Wickramasinghe, N. C.: 1975, *Monthly Notices Roy. Astron. Soc.* **170**, 11–16P.

Wickramasinghe, N. C. and Guillaume, C.: 1965, *Nature* **207**, 366.

Wickramasinghe, N. C., Lukes, T., and Dempsey, M. J.: 1974, *Astrophys. Space Sci.* **30**, 315.

Wickramasinghe, N. C. and Nandy, K.: 1972, *Rep. Prog. Phys.* **35**, 157.

Witt, A. N. and Lillie, C. F.: 1973, *Astron. Astrophys.* **25**, 397.

Woolf, N. J. and Ney, E. P.: 1969, *Astrophys. J. Letters* **155**, L181.

EFFECTS OF CHARGED DUST GRAINS

S. HAYAKAWA

Dept. of Physics, Nagoya University, Nagoya, Japan

Abstract. Dust grains expelled by radiation pressure of stars are charged to potentials in the range 30–40 V in H I clouds. These grains may be responsible for the following phenomena which are otherwise hardly explicable. (1) A considerable fraction of electrons knocked-out by charged grains of high speeds have energies around 15 eV and produce singly ionized ions but not doubly ionized ones in accord with an ultraviolet observation of interstellar atoms and ions. (2) Transverse momentum transferred to grains by Coulomb scattering of ambient electrons and protons is greater than that by multiple scattering of cosmic ray protons, thus the former being more effective for the grain alignment than the latter. (3) At a shock front charge separation due to a large inertial mass of grains produces an electric field, thus accelerating charged particles and causing a drift of interstellar matter.

1. Introduction

It has been suggested by Wickramasinghe (1972) that dust grains are expelled from cool stars into the interstellar medium at speeds comparable to or higher than 10^8 cm s^{-1}, and that such high speed grains may be responsible for heating the interstellar matter and other astrophysical processes. The high speed grains are positively charged to potentials in the range 10–40 V due mainly to secondary electron emission by collisions with ions and atoms (Wickramasinghe, 1974). A number of interesting effects associated with charged dust grains have been discussed by Wickramasinghe (1972, 1974).

There are several more effects of charged grains which seem to be extremely important: (1) Electrons recoiled by charged grains of high speeds may be responsible for ionization of interstellar matter. (2) Streaming grains without spherical symmetry may be aligned by collisions with ambient matter on an argument similar to the alignment by cosmic rays as proposed by Salpeter and Wickramasinghe (1969). (3) Charge separation due to a large inertial mass of grains may take part in the acceleration of particles.

These effects seem to explain relevant astronomical phenomena which are otherwise difficult to understand. However, the results depend rather critically on the assumptions that most dust grains are of high speeds and highly charged.

2. Coulomb Scattering

Let us consider that a particle (mostly an electron or a proton in practical cases) of mass m, charge e and velocity v collides with a spherical grain of charge q and radius a. Since the grain is much heavier than the incident particle, the velocity v is meant by that of the incident particle in the rest system of the grain when the latter is in motion.

For an impact parameter s the scattering angle θ is given by

$$\cot(\theta/2) = 2s/b, \qquad b \equiv 2eq/mv^2, \tag{2.1}$$

where b is the distance of closest approach for a positive particle with $mv^2 < 2eq/a$. Otherwise the minimum impact parameter is given by

$$s_0 = [1 - (b/a)]^{1/2}a, \tag{2.2}$$

and the distance of closest approach is equal to the grain radius.

After scattering the momentum components parallel and perpendicular to the incident direction become $mv\cos\theta$ and $mv\sin\theta$, respectively. Hence the components of momentum transfer are

$$\Delta P_{\parallel} = mv(1 - \cos\theta) = \frac{2mv}{1 + \cot^2(\theta/2)} = \frac{2mv}{1 + (2s/b)^2}, \tag{2.3a}$$

$$\Delta P_{\perp} = mv\sin\theta = \frac{2mv\cot(\theta/2)}{1 + \cot^2(\theta/2)} = \frac{2mv(2s/b)}{1 + (2s/b)^2}. \tag{2.3b}$$

The components of momentum transfer are the same if a grain moving with velocity v collides with a particle at rest. The particle is then recoiled with energy

$$E = \frac{1}{2m}(\Delta P_{\parallel}^2 + P_{\perp}^2) = \frac{2mv^2}{1 + (2s/b)^2}. \tag{2.4}$$

The recoil energy increases as the impact parameter decreases, and for $s = s_0$ it reaches the maximum value

$$E_{\max} = |eq|/2a \tag{2.5}$$

for a negative particle at $mv^2 = |eq|/a$. For a positive particle $E_{\max} = 2mv^2 = 4eq/a$.

The differential cross section for recoil energy E is

$$\frac{d\sigma}{dE} = \frac{2\pi e^2 q^2}{mv^2 E^2}, \tag{2.6}$$

in which the upper limit of E is $2mv^2/[1 + (2s_0/b)^2]$. Since the cross section is inversely proportional to the mass, the recoil effect is important for electrons.

3. Ionization

Direct collision of a fast grain with a neutral atom cannot produce ionization, since the maximum frequency of electric field available by the collision is v/a, which is much smaller than the frequency of orbiting outer electrons of atoms. It is also

difficult to break-up molecules, since v/a is hardly greater than the vibration frequency. Hence, the Coulomb collisions of grains with atoms are elastic, and those
with molecules are quasi-elastic in the sense that rotational levels may be excited.
It may be worth while to mention here that the spectrum of inner bremsstrahlung
associated with recoil electrons is extended only to a frequency comparable to v/a
and is accordingly limited to the infrared region.

Energetic electrons can be produced only by the Coulomb collisions with free
electrons. The maximum energy of the electrons is in the range 15 to 20 eV in H I
clouds, whereas it is lower than 10 eV in H II regions, if reference is made to the
grain potentials calculated by Wickramasinghe (1974). Hence recoil electrons produced in H I clouds may be responsible for ionization of interstellar atoms.

An electron recoiled with energy E ionizes atoms with probability $p(E)$ in competi-
tion with excitation of atoms and collision with free electrons. Since the ionization
cross section per hydrogen atom is a little less than a half of the total cross section
for all processes in H I clouds, we assume

$$p(E) = p_0 \left(\frac{I}{E} - \frac{I^2}{E^2} \right), \qquad p_0 \simeq 1, \tag{3.1}$$

where I is the ionization energy and the energy dependence simulates the classical
ionization cross section. Then the ionization per hydrogen atom is obtained to be

$$\zeta = \frac{n_e}{n_H} f n_g v \int_I^{Em} \frac{d\sigma}{dE} p(E) \, dE$$

$$\simeq \frac{n_e}{n_H} f n_g v \frac{3\pi e^2 q^2}{m_e v^2 I} p_0 \left(\frac{E_m}{I} - 1 \right)^2, \tag{3.2}$$

where n_e, n_H and $f n_g$ are the densities of free electrons, hydrogen atoms and high
speed grains, respectively, and $E_m - I \ll I$ is assumed. Its numerical value is

$$\zeta \simeq 3 \times 10^{-16} \left(\frac{n_e/n_H}{10^{-2}} \right) \left(\frac{f n_g}{10^{-12}} \right) \left(\frac{10^8}{v} \right) \left(\frac{eq/a}{30} \right)^2 \times$$

$$\times \left(\frac{a}{10^{-5}} \right)^2 \left(\frac{p_0}{1} \right) \frac{(E_m - I)^2/I^2}{0.2} \, s^{-1}, \tag{3.2'}$$

where numerical values except for eq/a are expressed in cgs units, whereas eq/a is
in eV. The value of ζ is comparable to the one required for interstellar ionization,
provided that the fraction f of high speed grains is considerable.

It should be emphasized that the ionization mechanism suggested above is incorporated with the interstellar abundances of atoms and ions (Hayakawa, 1974). An
ultraviolet observation by Morton et al. (1973) and by Rogerson et al. (1973) has
indicated that N II is as abundant as N I in spite of the fact that the ionization potential of N I is greater than that of H I, and that doubly ionized ions of metallic elements
are much less abundant than singly ionized ions whose ionization potentials are

above 30 eV. This can hardly result from ionization agencies of continuous energy spectra, such as cosmic rays and soft X-rays. The ionization by high speed grains just meet the requirement, since the recoil electrons have energies lower than 15–20 eV and can hardly produce doubly ionized ions.

4. Grain Alignment

Salpeter and Wickramasinghe (1969) have argued that charged grains with a net stream along the interstellar magnetic line of force may be aligned with their long axes preferentially perpendicular to the magnetic field, if grains are bombarded by isotropic cosmic rays of sufficiently large intensity. Without cosmic ray bombardment only a stream of matter along the magnetic field contributes to alignment, so that grains are preferentially aligned in a direction parallel to the magnetic field, since streaming in other directions is averaged out by gyration. Cosmic ray particles give transverse momentum to grains due to multiple scattering inside the grains, so that the spin axes of grains tend to be parallel to the magnetic field or the long axes are perpendicular to the magnetic field. However, the flux of low energy cosmic rays needed for such alignment is found too large compared with its upper limit obtained from other cosmic ray information.

Transverse momentum is given to grains also by Coulomb scattering. The average value of transverse momentum vanishes but the average square value increases with time as a grain moves with velocity v through interstellar matter. The rate of squared transverse momentum gain is calculated from Equation (2.3b) as

$$
\left(\frac{dP_\perp^2}{dt}\right)_c = nv(2mv)^2 \int_{s_0}^{D} \frac{(2s/b)^2}{[1 + (2s/b)^2]^2}\, 2\pi s\, ds
$$

$$
= 64\pi n \frac{e^2 q^2}{v}\left[\ln\frac{1 + (2D/b)^2}{1 + (2s_0/b)^2} - \frac{1}{1 + (2s_0/b)^2} + \right.
$$

$$
\left. + \frac{1}{1 + (2D/b)^2}\right], \tag{4.1}
$$

where D is the Debye length of an ambient plasma and n the density of charged particles of mass m. This depends on the particle mass mainly through b in such a way that the rate of squared transverse momentum gain increases with mass. Hence protons are more effective than electrons for transverse momentum gain.

The rate of squared transverse momentum gain due to multiple scattering of cosmic ray protons is given by

$$
\left(\frac{dP_\perp^2}{dt}\right)_{CR} = 4\pi r_e^2 \frac{Z^2}{A} Nm_g \frac{m_p c^2}{E} (m_e c)^2 \Lambda_{ms} F, \tag{4.2}
$$

where r_e is the classical electron radius, Z and A are the average atomic and mass numbers of elements in the grain, respectively, N is the Avogadro number, Λ_{ms} is the logarithmic factor in the formula of multiple scattering, and E and F are the average energy and the flux of cosmic rays, respectively.

The ratio of Equation (4.1) to Equation (4.2) is

$$\left(\frac{dP_\perp^2}{dt}\right)_c \bigg/ \left(\frac{dP^2}{dt}\right)_{CR} = \frac{4\pi a^2}{4\pi r_e^2} \frac{16}{Nm_g} \frac{A}{Z^2} \left(\frac{c}{v}\right)^2 \left(\frac{eq}{am_ec^2}\right)^2 \frac{\Lambda_c}{\Lambda_{ms}} \frac{E}{m_pc^2} \frac{nv}{F}, \tag{4.3}$$

where Λ_c represents the square bracket in Equation (4.1). Since this is on the order of nv/F, the Coulomb scattering of protons and electrons is far more effective than the multiple scattering of cosmic rays. Therefore, the Coulomb scattering of ambient charged particles must be responsible for the alignment mechanism proposed by Salpeter and Wickramasinghe (1969) rather than cosmic rays, if this mechanism works.

Since the alignment parameter due to high speed grains alone is about $-\frac{1}{2}$, the alignment mechanism discussed above is appreciable if about 10% of interstellar grains are of high speeds.

5. Electric Field Induced by Grains

If a stream of matter consisting of gas and grains collides with a cloud which may or may not have a magnetic field, the gaseous part and grains are separated because of large inertia of the grains, as indicated by the dark lane along spiral arms of galaxies. Since grains are positively charged, this causes charge separation and accordingly produces an electric field.

The field strength is estimated by considering that the electric field produced by grains is relaxed by the current flowing through an ambient plasma. If the charge separation is caused by grains of charge q, density n_g moving with velocity v perpendicular to a magnetic field, the field strength is given by

$$\mathscr{E} = qn_gv/\sigma_\perp, \tag{5.1}$$

where

$$\sigma_\perp = \sigma v_c^2/(\omega_g^2 + v_c^2) \tag{5.2}$$

is the electric conductivity in the direction perpendicular to the magnetic field but parallel to the electric field. The gyration frequency ω_g and the collision frequency v_c of electrons are given by

$$\omega_g = eB/m_ec = 1\cdot76 \times 10^7 B \text{ rad s}^{-1},$$

$$v_c = 0.90n_eT_e^{-3/2}A \text{ s}^{-1}, \tag{5.3}$$

where B is the magnetic field strength, n_e the electron density, T_e the electron temperature, and A the logarithmic factor of order 10. In ordinary conditions of H I clouds ω_g is much larger than ν_c, and therefore $\sigma_\perp$ is much smaller than σ. Since

$$\sigma\nu_c \simeq 1.3 \times 10^8 n_e,$$

we obtain $\sigma_\perp$ to be

$$\sigma_\perp \simeq 0.4\left(\frac{n_e}{3 \times 10^{-2}}\right)^2 \left(\frac{10^2}{T_e}\right)^{3/2} \text{s}^{-1}. \tag{5.4}$$

Substituting this value in Equation (5.1) we obtain

$$\mathscr{E} \sim 10^{-7}\left(\frac{q/a}{30 \text{ V}}\right)\left(\frac{n_g}{10^{-12}}\right)\left(\frac{v}{10^8}\right)\left(\frac{1}{\sigma_\perp}\right) \text{V cm}^{-1}. \tag{5.5}$$

If the field strength persists as long as the gyration radius of a charged grain, which is estimated to be about 3×10^{16} cm, the potential difference of about 10^9 V may result. This can accelerate charged particles and can be also responsible for a drift of interstellar gas in the northern or southern direction depending on the sense of magnetic field. The electric field produced by grains may therefore account for motions of the interstellar matter perpendicular to the galactic plane which are likely to be associated with interstellar shocks (Sofue and Tosa, 1974).

6. Discussion

Astrophysical significance of the high speed charged grains discussed above depends critically on how large is the fraction of such grains. The possibility that grains are expelled by radiation pressure of stars has been demonstrated by Salpeter (1974). Following his arguments we estimate the speed and flux of grains expelled.

There are three critical luminosities of stars for lift-off of material near the photosphere. For the luminosity L greater than a critical value $L_{cr,z}$ both the gas and the grains are expelled, whereas $L_{cr,p}$ is a critical luminosity above which pure grains can be expelled. If L is not much smaller than $L_{cr,z}$, the grains accelerated by radiation pressure are subjected to gas drag, so that the drift velocity of the grains is suppressed. If the luminosity is lower than $L_{cr,l}$, the grains are decoupled with the gas and escape with considerable speeds. These critical luminosities are estimated to be

$$L_{cr,p} \simeq z^{2/3}L_{cr,l} \simeq zL_{cr,z} \simeq L_\odot, \tag{6.1}$$

where $z \simeq 10^{-3}$ is the mass fraction of material condensable to grains. The grain speed at infinity is given by

$$v \simeq (L/L_{cr,p})v_e, \tag{6.2}$$

where v_e is the escape velocity. The rate of mass loss is estimated to be

$$\phi \simeq zQL/uc, \tag{6.3}$$

where Q is the momentum transfer efficiency factor of the grains for stellar radiation and u the thermal velocity of the gas.

Stars with luminosities of $L_{cr, l} \simeq 10^2 L_\odot$ are giants, and the escape velocity therefrom is about 300 km s^{-1}. Hence Equation (6.2) gives the grain speed exceeding 1000 km s^{-1}. For such giants the thermal velocity is $u \simeq 5$ km s^{-1}, and therefore the loss rate is estimated to be $\phi \approx 3 \times 10^{16} Q(L/L_{cr, l})$ g s^{-1}. The time scale for the high speed grains to be decelerated in an interstellar medium of hydrogen density n_H is as large as

$$t_d \simeq \frac{4}{3} \frac{a\varrho}{m_H n_H v} \simeq 6 \times 10^{11} \left(\frac{a}{10^{-5}}\right)\left(\frac{\varrho}{3}\right)\left(\frac{10^8}{v}\right)\left(\frac{0.5}{n_H}\right) \text{ s}, \tag{6.4}$$

where ϱ is the density of grain material. Since this is smaller by a factor of about 10^4 than the retention time of interstellar grains, only a minor fraction of total grains would have high speeds unless no acceleration of grains could take place. Even without acceleration most grains in the vicinity of the grain producing stars are of high speeds. The distance within which most grains are of high speeds is given by

$$r \simeq (3\phi t_d/4\pi\varrho_g)^{1/3} \simeq 10^{18} \text{ cm}, \tag{6.5}$$

where $\varrho_g \simeq 10^{-26}$ g cm^{-3} is the average mass density of grains in interstellar space. Hence the effects discussed in the present paper may be appreciable at least within 1 pc from the sources. If charged grains are accelerated in interstellar space (Wickramasinghe, 1974), the region in which they show appreciable effects may be larger.

Acknowledgement

I thank Prof. N. C. Wickramasinghe for his prompt information on his theory of grain charge and for enlightening discussions.

References

Hayakawa, S.: 1974, *Astrophys. Space Sci.* **31**, L13.
Morton, D. C., Drake, J. F., Jenkins, F. B., Rogerson, J. B., Spitzer, L., and York, D. G.: 1973, *Astrophys. J. Letters* **181**, L103.
Rogerson, J. B., York, D. C., Drake, J. F., Jenkins, F. B., Morton, D. C., and Spitzer, L.: 1973, *Astrophys. J. Letters* **181**, L110.
Salpeter, E. E.: 1974, *Proc. of the Symposium on Solid State Astrophysics*, held at the University College, Cardiff, Wales, between 9–12 July, 1974; printed in *Astrophys. J.* **193**, 579, 585.
Salpeter, E. E. and Wickramasinghe, N. C.: 1969, *Nature* **222**, 442.
Sofue, Y. and Tosa, M.: 1974, *Astron. Astrophys.*, in press.
Wickramasinghe, N. C.: 1972, *Monthly Notices Roy. Astron. Soc.* **159**, 269.
Wickramasinghe, N. C.: 1974, *Astrophys. Space Sci.* **28**, 525.

LARGE-SCALE IONIZATION FRONTS AND THE NATURE AND DISTRIBUTION OF LIGHT SCATTERING PARTICLES IN THE ORION NEBULA

MICHAEL A. DOPITA

Dept. of Astronomy, University of Manchester, England

SYUZO ISOBE*

Astronomisches Rechen-Institut, Heidelberg, Germany

and

JOHN MEABURN

Dept. of Astronomy, University of Manchester, England

Abstract. Image-tube filter photographs calibrated against photoelectric filter photometry have been used to give maps of M42 in absolute flux units over the central 15 arc min of the nebula in Hα, [N II] (λ 6584 Å), Hβ and continuum at λ 4700 Å. Maps of the ratios Hα/[N II] and (for the first time) of continuum/Hβ have been produced with unprecedented spatial resolution. These show that the gas to dust ratio is high near the exciting stars and falls strongly in the vicinity of large scale ionization fronts marked by minima in the Hα/[N II] ratio.

These results are interpreted in terms of detailed shell models containing either ice or graphite or silicate scattering particles. In all models there must be a central hole in the distribution of scattering particles. The effect of neutral globules and intrusions is investigated. It is found that all types of grain are trapped inside neutral intrusions near the centre of the nebula by the pressure of the Lα light surrounding the globule, but in the early evolution of the nebula particles can escape into the ionized medium when fronts are R-type. Ice grains escaping at this time will be destroyed for distances to the exciting stars less than 1 pc. These results can explain both the central hole in dust and the underabundance of oxygen in the ionized gas observed earlier. Arguments depending on colour index of the scattered light indicate that mixtures of scattered light from ice in the globules and from ice in the ionized medium can explain the observations, but that the graphite and silicate particles fail.

A schematic model of the Orion Nebula is presented to attempt to explain the large scale phenomena observed here. It demonstrates that simple shell models for this nebula are dubious.

1. Introduction

There have been many 'monochromatic' photographic studies of the Orion Nebula, notably that by Wurm and Rosino (1959). Unfortunately, much of the early work was done with plate-filter combinations which were at best 200 Å in effective band-width. This has meant that attempts at determining absolute fluxes by photographic means have been severely hampered by contamination by other lines and nebular continuum

---

* At present at Tokyo Astronomical Observatory, Mitaka, Osawa Z-21-1, Tokyo, Japan.

let through the filters (Schmitter and Recillas-Cruz, 1971; Isobe *et al.*, 1972). For accurate determinations we have had to rely on a limited number of photoelectric measurements made with either narrow-band interference filters (Reitmeyer, 1965, O'Dell and Hubbard, 1965) or else scanner photometers (Peimbert and Costero, 1969; Simpson, 1973).

The development of the image tube has led to some filter photographs (Elliott and Meaburn, 1974a, b) but these were only of the four arc minute core of M42.

The ratio of Hα/[N II] has been measured in M42 previously by Fabry-Perot techniques (Foukal, 1969) and later extensively by Baudel (1970) using a multislit spectrograph. This represented the first attempt to map a ratio of lines across the nebula, but unfortunately the technique meant that the image was badly undersampled, and many smaller scale features of the variation have been lost. Dopita (1973) at low angular resolution in M42 has measured this ratio using photoelectric filter photometry in a single strip.

Due to the extreme faintness of the continuum measurements of the continuum to Hβ ratio have been very limited in scope (five by O'Dell and Hubbard (1965) and six by Simpson (1973)). The present paper represents the first attempt to produce detailed maps at reasonable angular resolution of both these ratios across a large part of the nebula. These are interpreted in terms of a nebular model.

2. Experimental

The photographs in the light of Hα, Hβ, [N II] λ 6584 Å and nebular continuum centred at λ 4700 were taken with the device shown in Figure 1 at the Athens National Observatory.

The filters, whose parameters are shown in Table I, were placed non-classically (Meaburn, 1970) near the focal plane of a 25.4 cm aperture f/6.5 Fecker-Ross lens, and (because of the small bandwidth of the emission line filters) thermally tuned to their optimum position.

The nebular image was formed on the photocathode of a Westinghouse WL phosphor output image tube and finally on baked III a J emulsion pressed against the

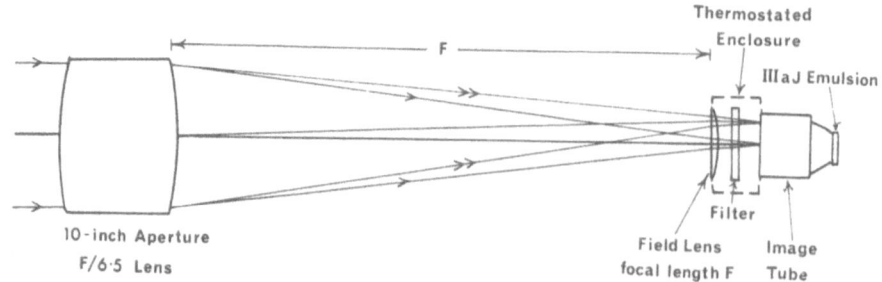

Fig. 1.   A schematic representation of the image tube camera system used in taking the photographs.

TABLE I

The filter and plate exposure parameters used in this study

Spectral feature	Hα	Hβ	[N II] λ 6584 Å	Continuum λ 4700 Å
Filter parameters:				
(1) Equivalent width	6.70Å±0.10Å	5.18Å±0.05Å	9.25Å±0.10Å	50.3Å±0.6Å
(2) Maximum transmission T	0.608±0.008	0.497±0.002	0.710±0.010	0.480±0.005
Exposure times	7 min	30 min	20 min	18 min

fibre optics of the output window. This filter and image tube system attached to a telescope has been fully described previously (Meaburn, 1975). The performance of such a device for nebular photometry has also been analysed (Elliott and Meaburn, 1974a).

The field of the camera in Figure 1 is 52′ and its resolution is 8″. The camera was pointed in the same direction in the sky for each of the photographs (the exposure times of which are given in Table I) to an accuracy of better than 8″. This eliminated (Elliott and Meaburn, 1974a) the large effects of the variations in the quantum efficiency of the photocathode in the measurement of ratios of brightnesses.

The field lens in Figure 1 images the entrance pupil of the system on infinity. This ensures that all parts of the field are considered equally by the filters (Meaburn, 1971).

The photoelectric observations were made using the same filters at the Pic-du-Midi Observatory with the same instrument, beamwidth (1′) positions on the sky and experimental precision as discussed by Dopita (1973). However, the reduction to absolute units was accomplished as follows. The planetary nebula IC 418 was considered as a standard emission line object and the flux given by Peimbert and Torres-Peimbert (1971) was used. For the continuum measurements the stars HD 35640 (B9V, $M_v = 6.23$) and HD 39927 (A0, $M_v = 6.28$) (Cousins, 1971) were used as standard continuum sources. These stars were also observed in the emission line measurements to monitor the sky transmission variation. The calibration of Oke and Schild (1970) for A0 stars was used to give absolute flux units. Thus systematic errors due to uncertainties in filter parameters are eliminated. Absolute flux measurements should be reliable to within five percent (three percent systematic, two percent random).

3. Reduction of the Observations

Isodensity maps of identical regions were made of all plates using a Joyce Loebl MK 3C isodensitracer at a 50× magnification. The following procedure, although not ideal, makes use of a considerable fraction of the data on the plates. The analysing aperture was set up at 50 μ on the plate which corresponded to 8″ on the sky (similar to limit imposed by resolution and guiding). Thus effective resolution is about 15″. Reduction of densities above fog to intensity using the photoelectric measurements is difficult because the lower resolution (1′) of the photoelectric measurements means that convolution effects cannot be neglected in regions where the brightness changes rapidly. To correct for this, the characteristic of the emulsion was assumed linear on a density log intensity plot (to first approximation). The isodensity map was then convolved up numerically in each of the beam positions by dividing the beam up into seven equal areas and finding the mean intensity in each of these areas using the above assumption. The mean intensity was found and this was reconverted to a corrected density figure.

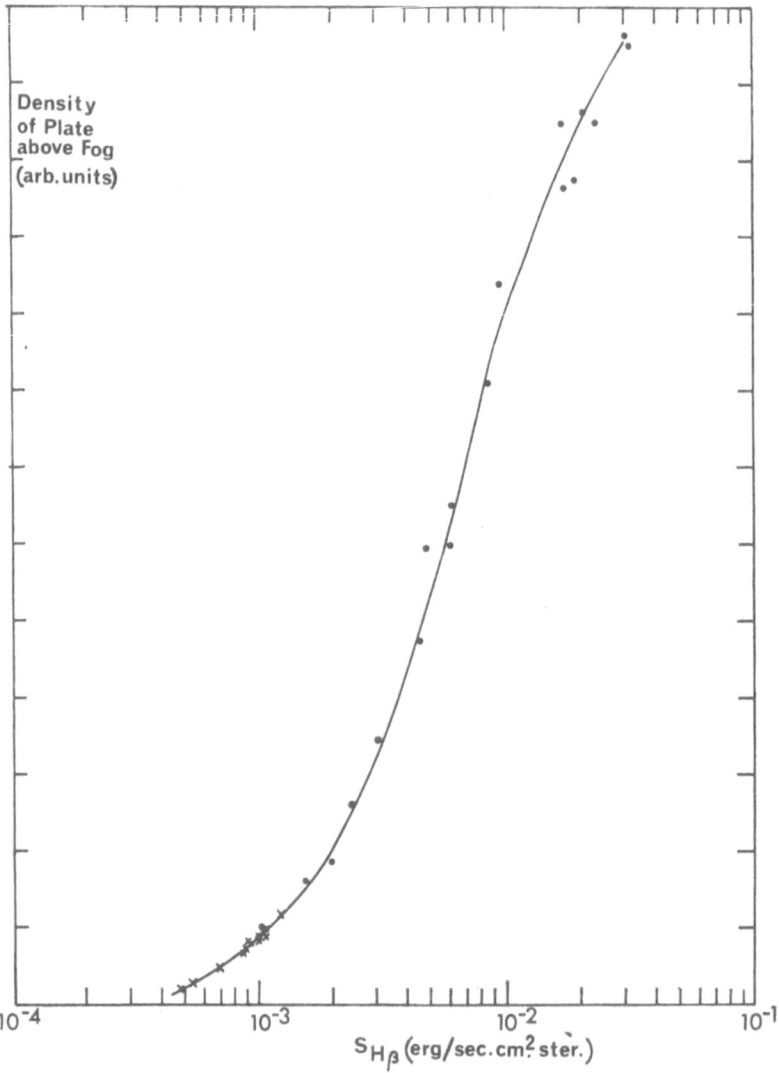

Fig. 2.   The plate characteristic for Hβ derived from direct calibration of the nebular photograph against photoelectric measurements made in the nebula.

This computed density was then plotted against the photoelectrically measured absolute intensity.

The characteristic resulting from this rather unusual reduction procedure is shown in Figure 2 for the case of the Hβ photograph. The scatter is less than 5% over the region of interest. The crosses at the lower end are measurements obtained over a region where the Hβ flux was changing slowly enough for convolution effects to be negligible.

Errors in the density-intensity relationship may be caused by errors in the convolution, and point to point variation of the background fog level, filter transmission and photocathode sensitivity of the image tube. Since the calibration is made over the whole nebula (approximately 8 mm on the plate), characteristics such as Figure 2 prove that all these effects give less than 5% error overall.

Hence, the total error in the assigned absolute fluxes of the density contours is less than 10%, consisting of 7% random and 3% systematic.

Ratios of fluxes could now be simply found by superimposing one isodensity map on top of another, and measuring the ratio wherever two density contours happened to cross. These values have errors due to photocathode sensitivity eliminated, because the nebular image was in exactly the same position on the photocathode for all the photographs. Thus errors in ratios are estimated at 18%, 12% and 6% systematic except in the region of stars where the continuum intensities have to be estimated. Near the central stars ratios were not measured due to image spreading of the stars in the continuum photograph, and saturation of the plate at Hα.

Maps of the measured ratios could be constructed to a high degree of accuracy because where the ratios change rapidly, points at which they are determined are crowded close together. In other regions, the determinations are spread out thinly and so the plate is undersampled. Undersampling leads to a greater degree of uncertainty in the exact position of the contour, but in these regions this is not so important.

Undersampling also becomes more severe in the faintest parts of the nebula due to the curvature of the plate characteristic here. However, the influence of plate noise is more severe in this region so that determinations have a low weight. In general, contours are only drawn out as far as the second contour above the fog level.

A total of more than 2300 points have been measured in the determination of the Hα/[N II] ratio map, and over 800 points in the continuum/Hβ ratio map. In the latter map, there was a small correction applied *a posteriori* to correct for contamination of the continuum filter by Hβ and vice versa.

If the Hβ filter has an equivalent width for transmission of continuum of $\Delta\lambda_H$, transmits a fraction T_H of Hβ, and if the corresponding values for the continuum filter are $\Delta\lambda_C$ and T_C respectively, the measured fluxes are

$$S_H^* = S_H + S_C(\Delta\lambda_H/T_H),$$
$$S_C^* = S_C + S_H(T_C/\Delta\lambda_C),$$

where S_H^*, S_H are the measured and true fluxes of Hβ respectively and S_C^*, S_C are, respectively, the measured and true fluxes of the continuum.

Provided that contamination effects are small, and letting A^* and A be the uncorrected and corrected values of the ratio of continuum flux to $H\beta$ (Å^{-1}) then

$$A^* = A[1 + (T_C/A \, \Delta\lambda_C) - (A \, \Delta\lambda_H/T_H)]$$

to first order.

For the filters used $\Delta\lambda_H = 5.18$ Å, $T_H = 0.497$, $\Delta\lambda_C = 50.3$ Å and $T_C = 0.0144$. Thus the corrections needed to be applied to A^* range from -17% at $A^* = 1.66 \times 10^{-3}$ to $+7.6\%$ at $A^* = 1.0 \times 10^{-2}$. From this point we define A to be equal to $1000 \, S_{4700 \, \text{Å cont}}/S_{H\beta}$ ($10^3 \, \text{Å}^{-1}$).

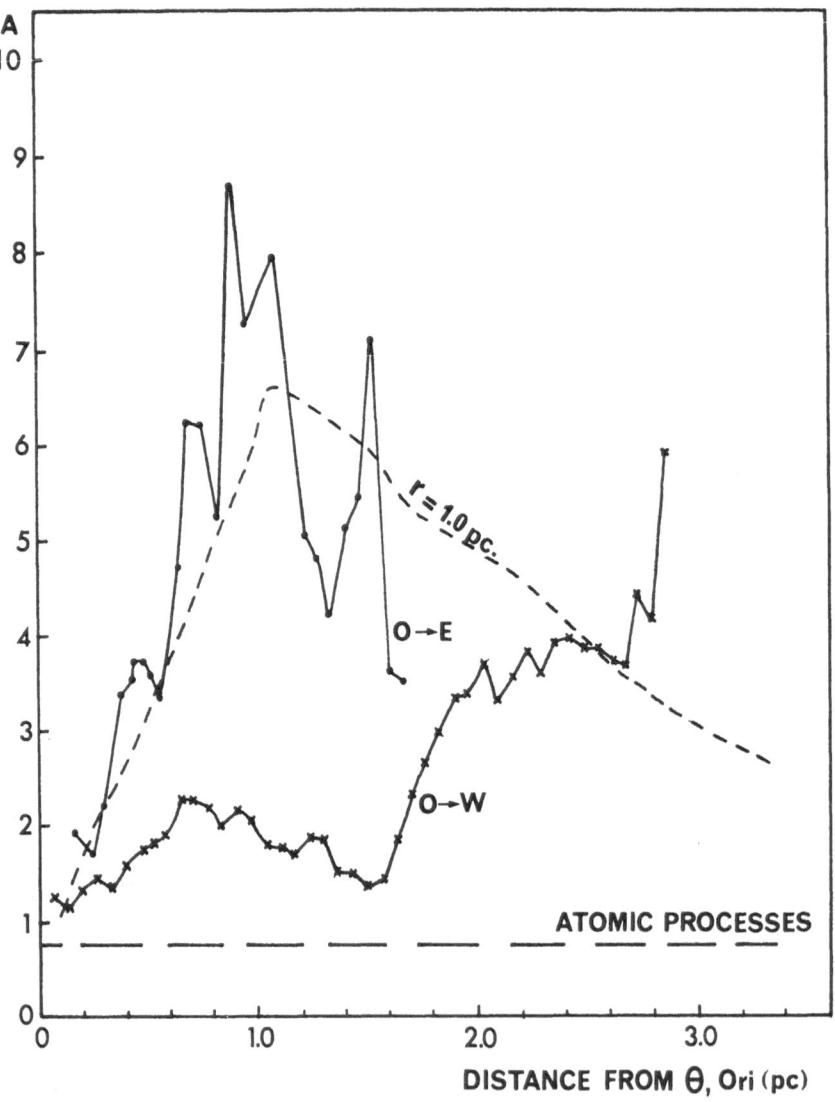

Fig. 3. The photoelectric measurements of A ($1000 \, S_{4700}/S_{H\beta}$) compared with a spherical model in which there is a central hole in composite grain distribution 1 pc in radius. The approximate level of atomic processes contributing to the continuum is also shown.

4. Results

The photoelectric results are shown in Figure 3. A clear minimum in the scattered light intensity is shown near the dominant exciting star θ^1 Ori. In the easterly direction there is a strong rise in A as the projected beam passes onto the dark lane. In the western part there is a similar rise in A, but this occurs further from the centre than the local rise in temperature and decrease in ionization that was taken (Dopita, 1973) to indicate the presence of an ionization front. This increase in the value of A in the outer regions agrees qualitatively with that of O'Dell and Hubbard (1965). The large majority of this scattered light must be caused by dust particles. At a temperature of 10 000 K the contribution of atomic processes to A is calculated at less than 0.7. Electron scattering does not become important even close to the centre where the density is above 10^4 cm^{-3} (Danks and Meaburn, 1971). Tenorio-Tagle (1973 and private communication)

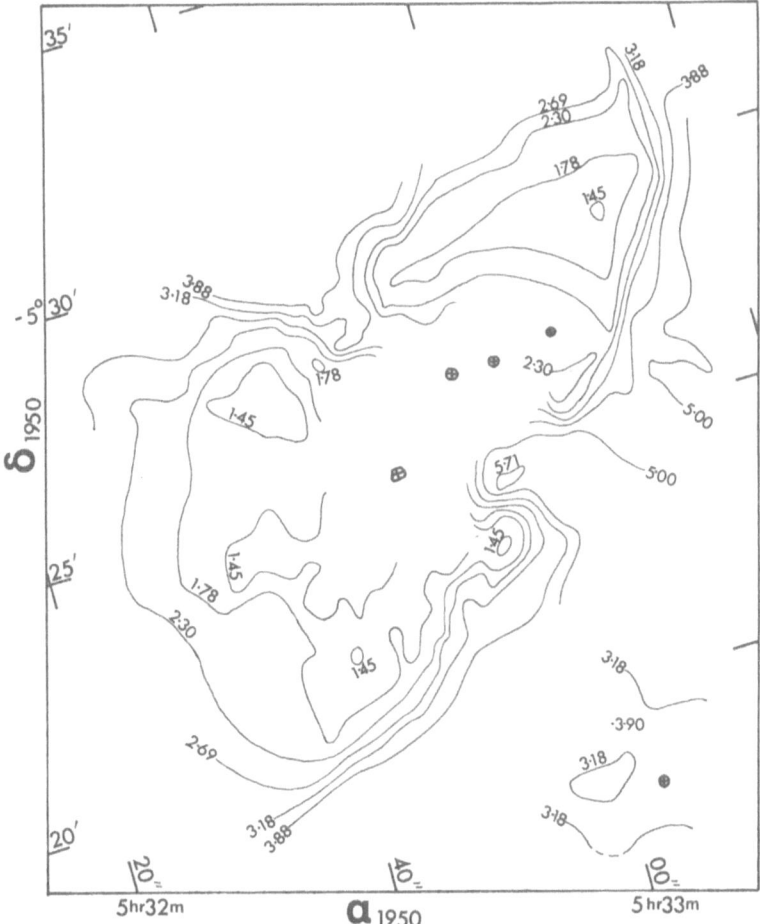

Fig. 4.   A map of A in absolute units derived from over 800 measurements on the photographs.

has shown that it is always less than 1 or 2% of the atomic processes contributing to continuum light.

The map of scattered light continuum A resulting from taking the ratio of the image tube photographs is shown in Figure 4. This clearly shows the irregularity of the zone within which A is strongly enhanced, and there is a strong correlation between enhancement of scattered light and the presence of dark lanes in the nebula.

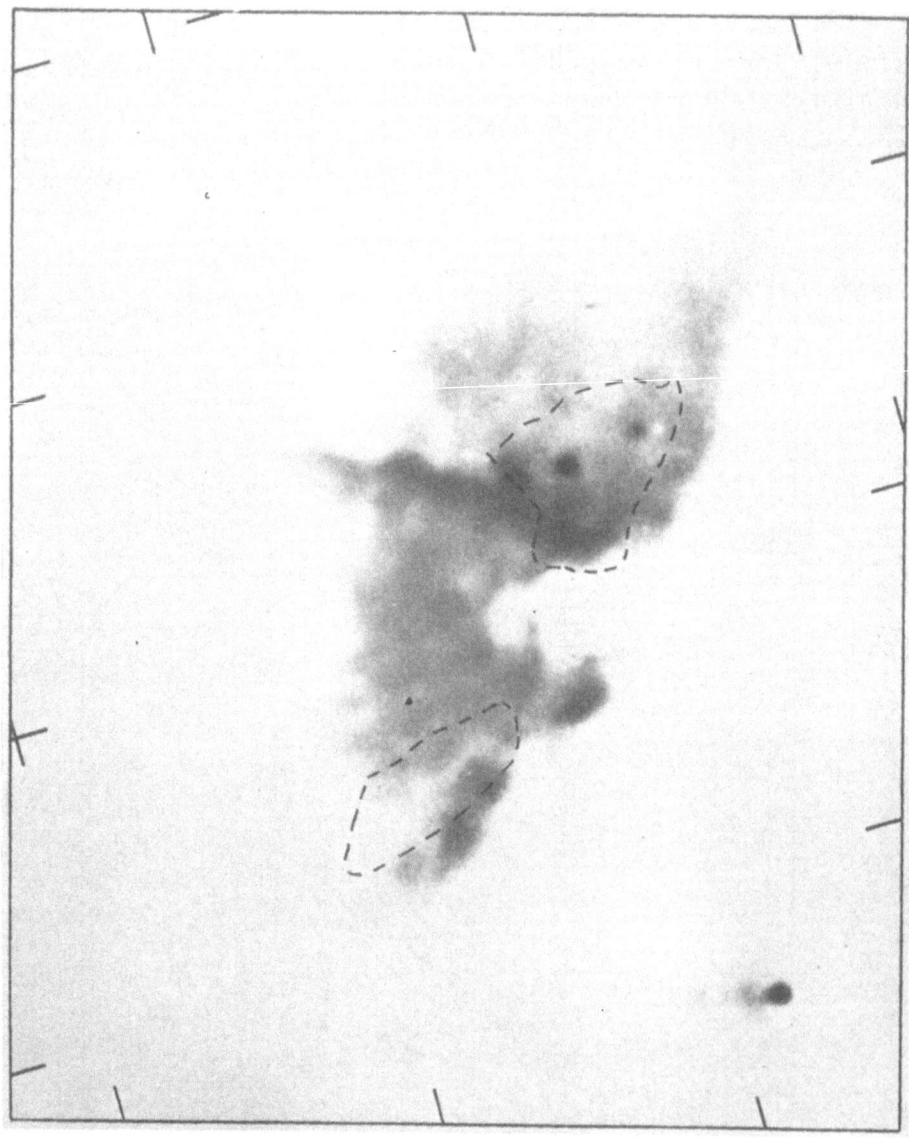

Fig. 5.   A reproduction of the [N II] λ 6584 photograph showing the bright knots and bars near ionization fronts. Dashed lines surround regions of [N II] line splitting (Deharveng, 1973).

The [N II] λ 6584 Å photograph on the other hand shows a series of bright knobs and bars on a spatial scale of about 1' (Figure 5). These appear on the map of Hα/[N II] ratio in Figure 6 as minima in this ratio. For M42, it has already been shown (Dopita, 1973; Elliott and Meaburn, 1974b) that variations in this ratio are caused more by variations in the degree of ionization than by changes in temperature. Since [N II] emission comes predominantly from those regions where hydrogen is ionized but helium is neutral, i.e. near ionization fronts, we regard these ridges of [N II] emission as delineating bright rim features. The knottiness of the [N II] radiation along these minima in the Hα/[N II] radiation probably reflects the irregularities in the density of the ionized gas near the ionization front and hence irregularities in the density of the neutral material. These globules of neutral material are presumably detached from the main mass as the ionization front eats in and become partially

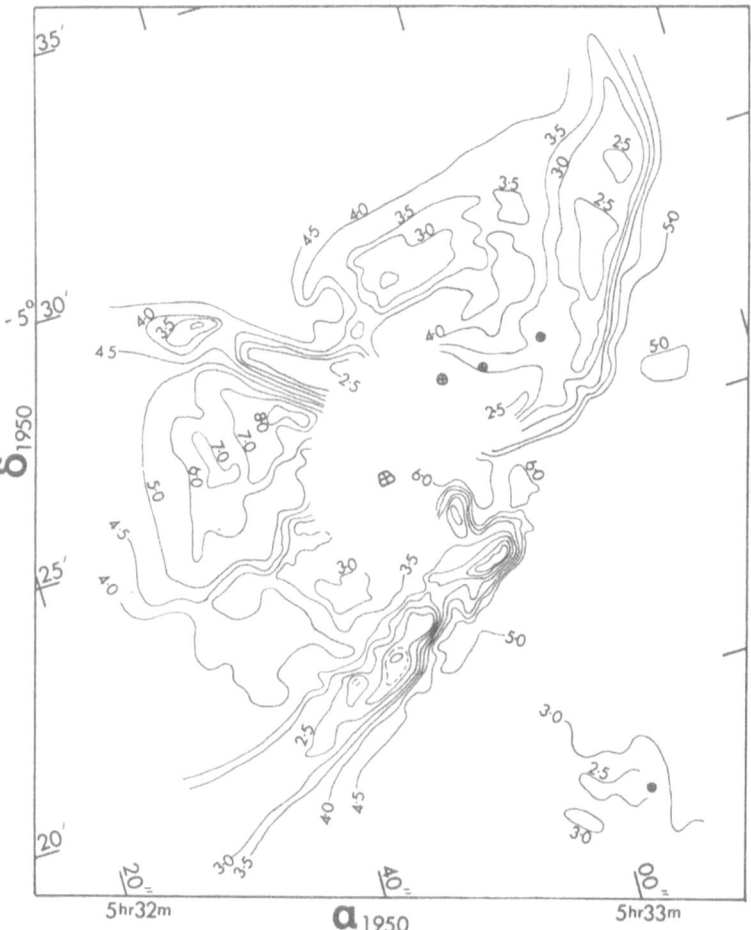

Fig. 6.   A map of the Hα/[N II] λ 6584 ratio in absolute units derived from over 2300 measurements on the photographs.

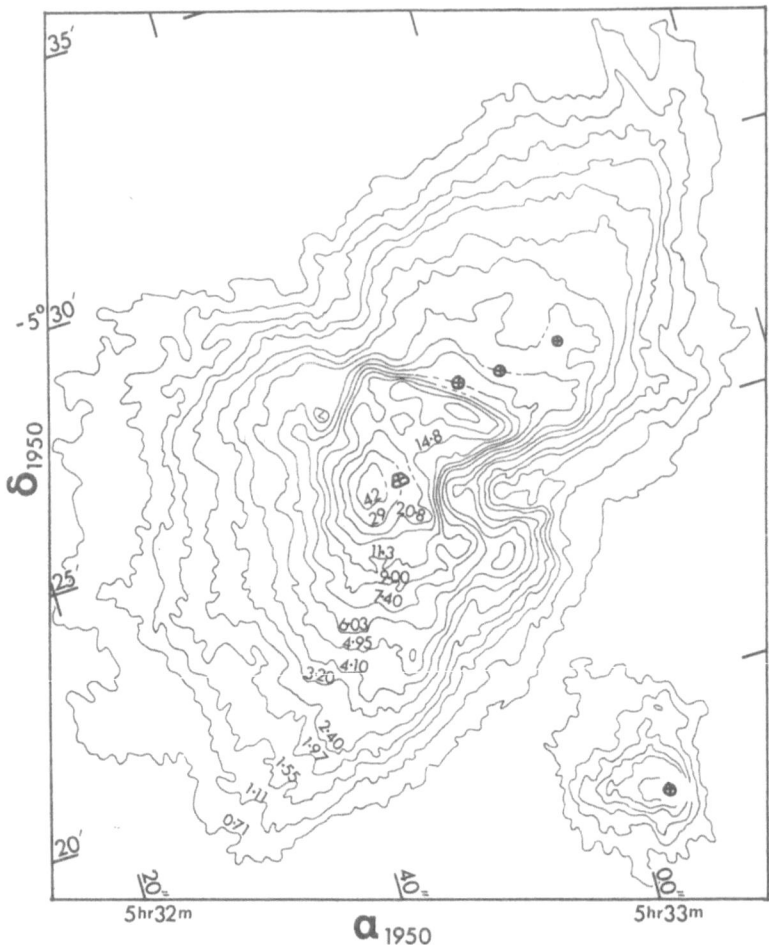

Fig. 7.   The absolute flux in Hβ derived from the photograph. For clarity, approximately half the
measured contours have been omitted. Contour units are erg sec^{-1} cm^{-2} ster^{-1} × 10^3.

ionized globules (PIG's). This is schematically represented in Figure 9. The photographs by Elliott and Meaburn (1974a) of the core regions of the nebula show that in
the centre the linear dimensions of these globules are smaller and their densities are
higher. Flow from ionization fronts in these central regions must account for the
vast majority of the ionized material in the nebula because the Hβ brightness map
(Figure 7) shows a very pronounced central maximum.

The whole complex of ionization fronts appears to form an interconnected structure
and this is taken as evidence to support the view that the exciting stars lie in front of a
neutral mass of a variable density (reaching a maximum at the point nearest to the
exciting stars) and with an extremely lumpy and folded surface, again represented in
Figure 9.

Large scale splitting of the [O II], [O III] and [N II] line has previously been found in many regions of the nebula (Wilson *et al.*, 1959; Meaburn, 1971; Dopita, 1971; Dopita *et al.*, 1973a; Dopita *et al.*, 1973b). Regions in which the N II splitting occurs have recently been mapped from superb interferograms by Mme. Deharveng (1973). These are shown as stippled areas in Figure 8. The minima in the $H\alpha/[N II]$ emission are also plotted on this figure as dotted lines. The splitting occurs inside but away from these minima in regions where the [N II] emission is comparatively faint. It is thought possible that this splitting is caused by the chance superposition along the line of sight of two separate gas flows streaming off ionization fronts. The idea of a simple Strömgren sphere cannot be supported on these observations since the state of ionization of the faint material further out from the exciting star than the ionization front is the same as the brighter material within the front. This suggests that faint material is foreground gas illuminated directly by the ionizing star – and the nebula is probably density bounded in this direction.

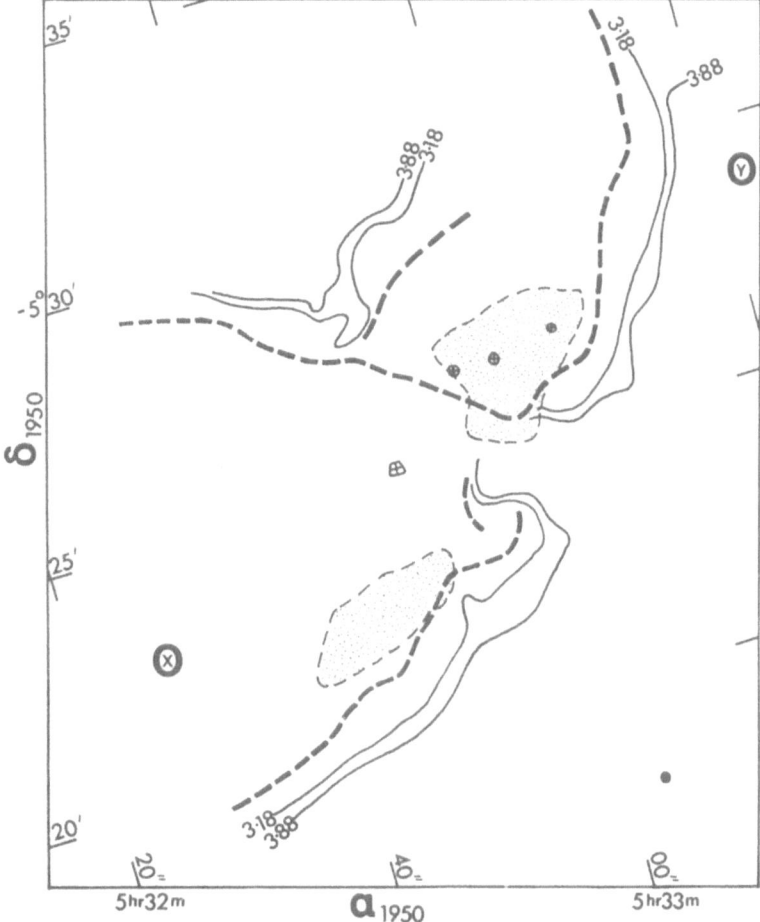

Fig. 8.   The correlation between [N II] line splitting (Deharveng, 1973), maximum [N II] emission relative to $H\alpha$ (dotted lines) and the measured A values (for which two contours are shown).

The intensity of the scattered light continuum increases strongly further out than the ionization fronts and its intensity is strongly correlated to the fronts. This is also shown on Figure 8. Hence the presence of dust is connected to the presence of the large scale fronts in the outer regions of the nebula. This could be caused by two effects, either the dust within the neutral clouds is directly scattering the light or else the scattered light is being produced by dust escaping into the ionized region from the neutral intrusions and is being pushed away from the star by radiation pressure and large-scale gas flows. We will attempt to distinguish between these alternatives in the following section. Figure 9 represents a highly simplistic picture of the large scale neutral intrusions across the nebula derived from these and other observations.

5. Discussion of Scattered Light Observations

There are several important candidates for interstellar grains, namely ice grains, graphite grains, silicate grains and iron grains. Various mixtures of these successfully reproduce observed interstellar extinction curves by adjusting abundance ratios and/or the size distribution of the grains. These include graphite-silicate-iron grain mixtures (Nandy and Wickramasinghe, 1971), enstatite silicate grains (Huffman and Stapp, 1971), graphite-silicate-silicon carbide (Gilra, 1971) and graphite-graphite core ice mantle (hereafter called composite) grains (Isobe, 1973).

However, all grains except ice (and ice mantle) grains are stable in the conditions encountered in interstellar space. On the other hand, the formation (and growth rate) and the destruction rates of ice and ice mantle grains depends strongly on local conditions (Mathews, 1967; Wickramasinghe and Williams, 1968; Isobe, 1972). It seems therefore more reasonable to explain the observed variation of interstellar extinction curves in terms of varying lines of sight abundance and size distribution of ice grains than in terms of variation in abundance and sizes of more stable grains (Isobe, 1973). Support for the existence of ice and of silicate grains comes from observations of infrared absorption features at $3.1\,\mu$ and $9.6\,\mu$ respectively (Gillett and Forrest, 1973) whilst graphite grains seem fairly essential to explain the feature at $\lambda^{-1} \approx 4.6\,\mu^{-1}$ (Isobe, 1973).

Evidence to indicate that the particles producing visible polarization are dielectric comes from the circular polarization data of Martin, Angel and Illing (1973) and of Kemp (1973). Greenburg (1968) has shown that cylindrical ice grains of mean radius $\approx 0.2\,\mu$ explain the observations of linear polarization better than other types of particle with the same shape.

In this section we will investigate to what extent the observed scattered light serves to limit the possible grain mixtures that may be present in the ionized gas and in the neutral inclusions in the Orion Nebula.·

The Hβ isophotes (Figure 7) show that a spherical shell model for the nebula is certainly not applicable. However, in the absence of any precise knowledge of the true situation (see Figure 9) a shell model represents the only theoretical model that we

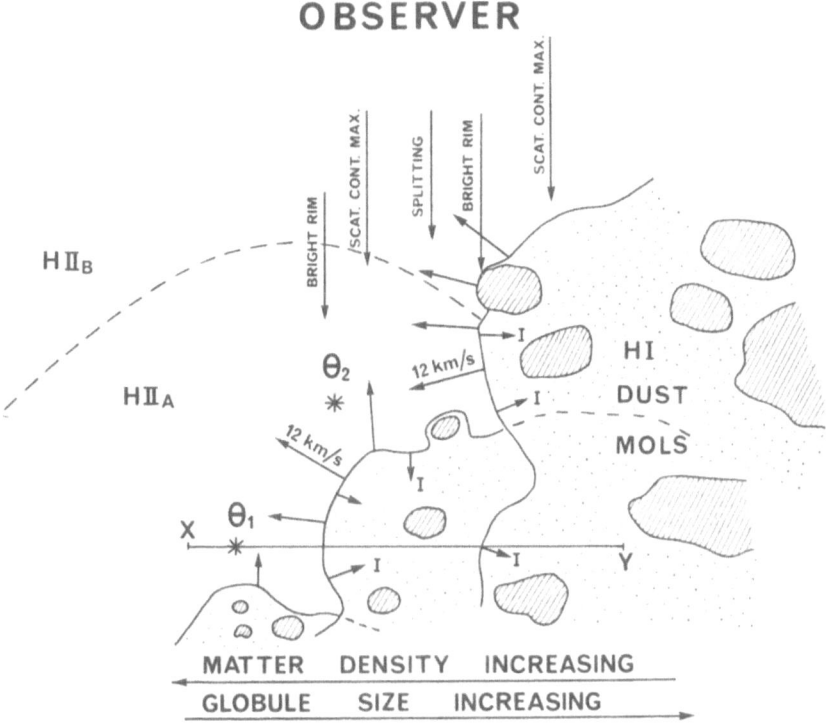

Fig. 9.   The derived schematic nebular model along the axis X, Y marked on Figure 8. H II$_A$ and H II$_B$ represent the regions repectively free of, and containing scattering particles in, the ionized gas.

can construct to correct local quantities to integrated (line of sight) quantities observed. It must be borne in mind that the results given by such an analysis will be valid only in a general way away from large scale neutral inclusions of matter.

We adopt as our density model a local root mean square density as follows:

$$N_e(r) = 262/r, \quad r \leqslant 0.45 \text{ pc};$$
$$N_e(r) = 130/r^2, \quad 0.45 \text{ pc} < r < 3.0 \text{ pc};$$

(1)

which satisfactorily fits Menon's (1961) detailed density distribution with the distance of M42 adopted as 450 pc (Mathews, 1967). To account for the extended envelope to the Orion Association we take

$$N_e(r) = 15, \quad 3.0 \text{ pc} < r \leqslant 5.0 \text{ pc}.$$

Using this density we calculate the surface brightness of Hα and Hβ and the intensity of the scattered continuum assuming various types and distribution of grains, on the assumption of negligible multiple scattering of light.

If r_p is the distance projected on the celestial sphere from the exciting star, then the surface brightness of Hα, Hβ and the scattered light of wavelength λ arising from the region r to $r+\Delta r$ along the line of sight are, respectively,

$$S_{H\alpha}(r_p, r) = \frac{1}{4\pi} 3.62 \times 10^{-21} N_e^2(r)T_e^{-1}(1{-}350\ T_e^{-1})\ \Delta y, \tag{2}$$

$$S_{H\beta}(r_p, r) = \frac{1}{4\pi} 1.36 \times 10^{-21} N_e^2(r)T_e^{-1}(1{-}792\ T_e^{-1})\ \Delta y, \tag{3}$$

$$S_\lambda(r_p, r) = \frac{L(\lambda)}{4\pi r^2} \int_0^\infty S(\theta, a, \lambda)n(a, r)\ da\ \Delta y, \tag{4}$$

where the expressions for $S_{H\beta}$ and $S_{H\alpha}$ are analytical expressions valid for an electron temperature, T_e, about 10^4 K and calculated from the computations of Pengelly (1964). $S(\theta, a, \lambda)$ is the amplitude function for axially symmetric scattering for grains with a radius a at wavelength λ to the direction θ (which is equal to arc sin (r_p/r)); and $n(a, r)$ is the number density of grains radius a at radial oordinate r in the nebula. $L(\lambda)$ is the luminosity of the central star at wavelength corrected for absorption by the dust out to a radial distance r. This elemental optical depth in this radial direction is given by

$$\tau_1(\lambda, r_p, r, \Delta r) = \int_0^\infty \pi a^2 Q_{ext}(a, \lambda)n(a, r)\ da\ \Delta r. \tag{5}$$

Whilst the elemental optical depth along the line of sight is given by

$$\tau_2(\lambda, r_p, r, \Delta y) = \int_0^\infty \pi a^2 Q_{ext}(a, \lambda)n(a, r)\ da\ \Delta y, \tag{6}$$

$$\Delta y = [(r + \Delta r)^2 - r_p^2]^{1/2} - [r^2 - r_p^2]^{1/2}; \tag{7}$$

$Q_{ext}(a, \lambda)$ being the extinction coefficient of the grains with radius a at wavelength λ.

The quantities observed at projected distance r_p can then be found for a given set of grain parameters by integrating equations (2) to (6) inclusive – i.e.,

$$S_{H\alpha}(r_p) = \int_{-y_0}^{y_0} S_{H\alpha}(r_p, r) \exp\left[-\tau_2(6563\ \text{Å}, r_p, r]\ dy, \tag{8}\right.$$

$$S_{H\beta}(r_p) = \int_{-y_0}^{y_0} S_{H\beta}(r_p, r) \exp\left[-\tau_2(4861\ \text{Å}, r_p, r)]\ dy, \tag{9}\right.$$

$$S_\lambda(r_p) = \int_{-y_0}^{y_0} S_\lambda(r_p, r) \exp\left[-\tau_1(\lambda, r_p, r)]\ \exp\left[-\tau_2(\lambda, r_p, r)]\ dy; \tag{10}\right.\right.$$

where

$$\tau_1(\lambda, r_p, r) = \int_0^r \tau_1(\lambda, r_p, r, \Delta r) \, \mathrm{d}r, \tag{11}$$

$$\tau_2(\lambda, r_p, r) = \int_y^{y_0} \tau_2(\lambda, r_p, r, \Delta y) \, \mathrm{d}y, \tag{12}$$

$$y_0 = (r_0^2 - r_p^2)^{1/2} \quad \text{and} \quad y = \pm(r^2 - r_p^2)^{1/2}. \tag{13}$$

We now discuss the solution of Equations (1) to (13) for various types of grain at the wavelengths of scattered light of 3000, 3500, 4000, 4500, 4700, 5000, 6000, 8000 and 10 000 Å.

A. ICE GRAINS

As shown in the diagrams of the scattering function at optical wavelengths (Isobe, 1974), the scattering by grains with radius smaller than 0.05 μ is nearly isotropic but becomes more and more forward throwing with increasing a. For example, the grains of radius 0.3 which are the effective scatterers in our model have $S(5°, 0.3\,\mu)/S(85°, 0.3\,\mu) = 88$ at a wavelength of 4700 Å. The presence of a graphite core makes no difference to the results for mantle radii more than five times the core radius.

Assuming that all oxygen atoms condense onto the ice grains, that the number ratio of oxygen atoms to hydrogen atoms (n_0/n_H) is 6.7×10^{-4} (Allen, 1973) and that the size distribution function for ice grains is given by $n(a) = n(0) \exp(-23.2a^3)$ (Isobe, 1973), the upper limit of ice grain scattered light intensity can be calculated. Figure 10 shows $S_{H\beta}(r_p, r)$ and $S_{\lambda 4700}(r_p, r)$ for a shell model with $\Delta r = 0.02$ pc at three values of r_p (0.1, 0.7 and 1.5 pc). The discontinuities in the scattered light curves arise because we used a discrete scattering function with a step of 10°. It is immediately clear from this figure that ice gives far too much scattered light in the inner region where A is in the region 20 to 50. We must therefore conclude that ice is not present in the ionized medium in significant quantities near the exciting stars. However, theory and observation come closer to agreement if we allow for the possibility of a central hole in the dust distribution. Figure 11 shows the dependence of scattered light due to ice grains with projected distance from centre as a function of dust-free sphere radii, r_c. Models with $r_c \approx 1.0$ pc give a fair fit to the photoelectric observations from the centre westwards (see Figure 7). However, it is clear that although ice may be important as a scattering particle in the outer regions, models relying on ice grains (Isobe, 1970) to explain the reddening curves of the Trapezium stars cannot be correct. There are two possibilities to explain this lack of ice grains in the central regions, either they have evaporated due to sputtering or else they were never present in the ionized medium. This second possibility will be discussed later.

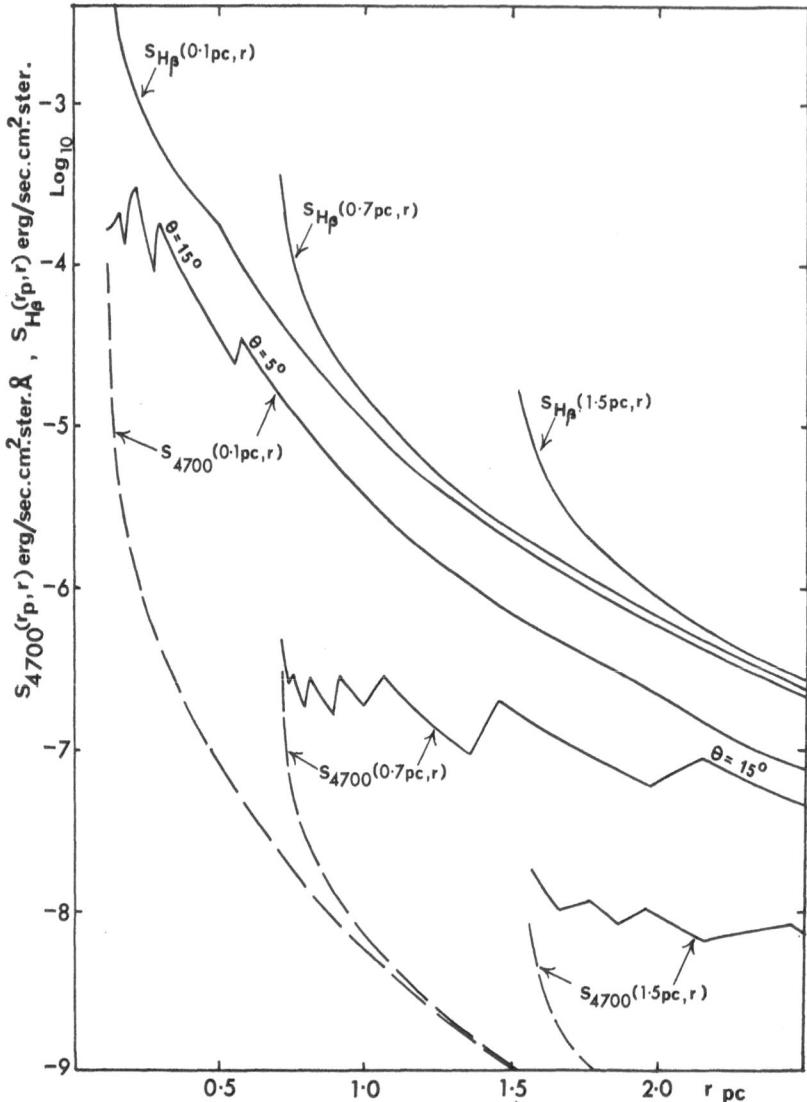

Fig. 10. Surface brightness distribution of Hβ and the scattered continuum light at λ 4700 from the
segment r=0.02 pc on the line of sight for r_p=0.1 pc, 0.7 pc, 1.5 pc.

Mathews (1969) has shown that the destruction rate by sputtering of the surrounding
ions is given by

$$da/dt = 6.5 \times 10^{-19} \, n_p e^\gamma \text{ cm s}^{-1},$$

where n_p is the proton density and γ is determined by the electrostatic potential of the
grains, V, from

$$\gamma = -eV/KT.$$

We shall adopt the destruction rate obtained by Mathews (although Barlow (1971) obtained a lower rate). Mathews' (1967) Equation (18) shows that the electrostatic potential of the grains is proportional to $1/n_p r^2$ and so is independent of r from $0.45 \text{ pc} < r < 3.0 \text{ pc}$ on our model. Recently, Feuerbacher *et al.* (1973) obtained the potential as functions of electron density and distance from an O5 and a B0 star considering the photoelectric properties of the grains, which gives results essentially consistent with Mathews. Adopting ice as their low yield material and graphite as their high yield material, we computed γ as a function of r (using their Figure 7) on our density distribution and on four times our density distribution (more appropriate to conditions near partially ionized globules). These cases are labelled 1.0 and 4.0 in Figure 12. Also plotted on this figure are the corresponding destruction times for

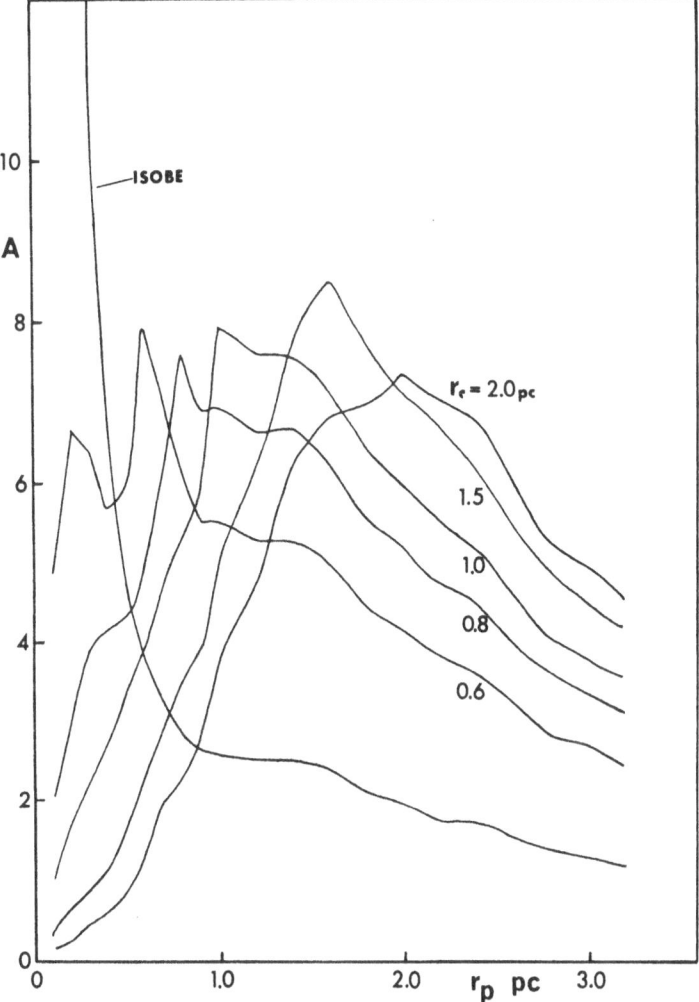

Fig. 11.   Ratio of the surface brightness $S_{\lambda 4700}(r_p)$ of the scattered continuum light to that $S_{H\beta}(r_p)$ of Hβ. 'Isobe' shows the ratio obtained from the grain distribution given by Isobe (1971). r_c is the critical distance within which all ice grains evaporate.

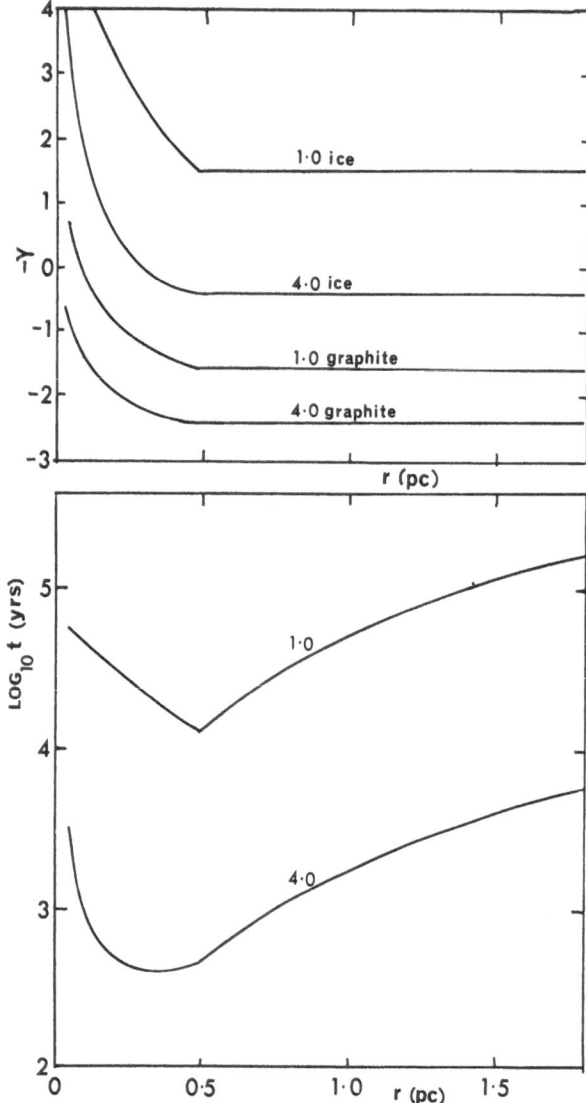

Fig. 12. Dependence of the grain charge ($\gamma = -\mathrm{eV}\,kT^{-1}$) and the destruction time scale of a grain with a radius $0.3\,\mu$ on the distance from the central star for two density cases, that is, one and four times $N(r)$.

grains of radius $0.3\,\mu$ which effectively contribute to the scattered continuum light on our model.

There are many estimates for the age of the Orion Nebula ranging from 10^4 yr to 10^6 yr (Johnson, 1957; Strand, 1958; Kahn and Menon, 1961; Vandervoort, 1963). However, an age of the order 10^5 yr since the turn-on of the exciting stars may be reasonable. In this case the possibility of evaporation of the grains nearer than about 1 pc is not excluded.

A further observational constraint on the grain model is provided by the wavelength dependence of the intensity of scattered light, although this is subject to much larger observational errors. We define the quantity

$$B(\lambda, r_p) = \log S_\lambda(r_p)/\log S_{\lambda\,4700\,\text{Å}}(r_p) - \log L(\lambda)/\log L(\lambda\,4700),$$

which corresponds to the $e(\lambda)/e(\lambda\,4630)$ of O'Dell and Hubbard (1965). This quantity was computed for each of the six regions observed by Simpson (1973), having corrected her continuum measurements for atomic processes. These are shown in Figure 13. In Figure 14 we show the computed wavelength dependence for a projected distance of 0.2 pc and 1.7 pc. It may be seen that there is fair agreement for Simpson's positions 3 and 4 but there appears to be a systematic difference from theory for the other positions. This point will be discussed later.

B. GRAPHITE AND SILICATE GRAINS

The scattering function and extinction coefficient for these grains as given by Isobe (1974) are used. Since the scattering function of the small radius grains considered here does not change much with varying θ, we can assume that all grains have a single radius. Thus Equation (4) simplifies to

$$S_\lambda(r_p, r) = \frac{L(\lambda)}{4\pi r^2} S(\theta, a, \lambda)n(r)\,\Delta y, \tag{14}$$

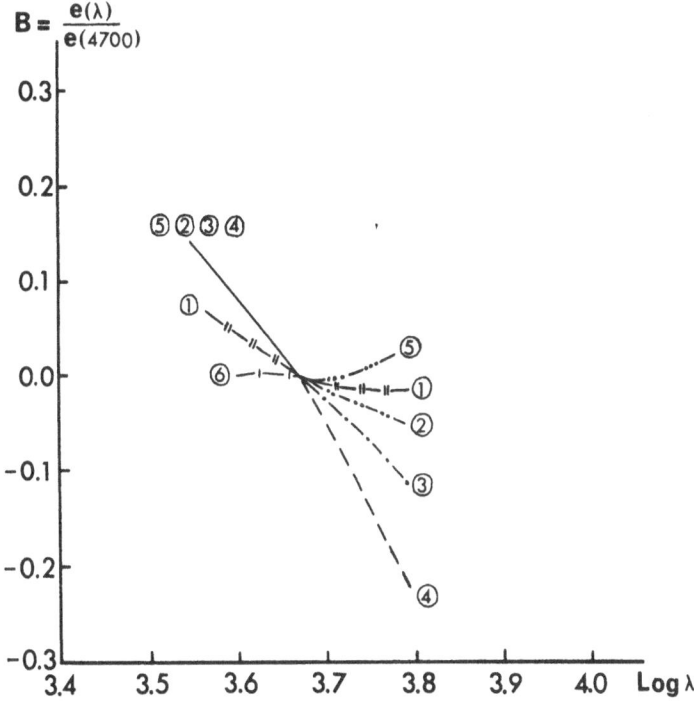

Fig. 13. The colour index $B(\lambda)$ of the scattered light derived from Simpson's (1973) measurements. (These are corrected for contribution by atomic processes.)

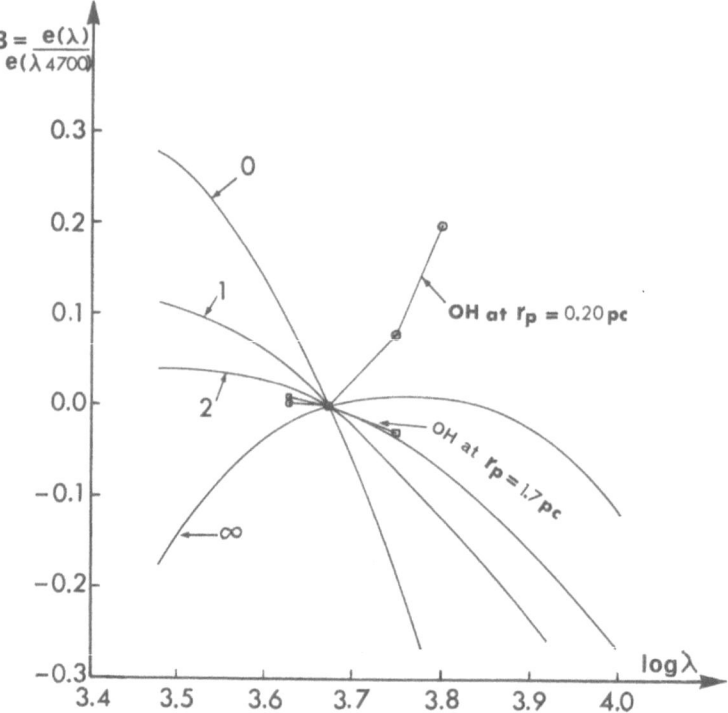

Fig. 14.   Theoretical wavelength dependence of the scattered continuum light normalized at 4700 at
$r_p = 0.2$ pc and 1.7 pc for the case of $r_c = 1.0$ pc.

where we compute for three grain distributions

$$n(r) = N_e(r)(n_d/n_H)(r/1 \text{ pc})^i \quad i = 0, 1, \text{ or } 2;$$ (15)

n_d/n_H is the dust to hydrogen number density.

The values of $S_{\lambda 4700}(r_p)$ are computed for the various sets of parameters given in Table II, and Figures 15 and 16 show the $A(r_p)$ resulting. Cases A0, A1 and A2 correspond to normalized surface brightness ratio given by O'Dell and Hubbard except in the inner regions where our curves are modified by the forward scattering of the graphite particles considered.

Only the cases C1, C2, D1, D2 for graphite grains and the cases H1 and H2 for silicate grains give enough scattered light to fit observed values. However, there is the following difficulty for all these models. The mass ratio of graphite and silicate grains to hydrogen atoms for a given n_d/n_H value is given by

$$f_m = \frac{n_d}{n_H} \frac{4}{3} \pi a^3 \varrho : \frac{1}{m_H},$$ (16)

where ϱ is the density of the grains (taken to be 2.2 gm cm^{-3}). The ratio is also shown on Table II for a radius of one parsec. The upper limit on the number ratios deter-

TABLE II

Parameters used in the calculations of $A(r_p)$ for graphite and silicate grains

a_μ	$\dfrac{n_d(1\,pc)}{n_H(1\,pc)}$	f_m	i	$\tau_{\lambda 4700}$	$A(0.5\,pc)$	$A(2.0\,pc)$	Case
				Graphite			
0.02	5×10^{-11}	2.2×10^{-3}	0	0.699	4.14×10^{-2}	4.58×10^{-2}	A0
			1	0.261	4.66×10^{-2}	2.14×10^{-1}	A1
			2	0.439	4.13×10^{-2}	6.63×10^{-1}	A2
	2.5×10^{-10}	1.1×10^{-2}	0	3.49	1.98×10^{-2}	1.52×10^{-2}	B0
			1	1.30	1.62×10^{-1}	5.06×10^{-1}	B1
			2	2.20	1.79×10^{-1}	2.00	B2
0.05	3.2×10^{-12}	2.2×10^{-3}	0	1.53	3.77×10^{-1}	3.69×10^{-1}	C0
			1	0.569	7.60×10^{-1}	3.11	C1
			2	0.960	7.19×10^{-1}	1.05×10	C2
	1.6×10^{-11}	1.1×10^{-2}	0	7.63	2.81	5.04×10^{-3}	D0
			1	2.85	1.85	3.36	D1
			2	4.80	3.07	2.15×10	D2
				Silicate			
0.05	6.4×10^{-11}	5.5×10^{-4}	0	0.015	4.68×10^{-2}	5.67×10^{-2}	E0
			1	0.005	3.26×10^{-2}	1.65×10^{-1}	E1
			2	0.009	2.80×10^{-2}	5.09×10^{-1}	E2
	3.2×10^{-12}	2.2×10^{-3}	0	0.073	2.23×10^{-1}	2.68×10^{-1}	F0
			1	0.027	1.62×10^{-1}	8.12×10^{-1}	F1
			2	0.046	1.39×10^{-1}	2.51	F2
0.10	8.0×10^{-14}	5.5×10^{-4}	0	0.091	2.67×10^{-1}	3.22×10^{-1}	G0
			1	0.034	2.03×10^{-1}	1.02	G1
			2	0.057	1.84×10^{-1}	3.26	G2
	4.0×10^{-13}	2.2×10^{-3}	0	0.454	1.01	1.14	H0
			1	0.169	9.85×10^{-1}	4.69	H1
			2	0.286	9.20×10^{-1}	1.58×10	H2

mined by the cosmic mass ratio abundances of carbon (4.0×10^{-3}) and silicate (9.3×10^{-4}) (Allen, 1973) are

$$n_d/n_H < 6\times10^{-12} \quad \text{for } 0.05\,\mu \text{ graphite grains (cases C and D)}$$

and

$$n_d/n_H < 1.7\times10^{-13} \quad \text{for } 0.1\,\mu \text{ silicate grains (case H)}.$$

Hence, all the cases which give a reasonable fit to the observations have abundances of grains greater than cosmic values in the outer regions.

This cannot be due to grains being forced out from the central regions because the terminal velocity of the grains due to radiation pressure is very low due to the coulomb

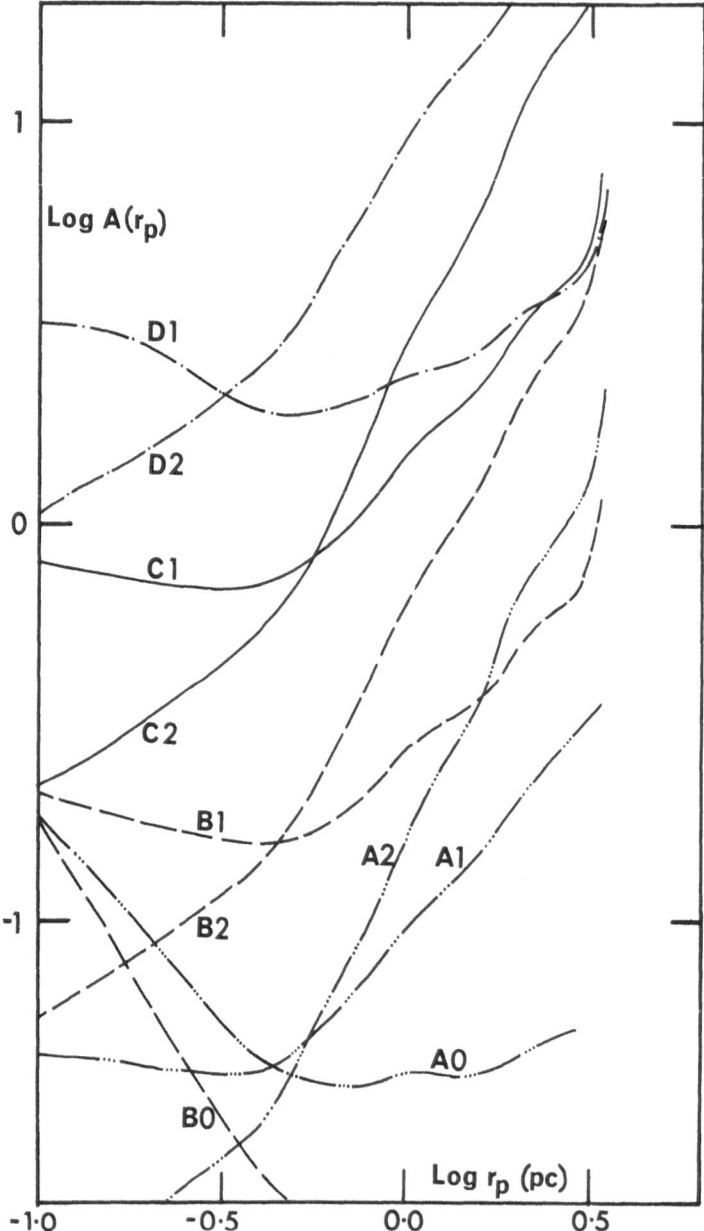

Fig. 15.   Ratio of the surface brightness $S_{\lambda 4700}(r_p)$ of the scattered continuum light to that $S_{H\beta}(r_p)$ of Hβ for the various distributions and sizes of graphite grains in the nebula. Assigned letters in the figure are referred in Table II.

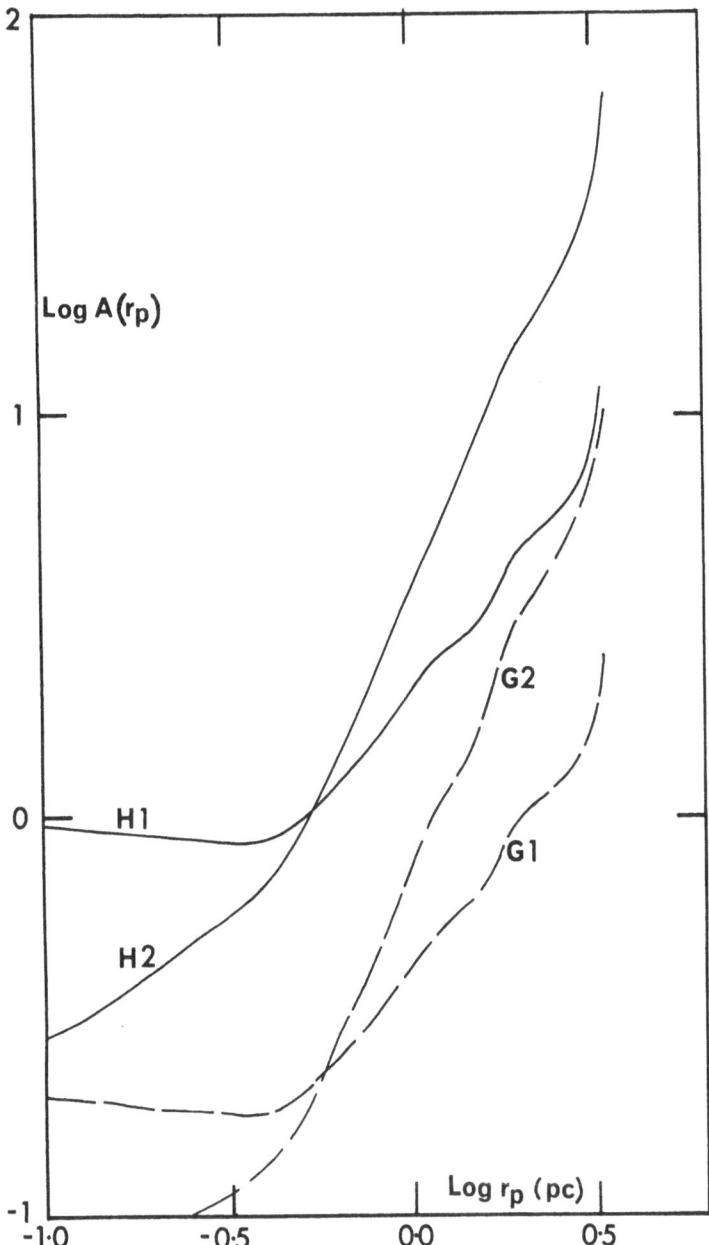

Fig. 16.   As Figure 15, but for silicate grains.

drag (≈ 0.4 to 0.6 km s^{-1}). Thus the grains need more than 10^6 yr to move a distance of about 1 pc. Even if they had no charge the time required would still be more than 10^5 yr.

However, the observations could be explained by light scattering off the grains inside neutral masses of material. This possibility will be discussed in the next section.

C. THE EFFECT OF NEUTRAL INCLUSIONS

The evidence for partially ionized globules (PIG's) and larger inclusions of neutral material in the Orion Nebula has been recently reviewed by Dopita *et al.* (1974a). These have been very clearly shown up in some remarkable filter photographs of the core region by Elliott and Meaburn (1974a). A large scale enlargement of one of these photographs is shown in Figure 17. The variability of extinction on a small scale is shown by the four stars of 0^1 Ori (A, B, C, D) which have different reddenings ($E_{B-V} =$ 0.30, 0.44, 0.38 and 0.41 mag respectively) although their projected distances from one another on the celestial sphere are less than 20″. Clearly, the presence of such globules must modify both the extinction and the amount of scattered continuum light in our models.

Firstly, inhomogeneities in the ionized gas will relax the conditions on the amount

Fig. 17.   A large scale enlargement of an [O III] filter photograph of the core regions of M42 showing evidence for globular structure. (After Elliot and Meaburn, 1974a.)

of dust present compared with the ionized material. We have used implicitly in our shell model a root mean square density and have taken the dust density to be correlated with this quantity. In fact, the condensations give a local density of ionized material much higher than that. If we define the condensation factor α to be $\langle N_e^2 \rangle^{1/2}/N_{loc}^2$ where α is the local density, then the dust density will be proportional to the quantity $\alpha^{-1/2}\langle N_e^2 \rangle^{1/2}$. Since α is observed to be of the order 0.05 (Danks and Meaburn, 1971), the true dust density in the presence of condensations could be a factor of four lower than computed. This would make ice even less abundant in the ionized region (15% of oxygen only locked up in the grains outside r equal to one parsec). If the grains are graphite or silicate, then these could (from the point of view of intensity of scattered light alone) be the particles involved. However, it will be argued that these give the wrong colour dependence of the scattered light.

The results of part (a) of this discussion showed that for ice grains, a central hole in the distribution was necessary to explain the scattered light, and this remains true.

However, this would make the total extinction in the nebula about 0.1 mag. Even accounting for foreground extinction of about 0.5 mag, this is too small for the values observed in the $\theta^{1,2}$Ori stars (about 1 to 2 mag). Furthermore, if the central hole was caused by evaporation of the ice, the oxygen abundance in the nebula would be normal in the central regions. Combining the temperatures given by the Hβ and [O III] line profiles (Dopita *et al.*, 1973a) with photoelectric measurements of the Hα, [O III], Hβ and [N II] lines, Dopita (1972a, b) has shown oxygen to be deficient throughout the nebula. This has been confirmed by Simpson (1973) using completely different methods.

The density distribution as given by the isophotes in Hβ in Figure 7 indicates that most of the ionized material is being produced from the flow from ionization fronts eating into these neutral inclusions, so if we suppose that the missing oxygen is locked up on ice grains, we must find a mechanism to keep ice grains trapped inside the neutral masses.

Such a mechanism is provided by radiation pressure from Lα photons surrounding the globule. The force due to Lα is given by

$$F_{rad}(r) = \text{Flux } (L\alpha)Q_{pre}(a, \lambda\ 1215)\pi a^2, \tag{17}$$

where $Q_{pre}(a, \lambda\ 1215)$ is the radiation pressure coefficient of the grains at $\lambda = 1215$ Å. Since the Orion Nebula is optically thick for the Lα radiation the flux $(L\alpha) \approx 2.6$ Flux (Lc) (Unno, 1952; Auer, 1968). Therefore

$$F_{rad}(r) = \frac{\pi a^2}{4\pi r^2 c} \int\limits_0^{912\ \text{Å}} 2.6L(\lambda)Q_{pre}(a, \lambda\ 1215)\ d\lambda. \tag{18}$$

The collisional (drag) force in the globule to the grains is given (cf. Mathews, 1967) by

$$F_{coll} = \tfrac{8}{3}\kappa(2\pi KTm_H)^{1/2}na^2W, \tag{19}$$

where κ is a factor of order unity and W is the relative grain-gas velocity. If we set

$F_{rad} = F_{coll}$, grains have a terminal velocity

$$W_t \propto 1/nr^2 T^{1/2}. \tag{20}$$

Consider a neutral mass with temperature 100 K at a distance of 0.5 pc from the centre. The density in the ionized hydrogen near to the ionization front is about 2000 cm^{-3} (Danks and Meaburn, 1971). The ratio of densities of ionized material to neutral material N_i/N_n is about 2.5×10^{-3} (Dyson, 1968), so the density in the globule is about 8×10^5 cm^{-3} which gives a terminal velocity reached in a time scale of about one year for the grain, W_t, of 1.8×10^4 cm s^{-1}. Since on the models of Dyson (1968) the ionization front eats into the globules at about $\frac{1}{10}$ the sound speed in the H I region ($\approx 7 \times 10^3$ cm s^{-1}), the grains are forced into the globules faster than the ionization front. Essentially the same argument applies to graphite and silicate grains (for which Q_{pre} is even larger than ice).

However, it is expected that some grains will be ejected into the ionized region near the inner region especially in the early evolution of the nebula. At this time the ionization front will be R-type (moving supersonically through the neutral medium). Thus grains are overtaken by the ionization front. In later stages, the ionization fronts will become well protected from the ionizing radiation of the stars and become D-type, in which case the above argument applies and grains are no longer released (except in the outer regions where the optical depth in $L\alpha$ becomes low). Since in Section 5A we showed that the mantles of composite grains ejected into the inner ionized region will have had time to evaporate but in Section 5B that any graphite cores remaining would be strongly coupled to the ionized gas, we expect that the grains in the ionized gas near the centre of the nebula to be graphite or possibly graphite-silicate mixtures. This may be sufficient to explain the anomalous extinction found by Lee (1968) in the central 3–5′ (Nandy and Wickramasinghe, 1971). Since graphite grains with small radius absorb effectively in the far UV (although Q_{ext} (visual) is small), these grains may be expected to efficiently absorb ionizing photons and possibly to preferentially absorb the helium ionizing photons. This would give rise to separation of the helium and hydrogen Strömgren spheres observed by Dopita et al. (1974b) in M42 and may well explain the correlation between IR excess and ionized helium abundance as measured by radio techniques in many nebulae (Churchwell et al., 1974; Mezger et al., 1974).

Since the rate at which grains are forced into the globule is not very much different from the ionization front velocity, local variations in the $L\alpha$ field could also let some grains out into the ionized material (in the shadow regions of partially ionized globules, for example). Composite grains are destroyed by sputtering near the ionization front because of the high densities in the ionized gas. The force exerted by radiation pressure radially outwards from the central star is given by

$$F'_{rad}(r) = \frac{\pi a^2}{4\pi r^2} \int\limits_0^{912\,\text{Å}} L(\lambda) Q_{pre}(a, \lambda)\, d\lambda +$$

$$+ \pi a^2 \, \text{Flux} \, (L\alpha) \cdot Q_{pre}(a, \lambda\, 1215), \tag{21}$$

and the frictional force on the grains due to surrounding ions is given by

$$F_{\text{fric}} = F_{\text{coll}} + F_{\text{coul}},\tag{22}$$

where the Coulomb force F_{coul} is given (cf. Spitzer, 1956) by

$$F_{\text{coul}} = \frac{m_H \log_a (\theta) n Z_g^2 \omega}{11.7 T_e^{3/2}}.\tag{23}$$

The terminal velocity, $W_t^1(r)$, is computed at about 5 km s^{-1} at $r > 0.45$ pc. Since the radius of the ionized condensation is of the order of the globule (10^{15-17} cm) which corresponds to the mass of the globule (0.3 $M_\odot$–5.0 $M_\odot$) (Dyson, 1968) the time needed for the escape of grains from the dense region close to the globule is of the order $7 \times 10^{1-3}$ yr considering radiation forces alone. This represents an upper limit as the grains will also share the gas velocity (≈ 10 km s^{-1}). The amount of radius decrease in the time $t \approx 10^1$ to 10^3 yr is 2×10^{-8} to 2×10^{-10} times n_p which is 10^{-5} to 10^{-7} cm. Therefore grains ejected from globules with mass less than 1 $M_\odot$ at r greater than 0.5 pc are hardly altered in their journey from the globule ionization front to the low density region. Thus we expect those grains escaping into the ionized medium outside about one parsec to survive their passage into the low density ionized medium and to survive there for longer than the age of the nebula ($\approx 10^5$ yr).

Let us now consider how grains within globules contribute to the scattered light. This will depend very critically on the geometry of the problem, in particular, scattered light which has to pass through a globule will be very efficiently absorbed from Dyson (1968) as

$$N_n R_i = 2 \times 10^{23} \, r^{-3/4},\tag{24}$$

where N_n is the number density of neutrals and R_1 is the globule radius. Since $n_H/m_V \approx 5 \times 10^{20}$ on a dirty ice model,

$$m_V = 200 \, r^{-3/4}.\tag{25}$$

Hence, extinction by dust grains in the globule is extremely high, and forward scattered light will be very quickly absorbed. (This result is important for the validity of the radiation pressure on grains argument above.) For composite grains, the majority of scattered light will come from globules which scatter light to the observer at an angle θ of about 90°. However, it is very difficult to estimate the total intensity of scattered light from neutral intrusions since we do not know the distribution in terms of the geometry of neutral material or in terms of grains in the globules. We will therefore compute the wavelength dependence on the basis of different proportions of scattered light from globules to scattered light from grains in the ionized medium with distances of the order 1 pc from centre (for which the scattering is forward throwing with θ of the order of 5°). In Table III there are shown the scattering efficiencies of several types of grains normalized to $M_{4700} = 1$ mag at $r = 1$ pc for $\theta = 5°$ and 85°. Assuming a value of $M_{4700} = 2$ mag for the scattered light from the grains in the globule, the wavelength

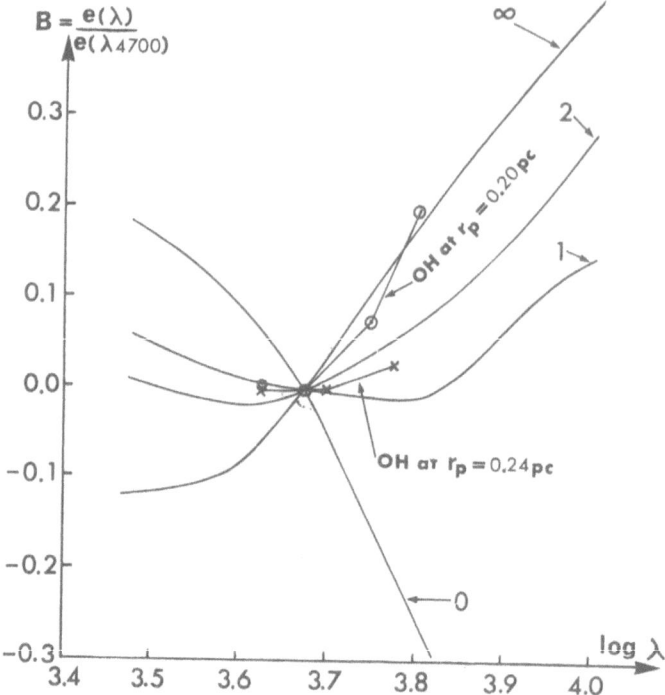

Fig. 18.   Wavelength dependences of the scattered continuum light at $r_p = 0.2$ pc normalized at λ 4700 for the four cases, that is, that the ratio of the light from the globules at $\theta = 85°$ to that from the grains at $r \approx 1.0$ pc are 0, 1, 2 and ∞. The dependences observed by O'Dell and Hubbard at $r_p = 0.20$ pc and 0.24 pc are also shown.

dependences were calculated for ratios of scattered light from the globules to that from grains in the outer regions of 0, 1, 2 and ∞. These are shown in Figure 18 for ice grains. Clearly the Simpson values fit these computations for contributions from globules ranging from about zero (region 4) to about equality with the outer regions (region 5).

Evidence that dielectric grains in globules do contribute to the scattered light also comes from the observation by Feibelmann (1973) of a polarized bright knot 1' from θ^1 Ori where presumably there is only a single globule in the line of sight.

Since graphite and silicate grains have a sharp decrease of $S(\theta, \lambda)$ depending on λ as shown in Table III we need unreasonably large line of sight reddening to give a concidence of the calculated wavelength dependence and the observed one. It is therefore unlikely that these grains contribute effectively to the scattered continuum light anywhere in the nebula.

Finally, we must consider the possibility that the many B and A type stars which are 3–7 mag fainter than θ^1 Ori (Parenago, 1954) may affect the scattered light measurements. Since on our composite grain model $S(0°–30°, \lambda)$ is some hundred times greater than $S(70°–90°, \lambda)$, even assuming the same dilution factor, the scattered light from such a star is sometimes stronger than that from θ^1 Ori at large r. This may

TABLE III

Wavelength and angle dependence of scattering for the different grain models

Wavelength (Å units)		3000	3500	4000	4500	4700	5000	6000	8000	10 000
Composite grains	$\int S(5°, a)n(a)\,da \times 10^{-4}$ erg (s cm² ster Å)⁻¹	6.72	5.83	4.86	3.91	3.56	3.09	1.94	0.81	0.38
	$\int S(85°, a)n(a)\,da \times 10^{-6}$ erg (s cm² ster Å)⁻¹	1.34	1.32	1.28	1.23	1.21	1.19	1.14	1.06	1.03
	m_λ	1.17	1.16	1.11	1.03	1.00	0.97	0.79	0.54	0.38
Graphite grains	$S(5°, 0.02\,\mu)n \times 10^{-6}$ erg (s cm² ster Å)⁻¹	7.75	3.53	1.83	1.12	0.94	0.73	0.35	0.12	0.05
	$S(85°, 0.02\,\mu)n \times 10^{-6}$ erg (s cm² ster Å)⁻¹	3.57	1.63	0.85	0.53	0.44	0.35	0.17	0.06	0.03
	m_λ	2.31	1.66	1.26	1.06	1.00	0.92	0.71	0.51	0.37
Silicate grains	$S(5°, 0.05\,\mu)n \times 10^{-5}$ erg (s cm² ster Å)⁻¹	24.76	12.44	6.81	4.02	3.33	2.54	1.15	0.35	0.12
	$S(85°, 0.05\,\mu)n \times 10^{-5}$ erg (s cm² ster Å)⁻¹	7.59	4.34	2.57	1.61	1.35	1.05	0.50	0.15	0.06
	m_λ	5.73	3.22	1.91	1.20	1.00	0.78	0.37	0.12	0.05

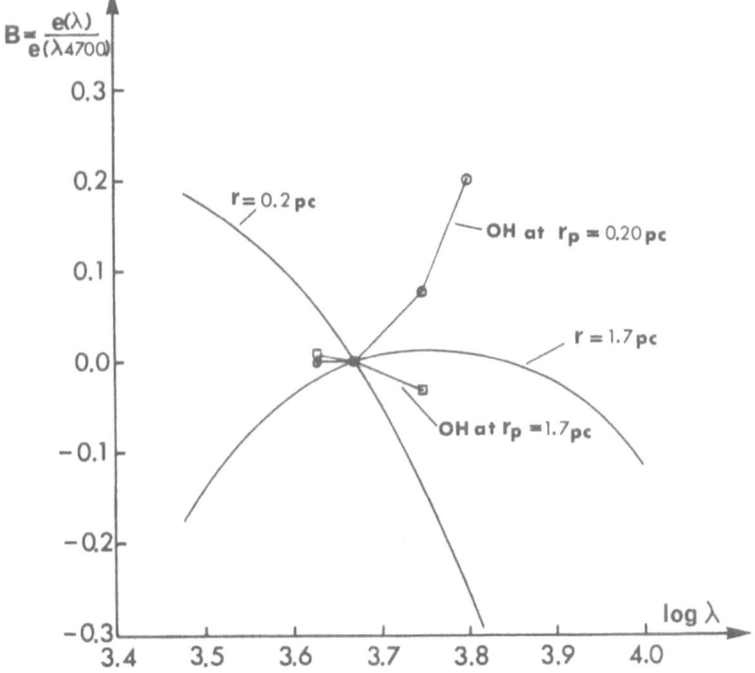

Fig. 19. Wavelength dependences of the scattered continuum light at $r_p=1.7$ pc normalized at λ 4700 for the four cases, that is, that the ratio of light from the central star to that from a star with $\theta=5°$ is 0, 1, 2 and ∞.

explain the strong fluctuations shown in A in our observations. In Figure 19 we plot the computed wavelength dependence of the scattered continuum light at $r_p = 1.7$ pc for several cases in which the ratio of scattered light from θ^1 Ori to that of a star with $\theta \approx 5°$ is 0, 1, 2 and ∞. The result can give a good fit to O'Dell and Hubbard's (1965) results for this distance.

6. Conclusions

A judiciously chosen distribution of ice grains in the ionized region and in the neutral intrusions can explain the scattered light observations whereas graphite and silicate grains alone would fail to do so. It seems clear that grains are escaping into the ionized material from the large scale ionizing fronts observed in this work, but at trapped in the smaller scale neutral masses closer to the centre. Observations of the colour index variations across globules and larger scale neutral masses are needed in order to verify some of the results of this paper.

A schematic model of M42 that attempts to explain the large scale phenomena demonstrated here has been constructed (Figure 9). This is considered to be a reasonable extension of those presented previously for the four arc minute core (Dopita *et al.*, 1974a; Elliott and Meaburn, 1974).

It appears certain that the whole of this H II region is located on the side of a very much larger inhomogeneous mass containing molecules, neutral hydrogen, dust and possibly protostars.

Acknowledgements

One of us (J.M.) wishes to thank the staff of the Athens National Observatory where the photographs analysed in this paper were taken. Likewise M.A.D. wishes to thank the staff of the Pio-du-Midi Observatory where the photoelectric work was carried out.

The instrumental and observational part of this work was generously sponsored by the Science Research Council.

References

Allen, C. W.: 1973, *Astrophysical Quantities*, 3rd ed., The Athlone Press, London, p. 31.
Auer, L. H.: 1968, *Astrophys. J.* **153**, 783.
Barlow, M. J.: 1971, *Nature Phys. Sci.* **232**, 152.
Baudel, L.: 1970, *Astron. Astrophys.* **8**, 65.
Churchwell, E., Mezger, P. G., and Huchtmeier, W.: 1974, *Astron. Astrophys.* **32**, 283.
Cousins, A. W. J.: 1971, *Roy. Obs. Ann.*, No. 7.
Danks, A. C. and Meaburn, J.: 1971, *Astrophys. Space Sci.* **11**, 398.
Deharveng, L.: 1973, *Astron. Astrophys.* **29**, 341.
Dopita, M. A.: 1972, *Astron. Astrophys.* **17**, 165.
Dopita, M. A.: 1973a, *Nature Phys. Sci.* **244**, 85.
Dopita, M. A.: 1973b, *Astron. Astrophys.* **29**, 387.
Dopita, M. A., Gibbons, A. H., and Meaburn, J.: 1973a, *Astron. Astrophys.* **22**, 33.
Dopita, M. A., Gibbons, A. H., Meaburn, J., and Taylor, K.: 1973b, *Astrophys. Letters* **13**, 55.
Dopita, M. A., Dyson, J. E., and Meaburn, J.: 1974a, *Astrophys. Space Sci.* **28**, 61.
Dopita, M. A., Elliott, K. H., and Meaburn, J.: 1974b, *Astrophys. Space Sci.* **28**, 163.
Dyson, J. E.: 1968, *Astrophys. Space Sci.* **1**, 388.

Dyson, J. E.: 1973, *Astron. Astrophys.* **27**, 459.

Elliott, K. H. and Meaburn, J.: 1974a, *Astrophys. Space Sci.* **28**, 351.

Elliott, K. H. and Meaburn, J.: 1974b, *Astron. Astrophys.* **34**, 473.

Feibelmann, W. A.: 1973, *Astron. Astrophys.* **27**, 317.

Feuerbacher, B., Willis, R. F., and Fitton, B.: 1973, *Astrophys. J.* **181**, 101.

Foukal, P. V.: 1969, *Astrophys. Space Sci.* **5**, 469.

Gillett, F. C. and Forrest, W. J.: 1973, *Astrophys. J.* **179**, 483.

Gilra, D. P.: 1971, *Nature* **229**, 239.

Greenberg, J. M.: 1968, in B. M. Middlehurst and L. H. Aller (eds.), *Nebulae and Interstellar Matter*, Univ. of Chicago Press, p. 247.

Huffman, D. R. and Stapp, J. L.: 1971, *Nature Phys. Sci.* **229**, 45.

Isobe, S.: 1970, *Publ. Astron. Soc. Japan* **22**, 429.

Isobe, S.: 1972, *Publ. Astron. Soc. Japan* **24**, 27.

Isobe, S.: 1973, *Publ. Astron. Soc. Japan* **25**, 253.

Isobe, S.: 1974, *Publ. Astron. Soc. Japan*, in press.

Isobe, S., Kawajiri, N., Ojima, T., Kawano, N., and Kurichara, H.: 1972, *Tokyo Astron. Bull. 2nd Ser.*, No. 218, 2549.

Johnson, H. L.: 1957, *Astrophys. J.* **126**, 134.

Kahn, F. D. and Menon, T. K.: 1961, *Proc. Nat. Acad. Sci.* **47**, 1712.

Kemp, J. C.: 1973, in J. M. Greenberg and H. C. van de Hulst (eds.), *Interstellar Dust and Related Topics*, D. Reidel Publishing Co., Dordrecht, Holland, p. 181.

Lee, T. A.: 1969, *Astrophys. J.* **152**, 913.

Martin, P. G., Angel, J. R. P., and Illig, R. M.: 1973, in J. M. Greenberg and H. C. van de Hulst (eds.), *Interstellar Dust and Related Topics*, D. Reidel Publishing Co., Dordrecht, Holland, p. 161.

Mathews, W. G.: 1967, *Astrophys. J.* **147**, 965.

Mathews, W. G.: 1969, *Astrophys. J.* **157**, 583.

Meaburn, J.: 1970, *Astrophys. Space Sci.* **9**, 206.

Meaburn, J.: 1975, *Appl. Opt.* **14**, 465.

Menon, T. K.: 1961, *Publ. Nat. Radio Astron. Obs.* **1**, 1.

Mezger, P. G., Smith, L. F., and Churchwell, E.: 1974, *Astron. Astrophys.* **32**, 269.

Nandy, K. and Wickramasinghe, N. C.: 1971, *Monthly Notices Roy. Astron. Soc.* **154**, 255.

Oke, J. B. and Schild, R. E.: 1970, *Astrophys. J.* **161**, 1015.

O'Dell, C. R. and Hubbard, W. B.: 1965, *Astrophys. J.* **142**, 591.

Parengo, P. P.: 1954, *Trudy Gas. Astron. Inst. Stemberga* **25**, 1.

Peimbert, M. and Costero, R.: 1969, *Bol. Obs. Ton. y. Tac.* **5**, 3.

Peimbert, M. and Torres-Peimbert, S.: 1971, *Bol. Obs. Ton. y. Tac.* **6**, 21.

Reitmeyer, N. L.: 1965, *Astrophys. J.* **141**, 1331.

Schmitter, E. F. and Recillas-Cruz, E.: 1971, *Bol. Obs. Ton. y. Tac.* **6**, 47.

Simpson, J. P.: 1973, *Publ. Astron. Soc. Pac.* **85**, 479.

Smith, M. J.: 1972, *Astron. Astrophys.* **16**, 482.

Spitzer, L.: 1956, *Physics of Fully Ionized Gases*, Interscience, New York.

Strand, K. A.: 1958, *Astrophys. J.* **128**, 14.

Tenorio-Tagle, G.: 1973, *Astrophys. Space Sci.* **23**, 221.

Unno, W.: 1952, *Publ. Astron. Soc. Japan* **3**, 158.

Vandervoort, P. O.: 1963, *Astrophys. J.* **138**, 294.

Wickramasinghe, N. C.: 1967, *Interstellar Grains*, Chapman and Hall Ltd., London.

Wickramasinghe, N. C. and Nandy, K.: 1971, *Nature Phys. Sci.* **230**, 16.

Wickramasinghe, N. C. and Williams, D. A.: 1968, *Observatory* **88**, 272.

Wilson, O. C., Münch, G., Flather, E. M., and Coffeen, M. F.: 1959, *Astrophys. J. Supp. Ser.*, No. 40, **4**, 199.

Witt, A. N.: 1973, in J. M. Greenberg and H. C. van de Hulst (eds.), *Interstellar Dust and Related Topics*, D. Reidel Publishing Co., Dordrecht, Holland, p. 53.

Witt, A. N. and Lillie, C. F.: 1973, *Astron. Astrophys.* **25**, 397.

Wurm, K. and Rosino, L.: 1959, *A Monochromatic Atlas of the Orion Nebula*, Asiago, Italy.

THERMAL EMISSION SPECTRA OF SILICATES
FROM PLANETARY NEBULAE

E. BUSSOLETTI

Istituto di Fisica, Università degli Studi, Lecce, Italy

and

J. P. BALUTEAU and N. EPCHTEIN

Observatoire de Meudon, France

Abstract. Calculations of the grain equilibrium temperature and of the expected infrared spectra of IC 418, BD+ 30° 3639, NGC 6572 and NGC 7027 have been performed using dielectric constants of lunar silicates. The results have been compared with previous work on pure graphite and ice-mantle grains. Lα heating of dust followed by thermal re-emission is consistent with the large infrared excesses detected in planetary nebulae. An extra source of heating is, nevertheless, necessary to fit correctly the experimental results. It appears from the calculations that, for each object, it is possible to define theoretically the most probable nature of the emitting dust.

1. Introduction

Recently Bussoletti *et al.* (1974) have calculated the expected infrared spectra of a number of planetary nebulae considering different grain models. The heating mechanism of Lα absorption by dust followed by thermal re-emission suggested by Krishna Swamy and O'Dell (1968) has been reconsidered.

The heat balance equation has been solved rigorously by taking into account the correct wavelength-dependence of the grain absorption efficiency $Q(\lambda)$.

In the previous work two grain models have been studied: (a) pure graphite, and (b) graphite-core ice-mantle.

This paper presents new calculations where different types of silicate s are considered in order to make the investigation on the dust nature more complete.

2. Calculations

In planetary nebulae which are optically thick to the Lyman continuum and Lyman lines, each stellar photon producing a hydrogen ionization is converted in a Lα photon and a Balmer photon.

The heat balance equation for the case of a single grain heated by a $L\alpha$ flux $F(L\alpha)$ can be written as

$$\pi\delta^2 U(L\alpha)F(L\alpha) = 4\pi\delta^2 \int_{IR} Q(\lambda)B(\lambda, T_g)\, d\lambda, \tag{1}$$

where $B(\lambda, T_g)$ is the black-body spectrum at a temperature T_g, while $U(L\alpha)$, δ and T_g are, respectively, the $L\alpha$ efficiency, the radius and the equilibrium temperature of the dust grain. The infrared flux F_{IR} emitted by the dust is given by

$$F_{IR} = \frac{F(L\alpha)B(\lambda, T_g)Q(\lambda)}{\int_{IR} B(\lambda, T_g)Q(\lambda)\, d\lambda}, \tag{2}$$

where we have assumed that all $L\alpha$ photons are absorbed by the dust. As was already pointed out by various authors, T_g is unaffected by the value of the grain size within the typical range $\delta=0.5\,\mu$–$0.01\,\mu$.

In what follows we shall assume spherical grains with $\delta=0.05\,\mu$ for graphite while $\delta=0.10\,\mu$ is the thickness of the ice mantle. The silicate dust radius is assumed to be equal to $\delta=0.2\,\mu$.

3. Grain Properties

The materials that we consider are lunar plagioclase and pyroxene silicates whose dielectric constants have been measured by Perry et al. (1972) in the frequency range 10–1600 cm^{-1}. These constants in form of tables have been kindly provided to us by Dr R. F. Knacke of Stony Brook University.

Table I reports, from the paper of Knacke and Thompson (1973), the percent mineral abundances of selected samples of silicates that we have considered in our study.

Figure 1 shows the infrared absorption efficiencies of these materials calculated from the Mie theory.

TABLE I

Major oxide in abundance in %

Composition	10058	14310	12009
SiO_2	41	50	39
Al_2O_3	11	20	12
MgO	12.5	8	9
FeO	20	8	22
CaO	10	11	12
TiO_2	3	1	4
(Fe/Ti)oxide	23	9	26

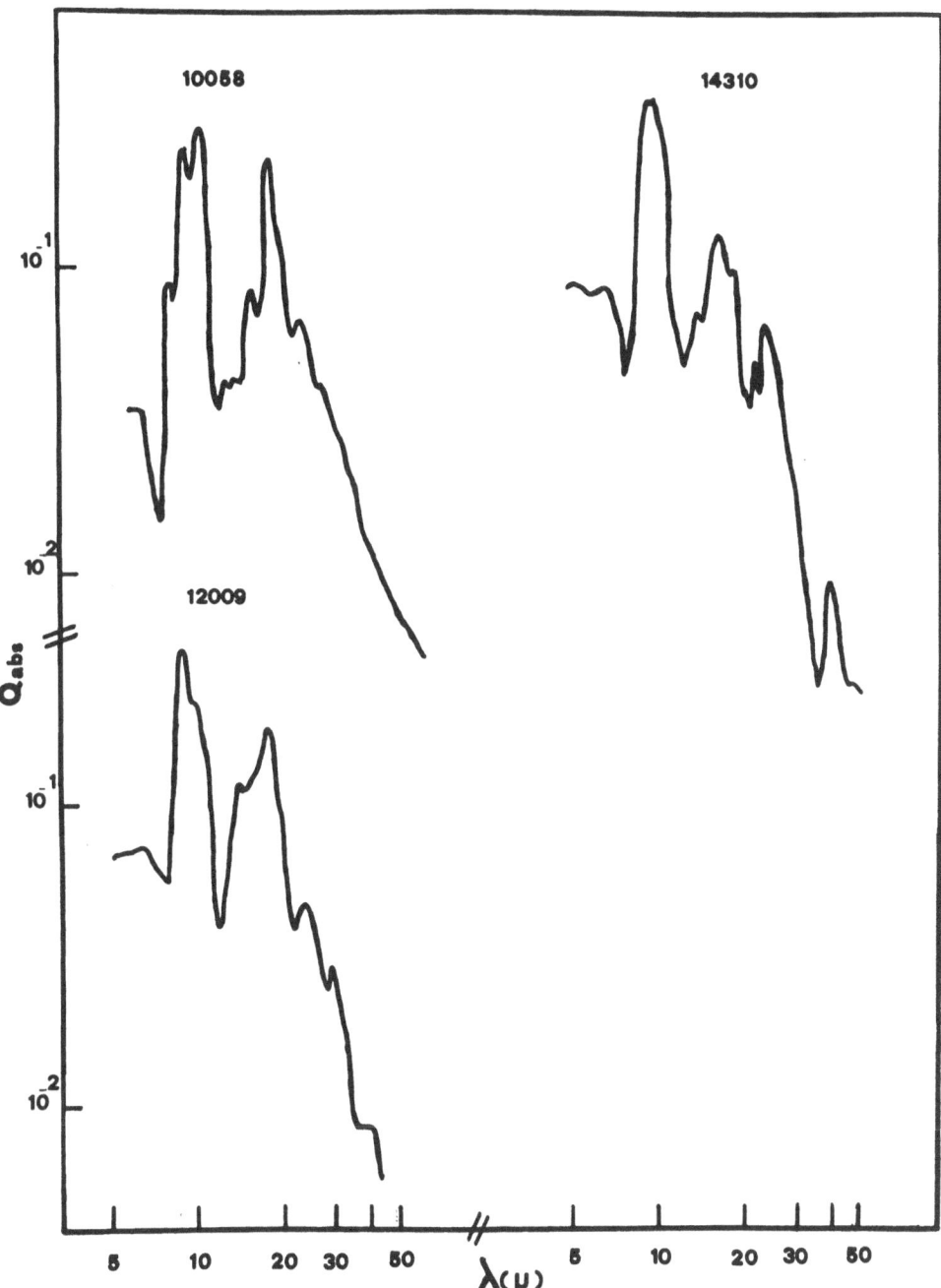

Fig. 1(a, b). Infrared absorption efficiencies for the three silicates whose characteristics are reported in Table I. Each sample is labelled according to the classification given in Knacke and Thompson (1973).

In our case, where $\lambda \gg \delta$ we can write, according to van de Hulst (1957) and Wick-ramasinghe (1967),

$$Q(\lambda) = -4\chi \operatorname{Im}\left[\frac{\varepsilon - 1}{\varepsilon + 2}\right] - \tfrac{4}{15}\chi^3 \operatorname{Im}\left[\left(\frac{\varepsilon - 1}{\varepsilon + 2}\right)^2 \left(\frac{\varepsilon^2 + 27\varepsilon + 38}{2\varepsilon + 3}\right)\right], \qquad (3)$$

where $\chi = 2\pi\delta/\lambda$ and $\varepsilon = m^2(\lambda)$ is the complex dielectric constant of the grains.

An examination of the figure shows that $Q(\lambda)$ is characterized by strong absorption bands peaked, respectively, at 9 μ, 11 μ, 18.5 μ for the silicate labeled 12009; at 9.5 μ and 16.5 μ for that 14310 and finally at 9 μ and 18 μ for that labeled 10058. These strong bands are due to Si–O stretching modes for longer wavelengths and to Si–O–Si bending modes for the shorter wavelengths. Less strong bands appearing beyond 22 μ are due to the different cation content of each material.

In addition to these lunar silicates, pure graphite particles and ice-mantle graphite-core particles have been considered.

In this case we have assumed the following dust properties: (1) *Graphite*, $m_g(\lambda) = 6.5 - i\,0.47\lambda$; (2) *Ice*, $m_i = 1.73 - i\,0.08$, according to Werner and Salpeter (1969).

4. Results and Discussion

As in our previous paper, we have considered two possible dust locations in the nebula: (1) dust only in the neutral region outside the ionized part of the planetary, (2) dust only within the ionized region of the nebula.

The mapping of NGC 7027 done by Knacke and Dressler (1973) and by Becklin *et al.* (1973) shows that regions of the nebula emitting the infrared excesses are generally correlated with those of radio emission indicating that hot grains coexist with the ionized gas.

Case 1 is considered here for completeness, because silicates have never been con-sidered previous to our present investigation.

We have analyzed the following objects: IC 418, BD+30° 3639, NGC 6572 and NGC 7027. Their physical characteristics are reported in Table II with the value of the Hβ flux corrected for the interstellar extinction according to Terzian and Sanders (1972). In the table we give the maximum and minimum diameters found in the litera-ture for each object.

TABLE II

Nebula	$\log F(\mathrm{H}\beta)$ $(\mathrm{erg\ cm^{-2}\ s^{-1}})$	Angular min. diam.	Angular max. diam.
IC 418	−9.20	6.2	–
BD+ 30° 3639	−9.41	2.5	5.5
NGC 6572	9.22	5.3	7.2
NGC 7027	8.66	4.9	7.3

Table III summarizes the values of the grain equilibrium temperatures that we find for the two dust locations in the nebulae. We note that for case 2 the ice mantle can no longer exist because the higher temperatures cause its evaporation.

Our calculations show that the temperatures for the silicates are much lower than those for pure graphite and than those for composite particles.

The infrared flux derived from Equation (2) in case (1) is too low to account for the experimental data of Woolf (1969), Neugebauer and Garmire (1970), Willner *et al.* (1972), Gillett *et al.* (1972), and Cohen and Barlow (1974).

We conclude then, as already found for graphite and ice-mantle grains, that also for silicates the infrared excesses in the 8 μ–22 μ range cannot be explained by the presence of cold dust surrounding the ionized region of the nebula.

In the case 2 model, the Lα radiation is resonantly trapped within this part of the planetary nebula. All grains, regardless of their position in the nebula, have a temperature uniform and higher than that expected for case 1. These values of T_g are reported in the second part of Table III.

The spectra that we have computed are, in general, quite different. A common result, nevertheless, is that the simple Lα heating of dust is no longer sufficient to account for the experimental data. This has been clearly shown by the measurements of Cohen and Barlow (1974). Among 113 planetaries that they have studied, only few were such that the infrared luminosity L_{IR} was lower than the Lα luminosity. On the contrary, it was usual that $L_{IR} \gg L_{L\alpha}$. The additional heating necessary to account for the experimental data can reasonably be attributed to the absorption of appreciable quantities of non-ionizing ultraviolet radiation by dust.

Figure 2(a, b) and Figure 3(a, b) show (in flux units vs microns) the calculated emission spectra of the four planetaries. The calculations have been performed for the minimum nebular diameter considering the two models of pure silicate and pure graphite. The spectra for silicates concern the mixture 12009. The results for the other mixtures do not differ in the band shape but only in the quantity of extra heating which is reduced of 30% at most.

The calculated spectra have been compared with all the experimental results available in the literature up to date. In the figures we have indicated the error bars for each point whenever reported in the original papers.

An inspection of Figures 2 and 3 leads to the following remarks:

IC 418. The spectrum which fits correctly the experimental results is that emitted by the graphite grains. Silicate particles can then be ruled out from the possible dust composition in this object. The photometric results up to 3.4 μ cannot be explained either by our model or by free-free emission and require the existence of further processes to account for them.

BD + 30° 3639. The spectrum computed for the silicate model are clearly in agreement with observations. In this case again the fluxes up to 3.4 μ are more intense than those expected by the simple free-free emission.

NGC 6572, NGC 7027. In both cases the spectra computed for the silicates are in

TABLE III[a]

Grain temperatures (deg K)

Objects	Case 1					Case 2			
	Silicates			Graphite	GC-IM	Silicates			Graphite
	12009	10058	14310			12009	10058	14310	
IC 418	18	23	26	86	53	97	101	104	188
BD+ 30° 3639	15	20	25	80	48	186	186	190	359
NGC 6572	17	23	27	85	51	141	144	148	275
NGC 7027	27	30	37	102	68	155	157	161	301

[a] Results for the minimum observed diameter of the object.

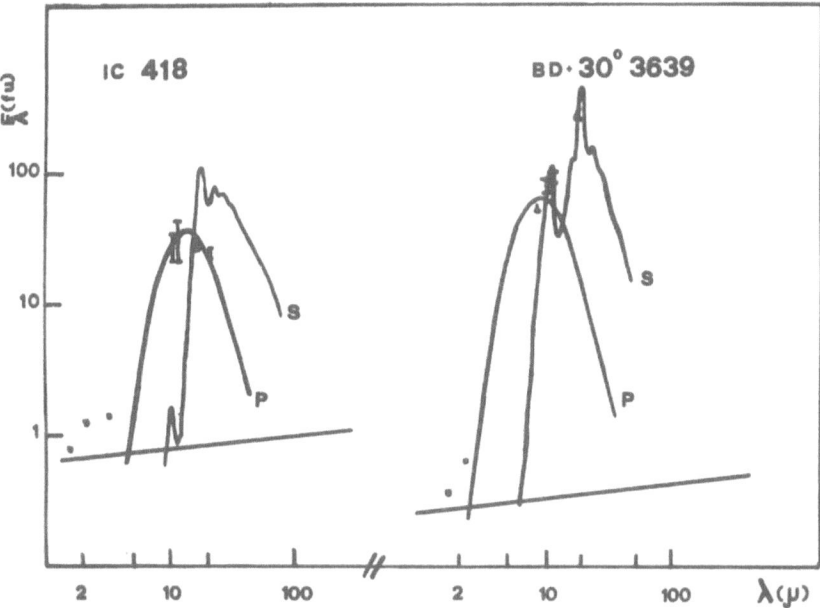

Fig. 2(a, b).   Calculated emission spectra, in flux units, for IC 418 and BD+ 30° 3639 for the case of dust mixed with the ionized gas. The expected free-free emission is given by the straight line at the bottom of the figures. The experimental points are reported, whenever possible, with their error bars. Experimental bars in the near infrared lie within the size of each point. P: pure graphite: S: silicate 12009.

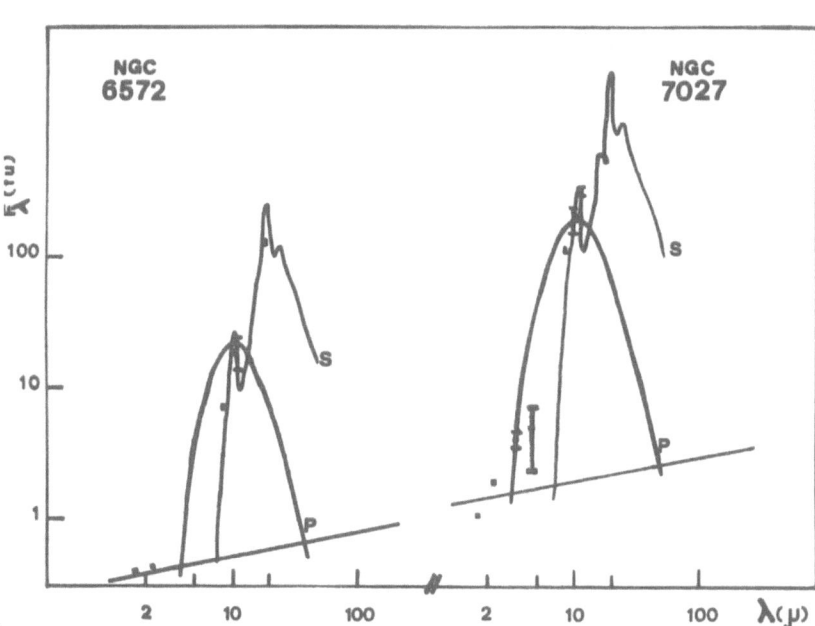

Fig. 3(a, b).   Calculated emission spectra, in flux units, for NGC 6572 and for NGC 7027 for the case 2 model of the dust location. The expected free-free emission and the observational data are also reported. P: pure graphite, S: silicate 12009.

very good agreement with the observations. The measurements in the near-infrared fit very well with the expected free-free emission.

5. Conclusions and Remarks

In this paper we have presented infrared spectra for planetary nebulae with higher accuracy than has been performed before, and we have compared the up to date observational data with predictions.

We have given a theoretical justification that the infrared excesses of planetary nebulae, in the wavelength range 8–22 μ are due to hot dust located within the ionized part of the nebula. It appears that the nature of dust is not unique in these objects. Graphite as well as silicates may be present.

This result does not seem to be, in our opinion, in contrast with the recent results of Cohen and Barlow (1975). These authors claim that some features appear in the 10 μ region but none clearly resembles the common silicate emission signature. Actually the multifilter narrow band photometry used in their investigation is centered at 8.6 μ, 10 μ and 18 μ. These wavelengths are outside, or on the edges of, the emission bands expected from the silicates that we have considered in our study.

In conclusion, we may affirm that dust heating by Lα radiation is consistent with large infrared excesses detected in the planetary nebulae. An extra heating, such as produced by absorption of nonionizing UV radiation, is nevertheless necessary for a correct fit with the observations.

The explanation of the entire infrared spectrum of a planetary nebula is still far from complete. Near-infrared data cannot always be interpreted as due to free-free emission, and require further physical processes – at present unknown – to explain them. The only certain feature is the fact that excesses in the region 8 μ–22 μ seem definitely to be due to hot dust mixed with gas. It is strongly suggested that continuous effort be made to obtain more observations. Low resolution spectroscopy is needed in the range 8–22 μ while, due to the nature of the dust itself, narrow-band photometry will be sufficient beyond 22 μ to obtain information on the emission mechanisms and on the nature of the dust.

Acknowledgements

The authors warmly thank Dr R. F. Knacke who provided tables of the dielectric constants of silicates. The staff of the C.E.C.U.S., the computing center of the University of Lecce, is thanked for having provided the maximum computing facilities to this work. E. Bussoletti was supported in this work under N.A.T.O. grant No. 861.

References

Becklin, E. E., Neugebauer, G., and Wynn-Williams, C. C.: 1973, *Astrophys. Letters* **15**, 87.
Bussoletti, E., Epchtein, N., and Baluteau, J. P.: 1974, *Astron. Astrophys.* **34**, 141.
Cohen, M. and Barlow, M. J.: 1975, *Astrophys. J.*, in press.

Gillett, F. C., Merrill, K. M., and Stein, W. A.: 1972, *Astrophys. J.* **172**, 367.

Knacke, R. F. and Thompson, R. K.: 1973, *Publ. Astron. Soc. Pacific* **85**, 341.

Knacke, R. F. and Dressler, A. M.: 1973, *Publ. Astron. Soc. Pacific* **85**, 100.

Krishna Swamy, K. S. and O'Dell, C. R.: 1968, *Astrophys. J.* **151**, L61.

Neugebauer, G. and Garmire, G.: 1970, *Astrophys. J.* **161**, L91.

Perry, C. H., Agrawal, D. K., Anastassakis, E., Lowndes, R. P., Rastogi, A., and Tornberg, N. E.: 1972, *The Moon* **4**, 315.

Terzian, Y. Y. and Sanders, D.: 1972, *Astron. J.* **77**, 350.

Van de Hulst, H. C.: 1957, *Light Scattering by Small Particles*, Chapman and Hall, London.

Werner, M. W. and Salpeter, E. E.: 1969, *Monthly Notices Roy. Astron. Soc.* **145**, 249.

Wickramasinghe, N. C.: 1967, *Interstellar Grains*, Chapman and Hall, London.

Willner, S. P., Beclin, E. E., and Visvananthan, N.: 1972, *Astrophys. J.* **175**, 699.

CONSIDERATIONS ABOUT THE ABSORPTION
EFFICIENCY OF DUST PARTICLES IN THE INFRARED

E. BUSSOLETTI, A. BORGHESI, G. LEGGIERI, and A. BLANCO

Istituto di Fisica, Università di Lecce, Lecce, Italy

Abstract. Analytical approximations used often in the literature for calculating energy rates emitted by dust grains in infrared are discussed. Comparisons with correct complete formulations are made for three grain models: (1) pure graphite, (2) ice mantle–graphite core, (3) silicates. λ^{-1} and λ^{-2} dependences for the average effective emissivity of such grains are used. We find that for silicate and graphite grains the simplified approximations are valid only when accuracies between 10% and 50% are required and only for grain temperatures higher than 80 K. At lower temperatures the validity of the approximations fails for the graphite particle while it is variable for the silicate dust grain. The ice core mantle particles can instead be treated with approximated formulae without introducing appreciable errors.

1. Introduction

The solid materials which have been considered in various models as constituents of the interstellar particles are of various kinds. Among these, ice, graphite and different silicates are the most commonly considered as emitting particles at wavelengths in the infrared.

The effective particle sizes are (in terms of a sphere radii δ) commonly assumed to be:

$$\delta_{ice} = 0.05\text{–}0.3\ \mu, \qquad \delta_{graph} = 0.05\text{–}0.1\ \mu, \qquad \delta_{sil} = 0.05\text{–}0.1\ \mu.$$

Under certain conditions, ice will condense on either graphite or silicate particles. Such core–ice mantle grains are generally characterized by a core with $\delta_c = 0.05\ \mu$ plus a mantle of $\delta_m = 0.1\text{–}0.2\ \mu$. The absorption efficiency Q_{abs} of such particles can be calculated by using the Mie theory once the dielectric constants of the materials are known over the IR wavelength range. Very often, in the literature, analytical approximations of Q_{abs} are used which are believed to approximate the real values within a few percent.

The aim of this work is to investigate the validity of this statement considering several grain models such as: (a) pure graphite, (b) ice mantle–graphite core, (c) silicate mixtures. The validity of these approximations and their limits will also be discussed.

2. Calculations

If the particle is small compared with the wavelength (which is the case in infrared), i.e., if $2\pi\delta/\lambda=x\ll1$, its absorption efficiency can be calculated by means of the approximated formula (Van de Hulst, 1957; Wickramasinghe, 1968)

$$Q_{abs}(\lambda) = -4x\,\mathrm{Im}\left\{\frac{m^2-1}{m^2+2}\right\} -$$
$$- x^3\,\mathrm{Im}\left\{\frac{4}{15}\left(\frac{m^2-1}{m^2+2}\right)^2\cdot\frac{m^4+27m^2+38}{2m^2+3}\right\}, \tag{1}$$

where $m=m'-im''$ is the complex refraction index of the material considered.

Equation (1), for $m'-1\ll1$ and $m''\ll1$ is often approximated by

$$Q_{abs}\simeq\tfrac{8}{3}xm''. \tag{2}$$

Actually the material absorptivity is far from being uniform in the infrared. Nevertheless, it is very common to consider a maximum possible emissivity or eventually an average effective emissivity of the grain (see, for example, Kaplan and Pikelner, 1970; Greenberg, 1971) of the form

$$\tilde{Q}_{abs}\simeq\frac{4\pi\delta}{\tilde{\lambda}}m'', \tag{3}$$

where $\tilde{\lambda}=\lambda_{eff}$ is the wavelength at which the emissivity is maximum. It is easy to see that $\lambda_{eff}=1.3\lambda_m$, where λ_m is the wavelength of maximum photon emission. On the basis of these assumptions, the particle emission, which should be correctly calculated as

$$\varepsilon_0 = 4\pi\delta^2\int Q(\lambda)\,B(\lambda,T_g)\,d\lambda, \tag{4}$$

is commonly evaluated with the approximation

$$\varepsilon_1 = 4\pi\delta^2\,\tilde{Q}_{abs}\,\sigma T_g^4, \tag{5}$$

where $B(\lambda,T_g)$ is the Planck function (in erg cm^{-2} s^{-1} μ^{-1}) at the grain temperature T_g, and σ is the Stefan constant.

The limits of this approach have been already discussed by Bussoletti et al. (1973) for planetary nebulae. These authors have pointed out the importance of considering correctly the slope of $Q(\lambda)$ within all the IR range. It is actually the crucial factor in determining the value of the energy rate or, inversely, of the grain temperature when the complete heat balance equation is considered.

In order to evaluate the size of the error when the approximate equation (5) is used instead of the correct equation (4), we have calculated the ratio

$$\varepsilon = \frac{\tilde{Q}_{abs}\sigma T_g^4}{\int Q(\lambda)B(\lambda,T_g)\,d\lambda}$$

for three grain models; (a) pure graphite, (b) ice mantle–graphite core, (c) silicate. We have assumed graphite particles with a standard radius $\delta=0.05\,\mu$ while the ice core has been taken to be $\delta_{\text{ice core}}=0.15\,\mu$.* $\delta=0.2\,\mu$ is assumed for the silicate. The complex refractive indices of graphite and ice have been taken from Werner and Salpeter (1968): $m_{\text{graph}}=6.5-i\,0.47\lambda$ and $m_{\text{ice}}=1.73-i\,0.08$. For the silicates we have used the dielectric constants measured from lunar dust by Perry et al. (1972) presented by Knacke and Thompson (1973). These constants, in a tabular form, have been kindly provided to one of us from Dr R. F. Knacke of Stony Brook University. For comparison, we have used in Equation (3) a value of $m''_{\text{sil}}=0.5$ which is a mean value for the imaginary part of the refractive index as considered, for example, by Greenberg (1971) in the infrared wavelength range. Figure 1 reports the plots of $Q_{\text{abs}}(\lambda)$ calculated by means of Equation (1) for the three grain models considered here. We note that the silicate mixture whose dielectric constants we have used is that labeled 14008 in the paper by Knacke and Thompson (1973).

In order to make our investigation more complete, we have performed calculations of Equation (6) considering for $Q_{\text{abs}}(\lambda)$ not only the usual λ^{-1} dependence, but also a λ^{-2} dependence which is probably more realistic when the imaginary part of the refractive index is linearly proportional to λ.

3. Discussion of the Results

The calculations have been performed for a temperature range from 3 K to 300 K. This interval covers a large range of temperatures which can be attained by dust grains in interstellar objects emitting in the infrared.

Figure 2 shows the values of ε calculated from Equation (6) for each of the grain models that we have considered here.

The results can be summarized as follows:

(1) *Graphite:* it appears clearly that the λ^{-1}-dependence is not acceptable. The errors range, in fact, from one to three orders of magnitude. More acceptable is the assumption $Q_{\text{abs}}(\lambda)\propto\lambda^{-2}$. In this case, the error introduced is less than 50% for temperatures greater than 40 K. At any rate, it is clear that, below this temperature, the approximation which is adopted by using Equations (3) and (5) is not acceptable.

(2) *Ice-core mantle:* For this composite particle the effect of the presence of absorbing mantles is noticeable. Large differences between this curve and the best for the graphite suggest that the presence of a core does not affect appreciably the emission properties of an absorbing mantle. In this case the usual approximations of Equations (3) and (5) are largely justified.

(3) *Silicates:* the behaviour of ε in this case is quite different compared to the previous grain models. We see that for $T_g>100$ K the error introduced with a simple λ^{-1}-dependence is less than or equal to 50%. In this case it is noticeable that the

* Actually ε is not dependent on the value of δ.

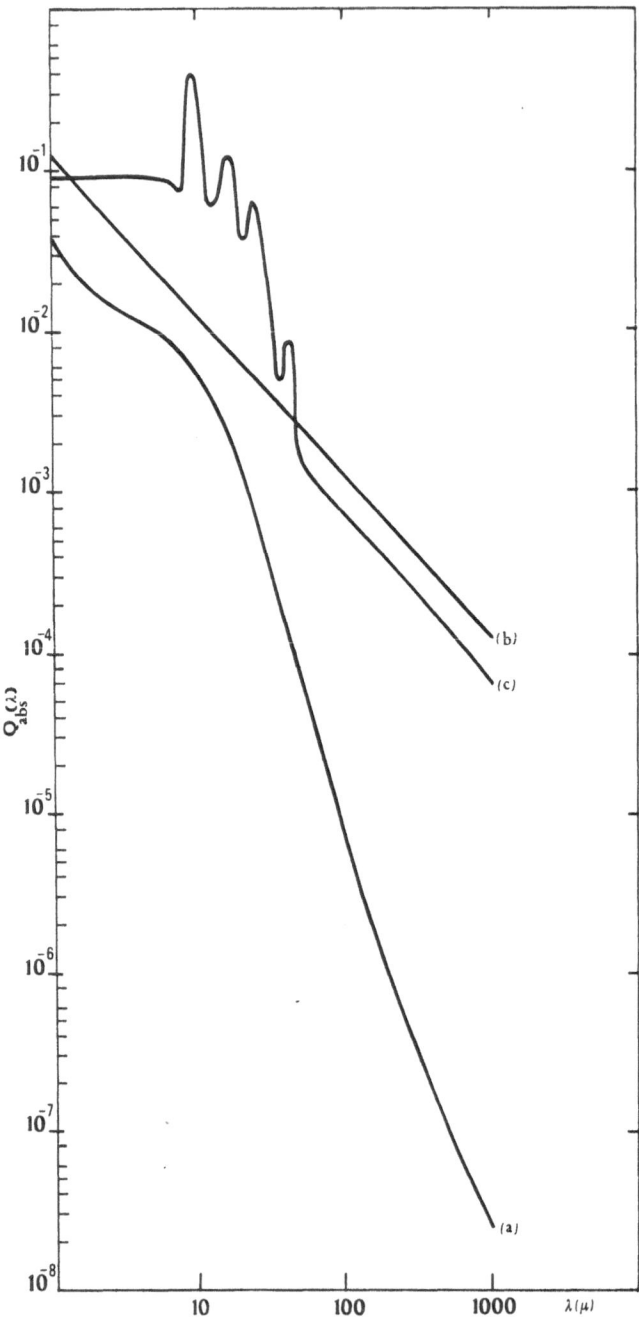

Fig. 1. Absorption efficiencies of respectively (a) pure graphite, (b) ice mantle–graphite core and (c) silicate grains calculated by means of Equation (1). The refractive indices of (a) and (b) particles have been taken from Werner and Salpeter (1968) while for the silicate the dielectric constants of Knacke and Thompson (1973) are used.

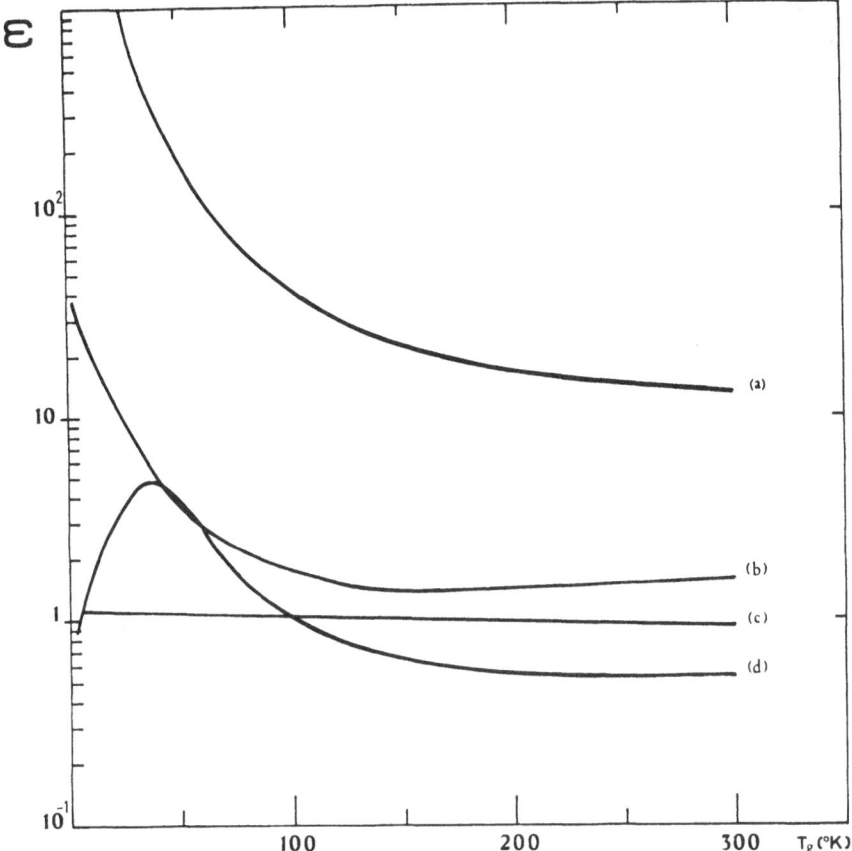

Fig. 2. Plot of the values of ε calculated according to Equation (6), see text, in the range of temperatures 3–300 K for each of the grain models considered in this study. The curves are respectively:
(a) pure graphite, $\tilde{Q}_{abs}(\lambda) \propto \lambda^{-1}$, (b) pure graphite, $\tilde{Q}_{abs}(\lambda) \propto \lambda^{-2}$, (c) ice mantle–graphite core,
$\tilde{Q}_{abs}(\lambda) \propto \lambda^{-1}$, (d) silicate, $\tilde{Q}_{abs}(\lambda) \propto \lambda^{-1}$.

approach of calculating the energy emitted by the dust simply as $\varepsilon_1 = 4\pi\delta^2 \tilde{Q}_{abs}\sigma T_g^4$
represents an underestimate for the energy rate (ε is actually lower than unity).

For temperatures lower than 100 K, ε is always greater than one and presents a
peak for grain temperatures around 35–40 K.

We note that these results represent actual lower limits to the values of ε. We
have in fact considered $\tilde{Q}_{abs} = \frac{8}{3}xm''$. Its maximum value is $(Q_{abs})_{max} = 2x$ for dielectric
particles and $(Q_{abs})_{max} = 3x$ for metallic particles. Because of the presence of the
imaginary part of the refractive index, we see that ε has to be increased by 50% when
$(Q_{abs})_{max}$ is considered.

4. Conclusions

The preceding results can be summarized as follows. For the silicate particles, the
assumption $\tilde{Q}_{abs} \propto \lambda^{-1}$ turns out to be good within 50% for grain temperatures
beyond 80 K–100 K. The same situation is met for the graphite when $\tilde{Q}_{abs} \propto \lambda^{-2}$ is

assumed. At temperatures lower than 80 K, the error for the graphite model diverges rapidly, and approximations for Q_{abs} produce unacceptable errors in the calculation of the energy rate emitted by dust. For silicates the error is maximum (a factor of five) at around 35–40 K. For lower temperatures ε decreases rapidly to acceptable values at temperatures $T_g < 20$ K commonly expected to exist inside dark clouds. The assumption of $\tilde{Q}_{abs} \propto \lambda^{-1}$ is instead completely justified in the case of the ice-mantle-graphite-core grain model. This result is a further indication of the low incidence of an absorbing core in composite particles when the mantles are also appreciably absorbing.

In conclusion, for silicate or graphite grains the simplified approach of considering the peak value for the absorption efficiency ($\tilde{Q}_{abs} \propto \lambda^{-\alpha}$, $\alpha = 1$ or $\alpha = 2$) is valid when accuracies between 10% and 50% are acceptable, and only for temperatures higher than 80 K.

At lower temperatures, the validity of the approximation is questionable and strongly dependent from the nature of the dust. This point must be stressed because it is very common to find in the literature interpretations of the experimental data on the assumption of $\tilde{Q}_{abs} \propto \lambda^{-1}$, regardless of the actual nature of the dust present in the object studied.

This is of great importance for the correct treatment of the heat-balance equation, or in the evaluation of the dust contents of dark clouds where the temperature is expected to be within the range 3–40 K (Krishna Swamy and Wickramasinghe, 1968; Werner and Salpeter, 1969; Field, 1969; Greenberg, 1971).

It is worth pointing out that our results seem to show, apart from the case of the ice-core mantle particles, the importance of a correct calculation of the absorption efficiency of dust in cold objects ($T_g \leqslant 80$ K). Actually the infrared experimental measurements become better and better (the present experimental errors are often lower than 30%) and cannot, therefore, be interpreted by means of theoretical formulae meriting less confidence than the experiments.

Acknowledgements

This work has been partly supported under NATO grant No. 861. E. Bussoletti acknowledges Dr R. F. Knacke for having provided to him the data used in this work. The staff of the C.E.C.U.S. (University of Lecce) is warmly thanked for enabling us to use their computing facilities.

References

Bussoletti, E., Epchtein, N., and Baluteau, J. P.: 1974, *Astron. Astrophys.* **34**, 141.
Field, G. B.: 1969, *Montly Notices Roy. Astron. Soc.* **144**, 411.
Greenberg, J. M.: 1971, *Astron. Astrophys.* **12**, 240.
Kaplan, S. A. and Pikelner, S. B.: 1970, *The Interstellar Medium*, Harvard Univ. Press, Cambridge, Mass.

Knacke, R. F. and Dressler, A. M.: 1973, *Publ. Astron. Soc. Pacific* **85**, 100.

Krishna Swamy, K. S. and Wickramasinghe, N. C.: 1968, *Monthly Notices Roy. Astron. Soc.* **139**, 283.

Perry, C. H., Agrawal, D. K., Anastassakis, E., Lowndes, R. P., Rastogi, A., and Tornberg, N. E.: 1972, *The Moon* **4**, 315.

Van de Hulst, H. C.: 1957, *Light Scattering by Small Particles*, John Wiley, New York.

Wickramasinghe, N. C.: 1968, *Interstellar Grains*, Chapman and Hall, London.

ON THE PRESENCE OF PHYLLOSILICATE MINERALS
IN THE INTERSTELLAR GRAINS

A. ZAIKOWSKI, R. F. KNACKE, and C. C. PORCO*

Dept. of Earth and Space Sciences, State University of New York, Stony Brook, N.Y., U.S.A.

Abstract. The composition of the interstellar silicate dust is investigated. Condensation or alteration of silicate grains at temperatures of a few hundred degrees, in the presence of H_2O, would result in hydrous or phyllosilicates, the silicate type most abundant in the type I carbonaceous chondrites. Infrared spectra of small particles ($\sim 0.1\,\mu$) of the high temperature condensates, olivine and pyroxene, at 300 K and 4 K do not give a good match to the interstellar absorption band near $9.8\,\mu$. Laboratory spectra of several phyllosilicates give better agreement as does the spectrum of a carbonaceous chondrite. We propose that the silicates in the interstellar grains are predominantly phyllosilicates and suggest additional spectral tests for this hypothesis.

1. Introduction

The detection of an interstellar absorption band near $10\,\mu$ has provided rather compelling evidence for the existence of silicate dust grains in the interstellar medium. The band has been observed in absorption and emission with indications that similar minerals may be responsible in both cases. There is also some evidence for an absorption band near $20\,\mu$ as would be expected for the silicon-oxygen bending mode near this wavelength (Woolf, 1973).

Absorption bands near 10 and $20\,\mu$ are characteristic of all silicates, and are due to near-neighbor (Si-O) vibrations. The composition and crystal structure of the silicates determines the details of the shapes and positions of the bands. These infrared properties have long been used in silicate mineralogy, but have not yet been extensively applied to the study of silicates in the interstellar grains. Olivine and orthopyroxene grains have been suggested by condensation studies and by analogy with common minerals, but most workers have not gone beyond the reasonable guess that a mixture of silicates is likely to be present.

In this paper we suggest that a particular class of silicates, the hydrous forms, are probable grain constituents. We will discuss below three lines of evidence, meteoritic composition, condensation processes, and the properties of the observed absorption bands, which indicate that these silicates may be an important component of the interstellar dust particles.

* Now at Division of Planetary Sciences, California Institute of Technology, Pasadena, Calif.

2. Meteoritic Composition and Condensation Processes

A number of theoretical and observational studies indicate that circumstellar processes play an important role in the formation of interstellar grains (Woolf, 1973). Gilman (1969) and Dorschner (1968) proposed that silicates could be ejected from stellar atmospheres, and observations of an apparent silicate feature in the spectra of M stars supports this suggestion. Herbig (1970) suggested that grains formed in primeval stellar nebulae could be expelled into the interstellar medium, and evidence for circumstellar T-Tauri dust shells is now strong.

If one grants the hypothesis that grains are either formed or processed in a circumstellar environment, then it is natural to look to the present Solar System content, especially meteorites, and to circumstellar or protostellar condensation processes for insight into the chemical composition and solid structure of the interstellar grains. Such a study has been carried out, for example, by Field (1973) who considered circumstellar and interstellar processes to construct models of grain composition.

With regard to the present content of the Solar System, there is much evidence that the type I carbonaceous chondrites contain unprocessed, apparently primitive material. These meteorites show little evidence of heating, are most similar in abundance to cosmic rays and the Sun, and are richest in depleted elements (Anders, 1971). Consequently, these meteorites may resemble an assemblage of interstellar material and thus provide direct insights into the interstellar dust (Wood, 1968). We remark, however, that this is by no means a universally accepted view, and conclusions based on this interpretation are correspondingly subject to debate.

The type I carbonaceous chrondrites are devoid of chondrules and consist primarily ($\sim 60\%$) of a fine-grained ($\sim 0.1\,\mu$) *hydrous silicate* assemblage with the silicate occurring as serpentine, chlorites, or montmorillonites (Nagy, 1966; Bass, 1971). Analysis of the type I meteorite, Orgueil, also indicates the presence of magnetite ($\sim 10\%$), troilite ($\sim 5\%$), sulfur ($\sim 2\%$), sulfates ($\sim 17\%$), and limonite ($\sim 0.5\%$). Water, organic compounds, and traces of olivine make up the remainder ($\sim 10\%$) of Orgueil (Kerridge, 1967; Jeffery and Anders, 1970). The type II's contain varying amounts of material in the form of chondrules distributed in a fine-grained matrix of hydrous silicate and fragments of olivine and pyroxene. Hydrous silicates are only a minor phase in type III's.

The predominance of the hydrous silicates in the most primitive meteorites suggests that these minerals deserve consideration as interstellar grain constituents. At temperatures below 1000 K, the stable phases of the ternary system, $MgO—SiO_2—H_2O$ shift away from forsterite and enstatite to silicates incorporating hydroxyls. First talc can coexist with enstatite and the vapor, and then serpentine becomes stable at lower temperatures (Bowen and Tuttle, 1949; Johannes, 1968). In general, processes at temperatures of a few hundred degrees in the presence of H_2O and other minerals tend to give silicate minerals incorporating the H_2O in the form of hydroxyls or water. The silicates found in carbonaceous chondrites include this type of *layer* or *phyllo-*

TABLE I

Phyllosilicates[a]

Name	Formula
Talc	$Mg_3(Si_2O_5)_4(OH)_2$
Chrysotile (Serpentine)	$Mg_6(Si_2O_5)_2(OH)_8$
Montmorillonite	$(\frac{1}{2}Ca, Na)_{0.7}(Al, Mg, Fe)_4(Si, Al)_8O_{20}(OH)_4 \cdot n\ H_2O$
Chlorite	$(Mg, Al, Fe)_{12}(Si, Al)_8O_{20}(OH)_{16}$

[a] From Frye (1974). The composition as well as nomenclature of the clay minerals varies. Details can be found in Frye and in Grim (1968).

silicates. Other examples of these minerals are given in Table I. Since major grain constituents should be composed of reasonably abundant elements, Table I is restricted accordingly. Details of the structure and chemistry of the phyllosilicates can be found in references such as Grim (1968) and Frye (1974).

There does not appear to be a generally accepted view regarding the origin and evolution of the low temperature hydrous silicates in the carbonaceous chondrites. In early work, DuFresne and Anders (1962) concluded that the hydrous silicates are the result of the alteration of olivine by the action of liquid water within a parent object, subsequent to accretion. More recently, Larimer and Anders (1967) have proposed that the hydrous silicates were formed prior to accretion from anhydrous phases (olivine and enstatite) by reactions with H_2O at temperatures of 300–400 K in the solar nebula. Cameron's (1973) models suggest that interstellar grains will be vaporized only within 2 AU of the early Sun. Consequently, accretion of cool, relatively unaltered interstellar grains may occur in the outer parts of the nebula. Further evidence for the formation of hydrous silicates in the presolar nebula is discussed by Lewis (1972) with reference to the low densities of the outer planets. Arrhenius and Alfvén (1971) favor direct condensation of phyllosilicates under non-equilibrium conditions in which the grains are cooler (because of their efficient infrared emission) than the gas. Notably, an experimental basis for the formation of phyllosilicates from the condensation of atoms has been indicated by the work of Meyer (1971) although precise characterization of the condensates was difficult because of poor crystallinity.

The grains observed surrounding M stars are at temperatures of a few hundred degrees. It is not known where condensation occurs, but if it takes place at some distance from the star, low-temperature processes leading to phyllosilicates could be important in M stars as well. Furthermore, in this case, as well as in prestellar nebulae, if high temperature condensates such as olivines and orthopyroxenes form first, they could later be converted to hydrous silicates if such grains spend any considerable time in the presence of H_2O.

We conclude that there are several possible mechanisms in the condensation processes that result in hydrous silicates. This suggestion differs from some previous studies which have assumed that silicates expelled into the interstellar medium would be high temperature condensates. In the next section we discuss the observational evidence for the mineral composition of the grains.

3. Comparison of Astronomical and Laboratory Spectra

3.1. ASTRONOMICAL SPECTRA

The silicate spectral feature near 10 μ, upon which the silicate identification is based, has been observed in emission in the circumstellar envelopes surrounding late-type stars as well as early-type stars and in the infrared spectra of H II regions (cf. Woolf, 1973). In absorption, the feature has been detected in the galactic center (Woolf, 1973), the Becklin-Neugebauer source in Orion (Gillett and Forrest, 1973), and in AFCRL No. 809–2992 (Merrill and Soifer, 1974). In both the Becklin-Neugebauer and the AFCRL sources the absorption can be approximately fitted by inverting the *emission* band in Orion, thus indicating that similar silicate materials may be present in many sources (Merrill and Soifer, 1974). Recent work by Forrest (1974) does, however, indicate real differences in the shape of the emission feature in some objects.

In Figure 1 we have replotted the absorption data from the three sources quoted above to facilitate intercomparison and comparison with laboratory data below. The absorption spectra are less affected by source temperature variations than the emission spectra, thus making comparison with laboratory spectra easier. Since the shape of the continuum in the three sources is unknown, we assumed that the points of minimum absorbance occur at 8 and 13 μ. This follows the procedure described by Lyon (1963) and seems justified because the data of Hunt *et al.* (1950) for some thirty silicates show transmissions over 95% at 8 μ for all silicates even when the band near 10 μ is obviously strongly saturated. Transmission at 13 μ is not as high but in most cases is in the range of 70 to 80%. Our conclusions below do not depend upon this normalization.

The three spectra are remarkably similar in band shape and position. All have a single absorption band centered near 9.8 μ. The band in the galactic center is somewhat narrower than the other two. Recently Gillett and Capps (1974) have found several more absorptions associated with heavily obscured regions. These reproduce rather closely the three spectra shown here with the exception of a band in the OH source, W33A, which is centered near 9.4 μ. Since the emission bands are apparently related to the absorptions, there are then a fair number of objects with a smooth band centered near 9.8 μ.

That the band shapes and positions do not agree in detail with laboratory spectra of silicates such as the pyroxenes or olivines has been noted previously by, among others, Gillett and Forrest (1973), Knacke and Thompson (1973), and Day (1974). The laboratory absorption bands of these minerals are too broad and contain too much structure. This sense of disagreement is, of course, harder to explain than if the con-

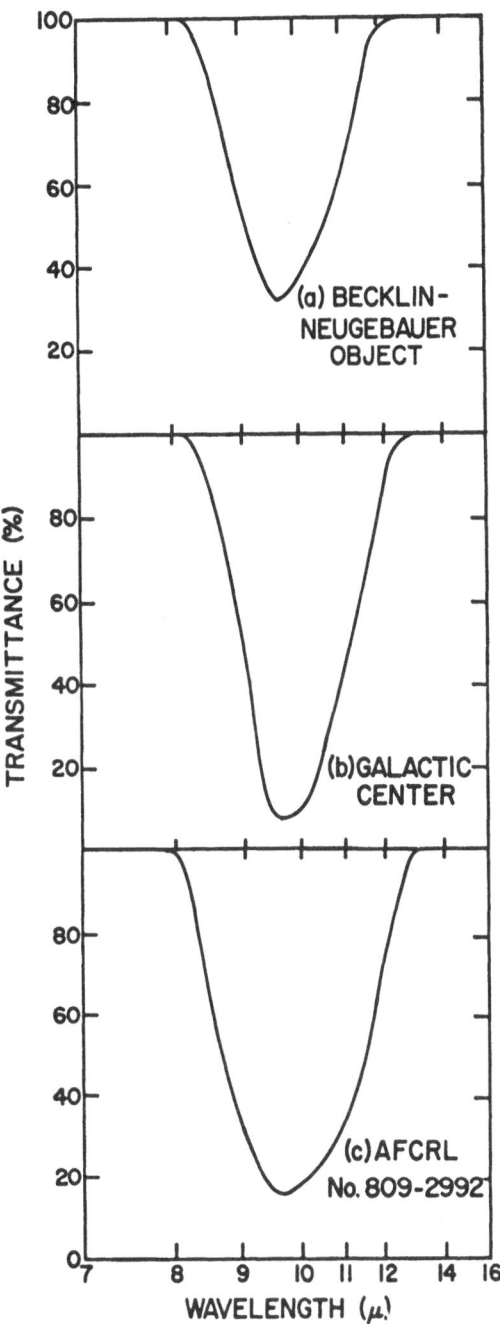

Fig. 1. Spectra of the Becklin-Neugebauer Object, the Galactic Center, and AFCRL 809-2992 obtained from references cited in the text and normalized to 100% transmittance at 8 μ and 13 μ.

verse were the case, since the expected mixture of different silicates in the interstellar grains would tend to broaden the interstellar band compared to the bands of individual minerals studied in the laboratory.

In the remainder of this section we will use these characteristics of the interstellar band to attempt to deduce the nature of the grain silicates.

3.2. LABORATORY SPECTRA, HIGH TEMPERATURE CONDENSATES

The minerals in this study were chosen on the basis of the arguments of Section 2. Of particular interest are the olivines and pyroxenes in their magnesium-rich forms, forsterite (Mg_2SiO_4) and enstatite ($MgSiO_3$), since these are the most abundant minerals expected from condensation (above 1000 K) of a gas of cosmic composition (Grossman, 1972). We also studied their low temperature alteration products, talc and serpentine, and the major silicate minerals which have been found in the carbonaceous chondrites, discussed above. A few other layer silicates of similar structure and composition were also investigated, as well as a type II carbonaceous chondrite. Mineral identifications were made on the basis of X-ray diffraction data.

While data on the infrared properties of enstatite and forsterite exist in the literature, Martin (1971) pointed out that much of the published data is not appropriate for quantitative comparison with interstellar absorption because of saturation in the bands of individual particles more than about 1 μ in diam. Therefore, we have obtained spectra of smaller particles in which the size was determined by inspection with an electron microscope. We have also obtained spectra at wavelengths between 15 and 40 μ because this region is becoming accessible to astronomical spectroscopy.

The particles were made by grinding natural and synthetic samples in an agate mortar with acetone. Typical grinding times were eight to twelve hours per gram of starting material. The particles were then allowed to settle in acetone for periods of time calculated from Stokes's law to yield upper limits on the particle size. The upper portion of the suspension was drawn off and centrifuged to collect the particles. They were then dried and mixed with KBr and pressed into a disk according to standard techniques. The only significant difference between our methods and those described in the literature (Hunt *et al.*, 1950; Lyon, 1963) appears to have been that a lower upper limit on particle size was used to determine settling times.

Olivine. Samples of magnesium-rich olivine (greater than 80% forsterite) which were prepared according to the procedures above are shown in the electron microscope photograph in Figure 2. The grinding and separating process resulted in particles with a mean diameter near 0.1 μ. This size should be small enough to avoid single particle saturation effects (cf. Day *et al.*, 1974); indeed it is of the order of estimated interstellar grain sizes. Note, however, the clumping of particles which may persist when the particles are mixed with KBr and pressed into a disk. If it does, the effective particle size would be larger than the individual particles in Figure 2. The figure shows that the grinding gave remarkably spherical olivine particles. While this is *not* necessarily desirable for comparison with interstellar particles, such particles might eventually provide good experimental tests of Mie calculations for silicates when good optical constants become available.

Spectra of forsterite are given in Figure 3. All the transmission spectra contain the absorption features of the silicates as well as scattering and reflection properties of the KBr disks. The latter influenced the continuum levels of the spectra and proved difficult to control. Thus the transmittance outside the bands, while usually high, is not a precise measure of the transmittance of the silicate particles. For quantitative estimates of

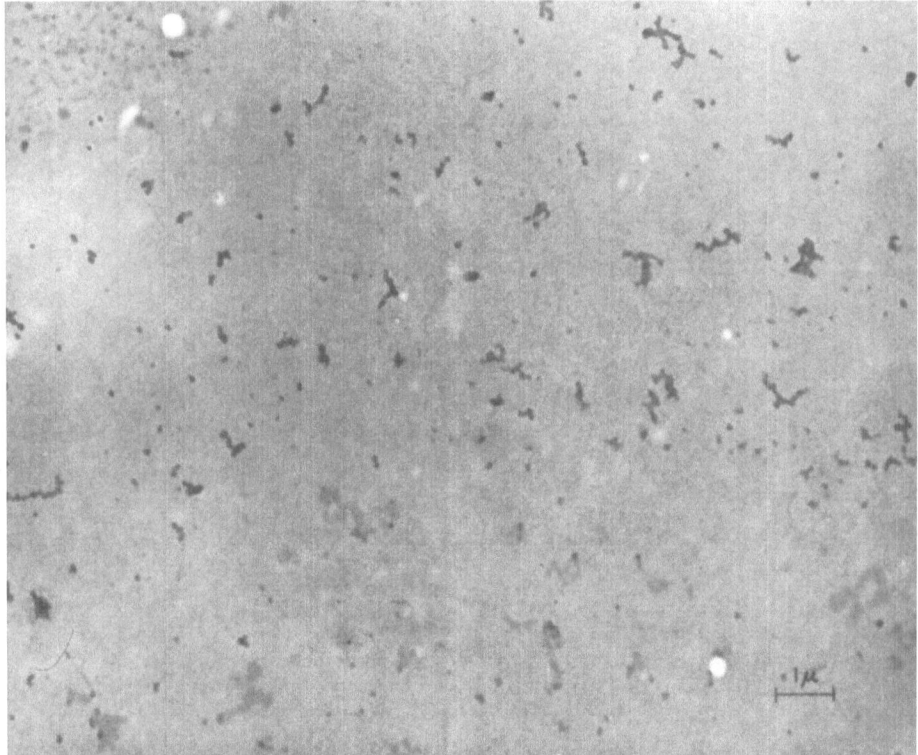

Fig. 2.   Electron microscope photograph of small forsterite particles.

the strength of the band near $10\,\mu$, it is usually more appropriate to draw a line through the 8 and $13\,\mu$ points to estimate the 100% level as was done for the astronomical spectra of Figure 1. We have not chosen to artificially renormalize the laboratory data in this way in the figures even though such renormalization makes comparison with the astronomical spectra easier.

The spectrum of the small-particle forsterite is shown in Figure 3a. The spectrum has absorption maxima near 10.2 and 11.3 μ, in disagreement with the observations. Note that the relative strength of the absorptions is such as to always shift the peak absorption far from 9.8 μ, the center of the astronomical absorption, even for spectra

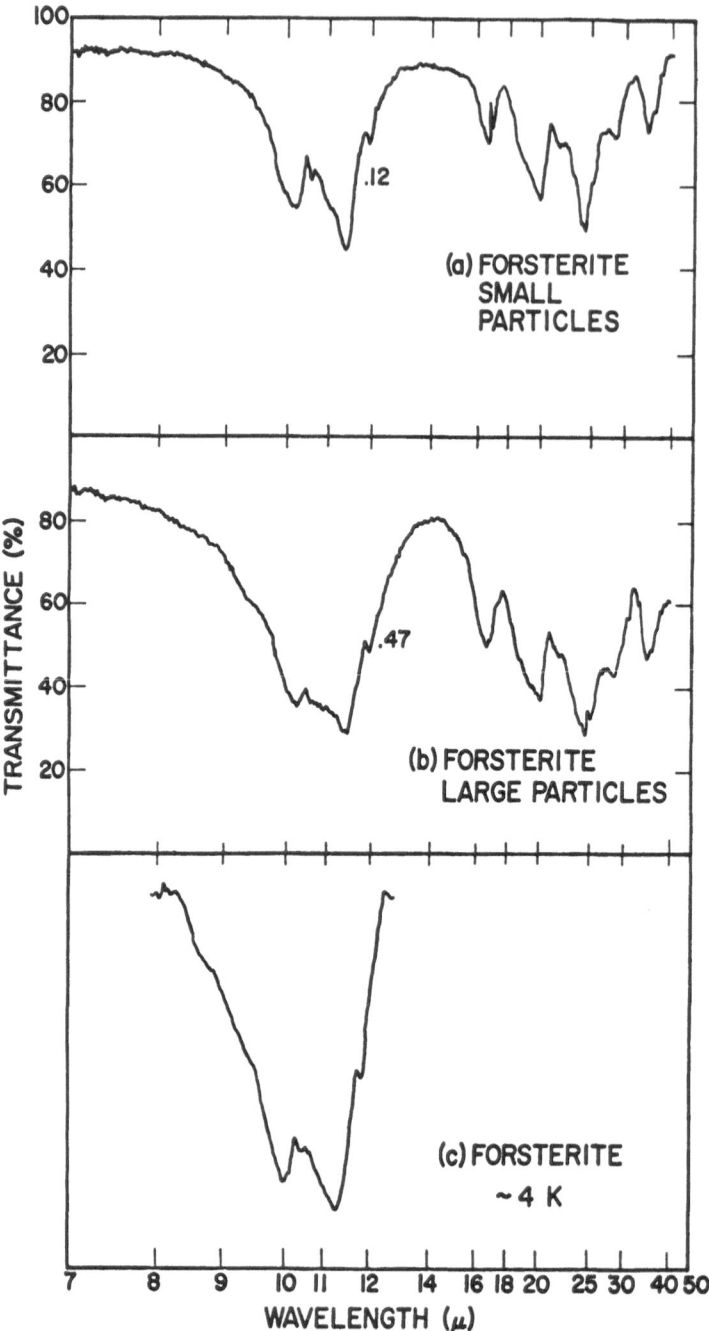

Fig. 3. (a) Spectrum of forsterite particles with mean diameter $\sim 0.1\ \mu$. The mass density (mg cm^{-2}) is given in the figures. (b) Spectrum of forsterite particles with mean diameter estimated to be $\sim 5\ \mu$. (c) Spectrum of small particle forsterite at ~ 4 K. The mass density could not be accurately determined but is estimated to be between 0.1 and 0.6 mg cm^{-2}. Uncertainty in the wavelength scale in the low temperature spectrum is about $0.2\ \mu$.

more strongly saturated than Figure 3a. Lyon's (1963) spectra and the recent spectra of Burns and Huggins (1972) are similar to Figure 3a. Burns and Huggins found a slight shift to longer wavelength as the iron content in the olivines is increased. However, Launer (1952) found the absorption at 10.2 μ to be stronger than the one at 11.3 μ. We do not know the reason for the disagreement with other spectra, but even in this case, forsterite gives a poor fit to the astronomical data.

In Figure 3b the spectrum of a sample containing a distribution of particles with sizes up to 40 μ is shown. The average size is estimated to be near 5 μ. The effects of saturation seem to be indicated in the band shape. Note also that there is almost four

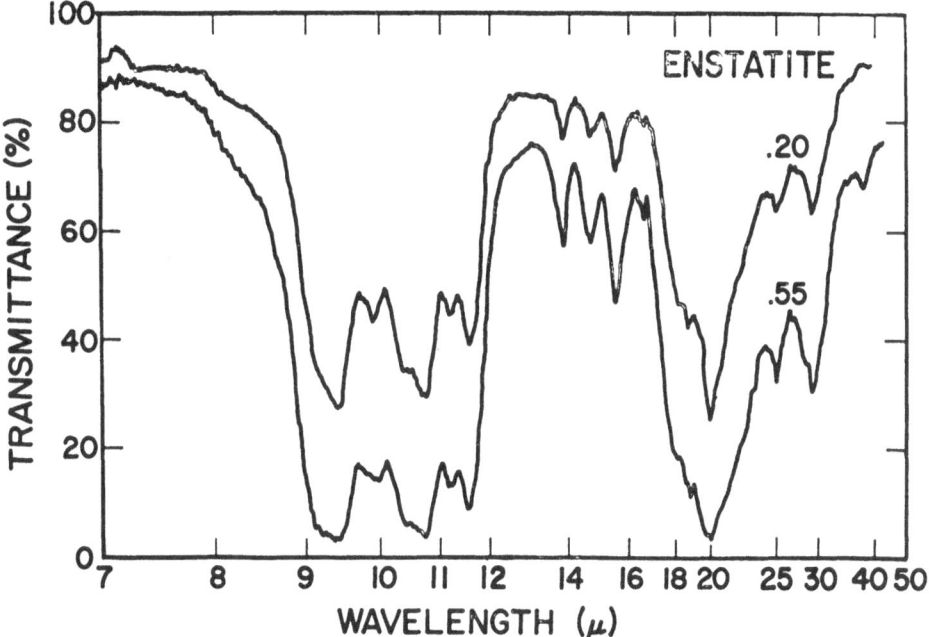

Fig. 4.   Spectra of enstatite.

times as much material in this sample as in the sample of Figure 3a although the absorption is not even twice as deep. It seems likely that these effects are due to single particle saturation as suggested by Martin (1971).

Since the interstellar grains giving the absorptions of Figure 1 are almost certainly colder than 100 K, comparison with laboratory spectra taken at ambient temperature might be questioned. In these materials the bandwidths probably reflect structural properties of the crystal and differences in the environment of individual silicate tetrahedra. Since these properties are not strongly temperature dependent at low temperatures, the absorption band shapes are not expected to be very temperature dependent either. To check this point, a spectrum (8–14 μ) of forsterite at 4 K was taken and is shown in Figure 3c. This spectrum is of lower resolution and quality than the room temperature spectra, and is not useful for quantitative analysis. The uncertainty in the wavelength calibration is about 0.2 μ. However, it clearly shows that neither the width, position, or relative intensities of the two peaks are strong functions of temperature. In particular, temperature effects do not significantly change the disagreement between forsterite and the astronomical spectra.

Enstatite. Enstatite particles were prepared in the same way as the forsterite particles were, so we estimate particle sizes of order 0.1 μ. Absorption spectra for two mass densities of these particles are shown in Figure 4. The disagreement with the astronomical observations is apparent. Enstatite has several distinct bands between 9.2 and 11.7 μ. The strong absorptions are at 9.4 and 10.8 μ in contrast with the single peak at 9.8 μ in the observations. Saturation of an interstellar enstatite band would give a flat-bottomed shape from 9 to almost 12 μ which would not reduce the disparity since the intensity in the spectra of Figure 1 is almost a factor of two higher at 9 and 11 μ than at band center.

Mass absorption coefficients may be derived from the data if 100% transmission is defined from the 8 and 13 μ points according to the procedure described above. For small particle forsterite we find 5800 cm^2 cm^{-1} at 11.2 μ compared to 3900 cm^2 gm^{-1} found by Day *et al.* (1974). A much smaller value would be derived from the large-particle sample (Figure 3b), again probably due to single-particle saturation. Similarly the disagreement with Day *et al.* is most probably due to the presence of larger particles in their spectra. We derive 5900 cm^2 gm^{-1} at 10.6 μ for enstatite. This is larger than would be estimated from Hunt *et al.*'s (1950) spectra.

3.3. LABORATORY SPECTRA, LAYER SILICATES

A selection of phyllosilicates were investigated following generally similar experimental procedures as were used for the forsterite and enstatite studies. Reviews of earlier infrared work on layer silicates are found in Lazarev (1972), Farmer and Russell (1964), and Stubican and Roy (1961).

Talc. Talc has one of the simpler chemical formulas of the phyllosilicates (Table I). Its structure consists of layers of SiO$_4$ tetrahedra with layers of Mg(OH)$_2$ between (Frye, 1974).

Because the layer structure is very different from the independent tetrahedra structure of olivine or the chain structure of pyroxene, the results of the grinding were also investigated with the electron microscope. Talc particles are shown in Figure 5. The flat platelets of talc did not settle out of the suspension according to Stoke's law, and the result was that particles and aggregates of particles up to several microns in size were present in our samples. This is probably also true of the other layer silicates studied below. We do not know if the particles in Figure 5 are solely the remainder of the original many-layered structures or if small particles cling to each other. The particles may be a result of both effects. Note that the layer structure of the bulk samples is maintained down to micron sizes.

Spectra of talc are shown in Figure 6. It is apparent that the 9–12 μ region resembles the spectra of Figure 1 much more closely than do the forsterite or enstatite spectra.

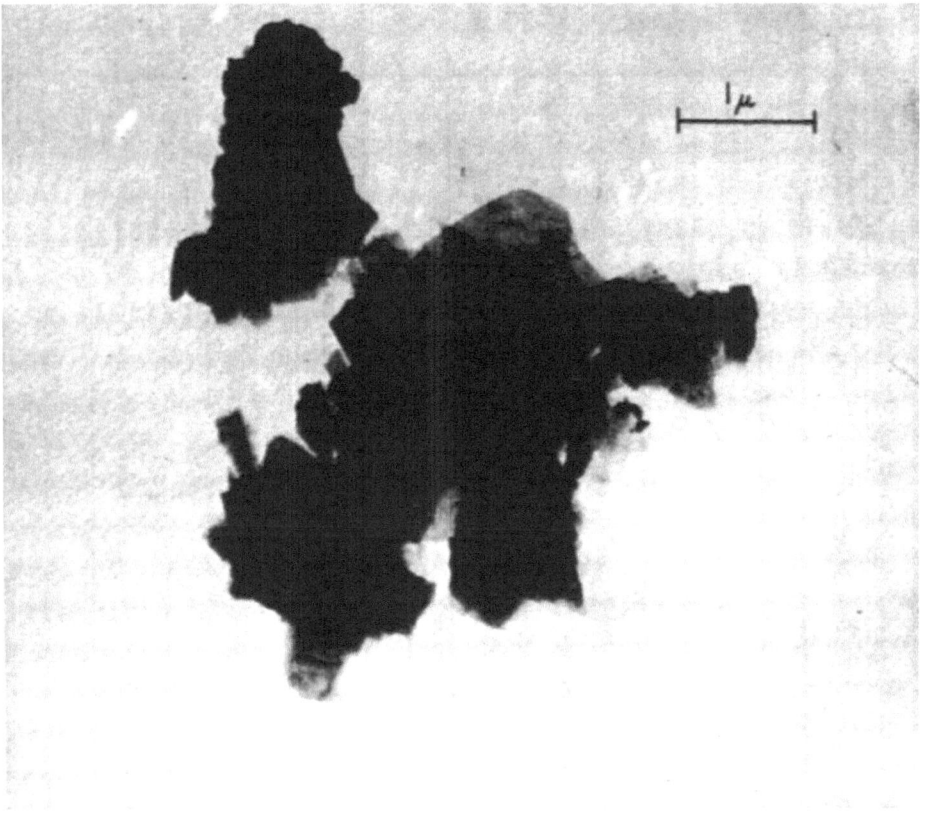

Fig. 5.   Electron microscope photograph of talc particles.

The band is centered near 9.8 μ, close to the position of the interstellar band. It is somewhat narrower than the interstellar band, but a *narrower* feature is not a problem because slight differences in composition or a mixture of silicates should give a better fit. Note the relatively strong absorption of talc (11 500 cm^2 gm^{-1}) at 9.8 μ. The large particles in talc and all the layer silicates below could give saturation effects at this wavelength.

A low temperature (~ 4 K) spectrum is shown in Figure 6b. Comparison of Figures 6a and 6b indicates that there may be some evidence for a small amount of temperature narrowing, but this is not certain. The band center does not shift as the mineral is cooled.

Farmer (1958) has published band assignments for talc.

Serpentine. The relationship of this silicate to talc was discussed in Section 2. It occurs naturally as fibers or rods resulting from the curling of the layer sheets. The spectrum is shown in Figure 6c. This mineral, with its strong absorptions at 9.3 and 10.2 μ, gives a rather poorer fit than talc to the interstellar curves. It seems unlikely that it could be the dominant silicate in the interstellar medium, but could be present with other silicates.

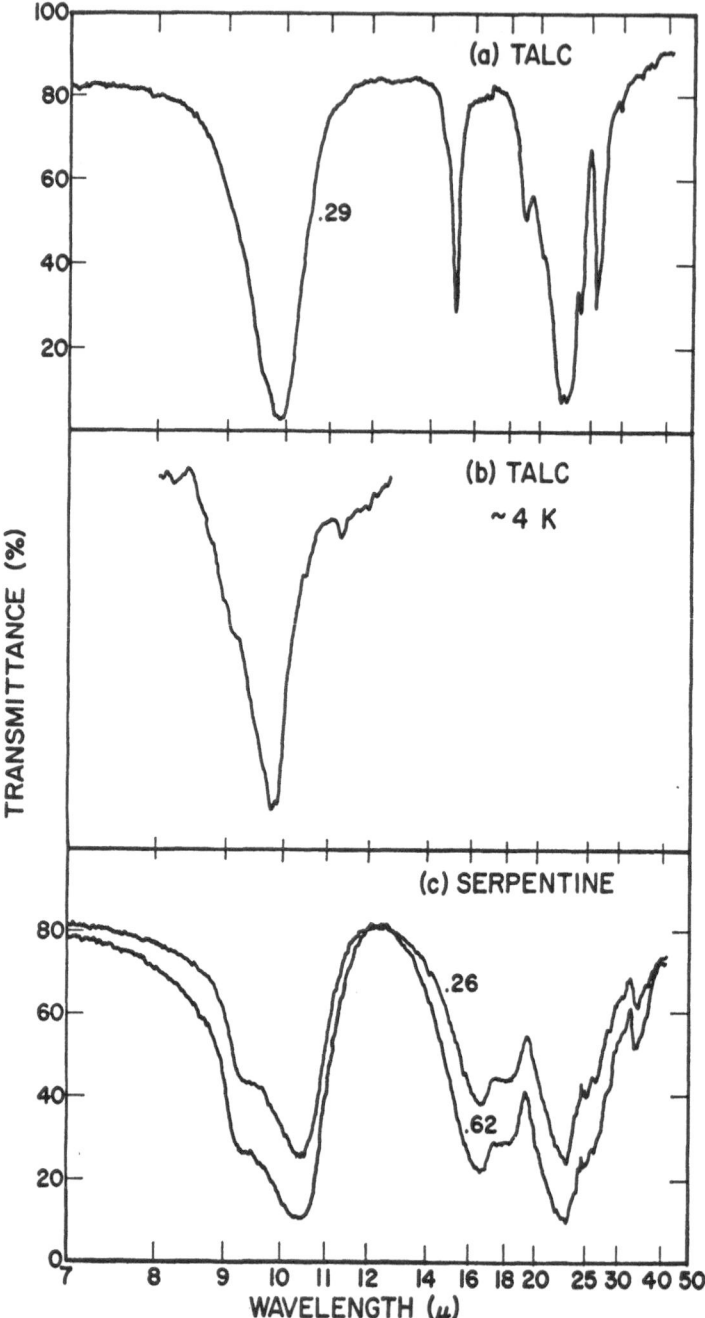

Fig. 6. Spectra of talc and serpentine. See caption to Figure 3 for explanation of uncertainties in the low temperature spectrum.

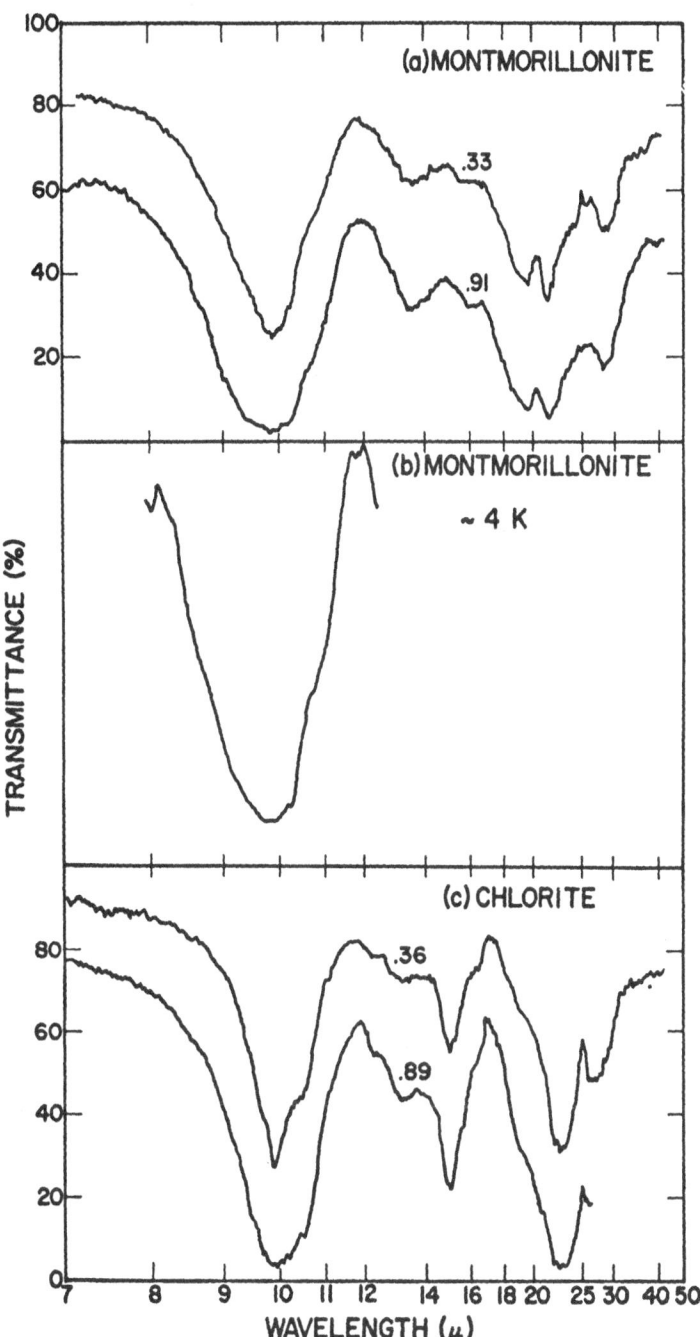

Fig. 7. Spectra of montmorillonite and chlorite. See caption to Figure 3 for explanation of uncertainties in the low temperature spectrum.

Montmorillonite. The structure of this layer silicate is similar to talc, but with the substitution of Al and Mg for some of the Si ions in the silicate tetrahedra. As a clay mineral it can contain variable amounts of water in between the silicate layers. It is of particular interest here because montmorillonite-like minerals are a major phase in type I carbonaceous chondrites (Bass, 1971). The spectrum, shown in Figure 7a, gives quite a good fit to the interstellar curves.

A spectrum of montmorillonite at ~ 4 K is shown in Figure 7b. The continuum (100% transmittance) level is not well located in this spectrum. Again, despite the lack of accurate quantitative data, there is no evidence for major changes in position or shape of the band at low temperatures.

In this mineral, as well as in the clay minerals, chlorite, chamosite, and hectorite, described below, broad absorptions occur between 2.6 and 3.3 μ in laboratory spectra (not shown). These are due to OH bonded to silicate tetrahedra and to layers of H_2O between the silicate layers. Preliminary measurements of spectra at these wavelengths indicate the absorptions in some of the minerals do not fit the interstellar 'ice' band at 3.1 μ (in the infrared spectra of the Becklin-Neugebauer and AFCRL sources); others are consistent with the astonomical data, especially if an ice mantle surrounds the grain. These shorter wavelength bands, in contrast to the 9.8 μ band, may be temperature dependent at low temperatures because of the association of water molecules in the crystal and the known temperature dependence of the water band as water freezes. The detailed comparison of hydrous silicate spectra with interstellar spectra at these short wavelengths is beyond the scope of the present work.

Chlorite. In Figure 7c we show the spectra of a chlorite mineral. Chlorites are similar to the talc structure but with an additional layer of $Mg(OH)_2$ between the talc sheets and with variable Fe, Mg, and Al content. A band near 9.9 μ is present and also a weaker absorption near 10.5 μ. The astronomical data are not good enough to exclude the weak feature.

Chamosite. This layer silicate has been identified in the Murchison (type II) meteorite by Fuchs *et al.* (1973). Its structure and chemistry is discussed by Deer *et al.* (1962). The band at 9.8 μ (Figure 8a) gives good agreement with the interstellar spectra. The strong absorption at 6.9 μ would be unobservable from the ground due to interference by atmospheric water bands.

Hectorite. This clay mineral was investigated because of its high magnesium oxide content (25% by weight) as compared to terrestrial montmorillonites which generally have high aluminum oxide content (Grim, 1968). Thus this mineral may be favored on abundance grounds. The spectrum (Figure 8b) exhibits the typical absorption near 9.8 μ but also a weaker feature near 9.5 μ which is not inconsistent with the astronomical data. This mineral has also been studied by Farmer (1958) who made band assignments.

Meteorites. Since hydrous silicates are important in the composition of type I and type II carbonaceous chondrites, it is natural to compare spectra of meteorites with the interstellar spectra.

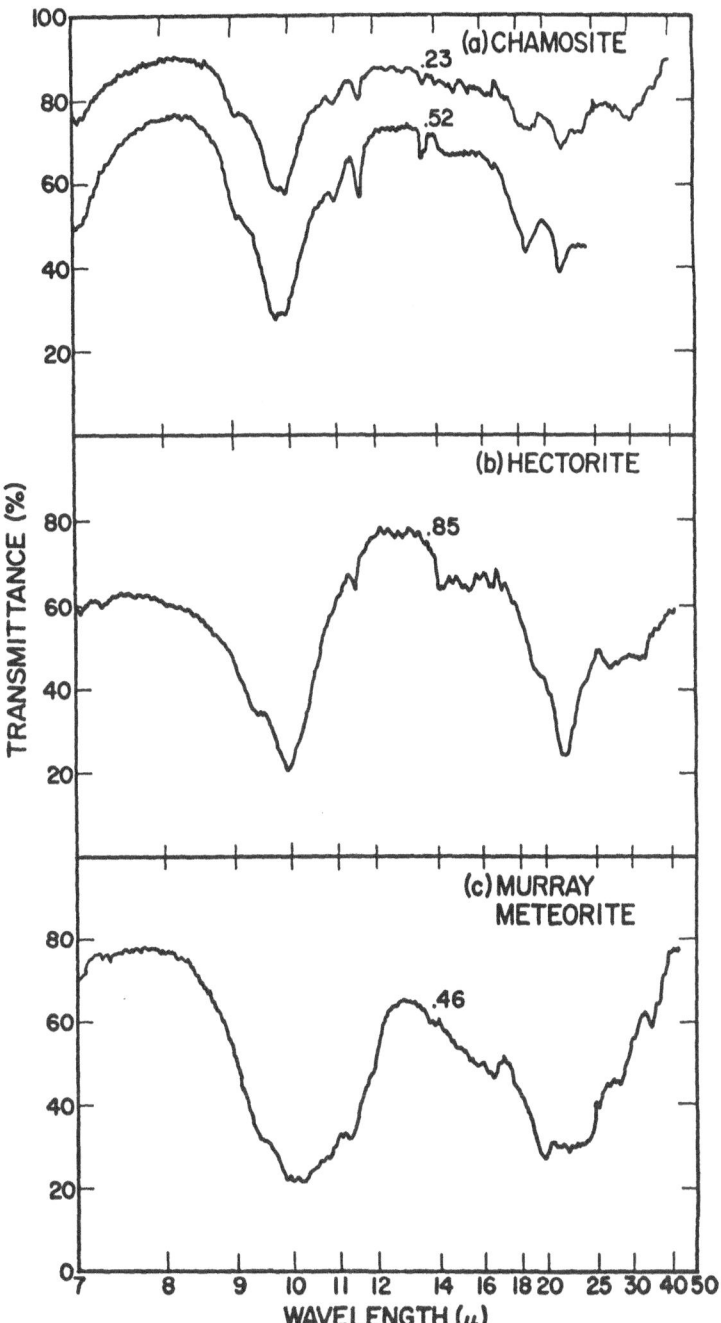

Fig. 8. Spectra of chamosite, hectorite, and the Murray meteorite.

We first obtained a spectrum of the type III carbonaceous chondrite, Allende. The spectra (not shown) of the matrix and a dark chondrule were quite similar to the spectra of olivine and enstatite, respectively. A white inclusion was very similar to olivine with slightly broader absorption features at 10.2, 11.3,.and 20 μ. A second white inclusion exhibited a very broad featureless absorption from 8 to 14 μ with a peak at 11.5 μ. We conclude that this meteorite is difficult to relate to interstellar matter as expected from the known high olivine content of the type III's.

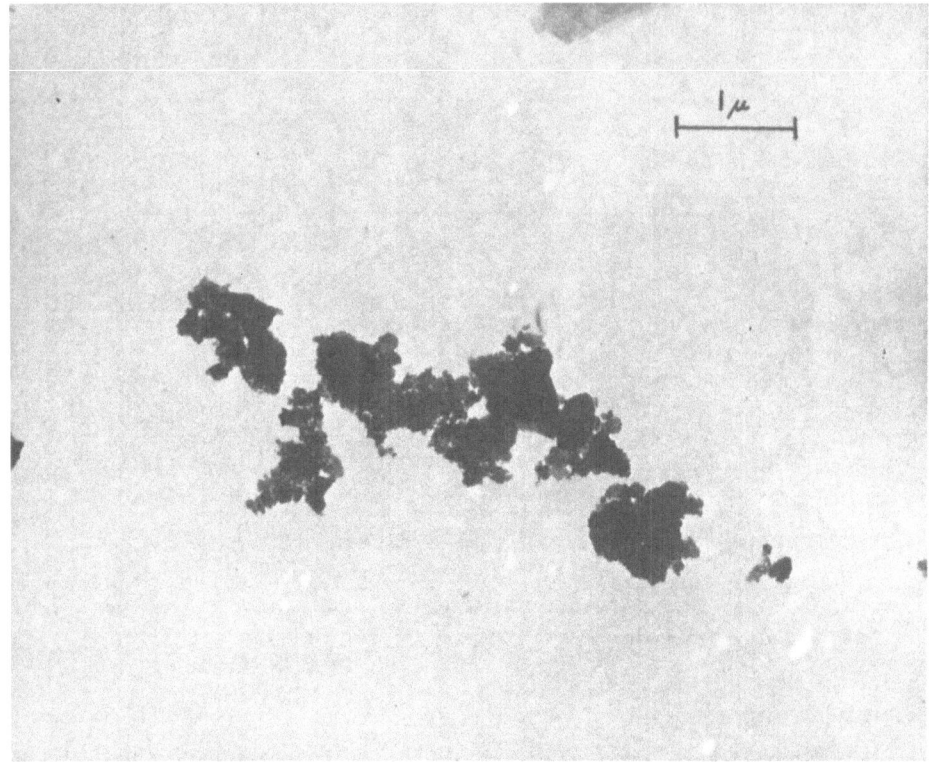

Fig. 9.   Electron microscope photograph of the Murray meteorite particles.

The type II carbonaceous chondrite, Murray, is approximately 90% equal amounts of olivine and hydrous silicate and about 10% pyroxene (DuFresne and Anders, 1962). Examination of the particles in the electron microscope (Figure 9) shows that the Murray particles were finer than the talc particles. This reflects the fine-grain structure of the Murray matrix. The absorption feature (Figure 8c) is fairly flat between 9.8 and 10.2 μ and is broader than the individual hydrous silicates. The olivine has the effect of broadening the absorption as well as possibly shifting the center to longer wavelength than typical for the hydrous silicates. In addition, the feature at 11.2 μ may be due to olivine (compare with Figure 3). Thus the spectrum seems interpretable as a mixture of olivine and hydrous silicates.

Recently Day (1974) has shown that an amorphous magnesium silicate gel also has a peak absorption near 9.8 μ and compared it to a carbonaceous chondrite. The Murray meteorite spectrum (Figure 8c) is similar to Day's. The hydrous silicates considered in the present work are crystalline as are the silicates in the carbonaceous chondrites (Bass, 1971). The known elongation of the interstellar grains (from polarization) indicates that there is definite internal organization in the particles.

The general similarity of the phyllosilicate spectra between 9.5 and 10.5 μ is shown in the figures. This similarity, due, it seems, to the similar layer structure of these minerals, was noticed earlier by Launer (1952). Their common tendency to a single, strong band usually near 9.8 μ is the property in which they most resemble the interstellar silicate feature. Thus no *particular* hydrous silicate can be identified with the interstellar bands. On the contrary, it is significant that a *class* of silicates has been identified whose spectra are similar to the astronomical spectra. Thus the suggestion that the interstellar silicates are hydrous does not rest on comparison with only a single silicate which is more likely to give accidental agreement, nor does it require that the interstellar silicates should be homogeneous in composition or structure. The observed absorption band could be the superposition of different hydrous silicates in the grains. Weaker neighboring absorptions might be washed out in the superposition, also broadening the band. Furthermore, even if none of the silicates in this study occur in precisely the same form in the interstellar medium, it is still likely that a real interstellar hydrous silicate would have a band very near 9.8 μ.

The band near 20 μ, present in all the silicates, might be useful to test the suggestion of phyllosilicates. A band near this wavelength has apparently been observed at low resolution by Woolf (1973). In most of the phyllosilicates the maximum absorption occurs near 22 μ. Enstatite has the peak absorption at 20 μ and another at 24.5 μ. Talc also has a sharp feature near 15 μ which might be observable at high altitude. Spectral signatures like these, and detailed comparison of laboratory and astronomical spectra, could provide further insights into the composition of the grains. Such information could be obtained from intermediate resolution ($\Delta\lambda/\lambda \sim 0.05$) spectroscopy.

4. Conclusions

From the comparison with laboratory spectra, we conclude that neither of the high temperature silicate condensates, forsterite or enstatite, nor the two in combination, give good fits to the astronomical data. The effects of particle size or temperature do not seem sufficient to significantly reduce the disagreement.

The interstellar absorption band near 9.8 μ resembles the spectra of hydrous silicate minerals which are predicted by condensation models and are found in carbonaceous chondrites. Direct comparison of the infrared spectra with a type II carbonaceous chondrite is consistent with this picture if allowance is made for an admixture of olivine. No individual phyllosilicate is singled out as a candidate for grain material. The observed bands could arise from absorption by an assembly of hydrous silicates with considerable latitude in the structure composition, size, and temperature of the particles.

We have not investigated other types of silicates in detail; however, some comments can be made based on published spectra (Launer, 1952; Lazarev, 1972). The silicas (e.g. quartz, vitreous silica) have a narrow band near 9.2 μ in disagreement with the observations. Silicates containing isolated SiO_4 tetrahedra (orthosilicates including olivine) have complicated spectra which do not, as a rule, give good fits to the astronomical data. The same is true for single and double chain silicates (clinopyroxenes including enstatite and the clinoamphiboles), although examples can be found which do not disagree badly. Some of the framework silicates have a band near the correct wavelength and might be investigated in more detail.

The phyllosilicate model is entirely consistent with the viewpoint of the 'dirty ice' or 'dusty snowball' models of interstellar grains as well as objects in the Solar System. Indeed the now established association of silicates (Maas et al., 1970) and H_2O (Wehinger et al., 1974) in comets suggests that phyllosilicates may be found in these objects also.

Acknowledgements

We gratefully acknowledge the contribution of B. Flaig, H. Rockefeller, and C. S. Springer of the SUSB Chemistry Department in making their infrared spectroscopic equipment available to us. E. Drummond and B. P. Lane of the SUSB Health Science Pathology Department obtained the electron microscope photographs. Meteorite samples were provided by T. Owen and O. A. Schaeffer. We thank D. H. Lindsley and J. J. Papike for assistance and valuable comments. A. Z. acknowledges support from AEC grant AT(11-1)3080. C. C. P. acknowledges support from NASA grant NGR 33-015-165.

References

Anders, E.: 1971, *Geochim. Cosmochim. Acta* **35**, 516.
Arrhenius, G. and Alfvén, H.: 1971, *Earth Planetary Sci. Letters* **10**, 253.
Bass, M. N.: 1971, *Geochim. Cosmochim. Acta* **35**, 139.
Bowen, N. L. and Tuttle, O. F.: 1949, *Geol. Soc. Am. Bull.* **60**, 439.
Burns, R. G. and Huggins, F. E.: 1972, *Am. Miner.* **57**, 967.
Cameron, A. G. W.: 1973, *The Role of Dust in Cosmogony*, paper presented at the Symposium on the Dusty Universe, Cambridge, Massachusetts.
Day, K. L.: 1974, *Astrophys. J. Letters* **192**, L15.
Day, K. L., Steyer, T. R., and Huffman, D. R.: 1974, *Astrophys. J.* **191**, 415.
Deer, W. A., Howie, R. A., and Zussman, J.: 1962, *Rock Forming Minerals*, Vol. 3, John Wiley and Sons, New York.
Dorschner, J.: 1968, *Astron. Nachr.* **290**, 171.
DuFresne, E. R. and Anders, E.: 1962, *Geochim. Cosmochim. Acta* **26**, 1085.
Farmer, V. C.: 1958, *Miner. Mag.* **31**, 829.
Farmer, V. C. and Russell, J. D.: 1964, *Spectrochim. Acta* **20**, 1149.
Field, G. B.: 1973, *The Composition of Interstellar Dust*, paper presented at the Symposium on the Dusty Universe, Cambridge, Massachusetts.
Forrest, W. J.: 1974, Private communication.

Frye, K.: 1974, *Modern Mineralogy*, Prentice Hall, Englewood Cliffs, New Jersey.

Fuchs, L. H., Olsen, E., and Jensen, K. J.: 1973, *Smithsonian Contrib. Earth Sci.* **10**, Smithsonian Institution Press, Washington, D.C.

Gillett, F. C. and Capps, R. W.: 1974, Private communication.

Gillett, F. C. and Forrest, W. J.: 1973, *Astrophys. J.* **179**, 483.

Gilman, R. C.: 1969, *Astrophys. J. Letters* **155**, L185.

Grim, R. E.: 1968, *Clay Mineralogy*, McGraw-Hill, New York.

Grossman, L.: 1972, *Geochim. Cosmochim. Acta* **36**, 597.

Herbig, G. H.: 1970, *Mem Soc. Roy. Sci. Liège*, Ser. 5, **19**, 13.

Hunt, J. M., Wisherd, M. P., and Bonham, L. C.: 1950, *Anal. Chem.* **22**, 1478.

Jeffery, P. M. and Anders, E.: 1970, *Geochim. Cosmochim. Acta* **34**, 1175.

Johannes, W.: 1968, *Cont. Miner. Petrol.* **19**, 309.

Kerridge, J. F.: 1967, in S. K. Runcorn (ed.), *Mantles of the Earth and Terrestrial Planets*, Chapt. 3, pp. 35–47, Interscience.

Knacke, R. F. and Thomson, R. K.: 1973, *Publ. Astron. Soc. Pacific* **85**, 341.

Larimer, J. W. and Anders, E.: 1967, *Geochim. Cosmochim. Acta* **31**, 1239.

Launer, P. J.: 1952, *Am. Miner.* **37**, 764.

Lazarev, A. N.: 1972, *Vibrational Spectra and Structure of Silicates*, Consultants Bureau, New York and London.

Lewis, J. S.: 1972, *Icarus* **16**, 241.

Lyon, R. J. P.: 1963, *N.A.S.A. Tech. Note*, TND-1871.

Maas, R. W., Ney, E. P., and Woolf, N. J.: 1970, *Astrophys. J. Letters* **160**, L101.

Martin, P. G.: 1971, *Astrophys. Letters* **7**, 193.

Merrill, K. M. and Soifer, B. T.: 1974, *Astrophys. J. Letters* **189**, L27.

Meyer, C., Jr.: 1971, *Geochim. Cosmochim. Acta* **35**, 551.

Nagy, B.: 1966, *Geol. Fören. Stockholm Förh.* **88**, 235.

Stubican, V. and Roy, R.: 1961, *Am. Miner.* **46**, 32.

Wehinger, P. A., Wyckott, S., Herbig, G. H., Herzberg, G., and Lew, H.: 1974, *Astrophys. J. Letters* **190**, L43.

Wood, J. A.: 1968, *Meteorites and the Origin of Planets*, McGraw-Hill, New York.

Woolf, N. J.: 1973, in J. M. Greenberg and H. C. van de Hulst (eds.), 'Interstellar Dust and Related Topics', *IAU Symp.* **52**, 485.

THE INFLUENCE OF GRAIN MANTLES ON THE FORMATION OF HYDROGEN MOLECULES ON GRAIN SURFACES

T. J. LEE

Royal Observatory, Edinburgh, Scotland

Abstract. The physical adsorption energy, E, of hydrogen molecules on various substrates at temperatures between 5 and 30 K and at the lowest practicable gas densities has been measured. Values of E/k are for condensed CO 340 K, CO_2 800 K, H_2O 850 K and for 'dirty' graphite 980 K and 'dirty' copper 800 K. From these measurements temperature ranges in which H atoms might combine on the surface to form H_2 molecules are estimated.

Duley has discussed the formation and composition of condensed gas mantles on interstellar grains. The effects of such mantles in promoting and poisoning hydrogen molecule formation are discussed.

1. Introduction

A number of authors have discussed the formation of hydrogen molecules by the combination of two hydrogen atoms on the surface of a grain of interstellar matter. Though chemisorption may catalyse molecule formation on certain surfaces, especially metals, at higher temperature, physical adsorption processes operate at the lower temperatures which grains are thought to have when molecule formation takes place. The key parameter in determining the rate of molecule formation on a surface is the physical adsorption or desorption energy of hydrogen atoms on the grain surface since this fixes the mean time a hydrogen atom resides on a grain. Desorption energies have not been measured for hydrogen atoms on cold surfaces, such experiments being difficult. Theoretical calculations have been made but confidence cannot be placed in the resultant values of adsorption energy within a factor of two. The physical situation is complex since to obtain the potential energy, the interaction of the adsorbed atom with all other atoms of the grain must be summed. Further, the zero point energy is significant and must be calculated. In a previous paper values of the physisorption energy of H atoms were estimated from measurements of the adsorption energy of hydrogen molecules on solid CO and H_2O surfaces. This paper presents physisorption energies for other substrates and discusses all the results in relation to grain mantle compositions proposed by different authors.

2. Adsorption-Desorption Processes

For equilibrium between a species adsorbed on a surface and the same species in the gas phase, first order kinetics can be used to describe the rate of absorption and desorption provided that the surface and gas phase species are the same, for example hydrogen atoms on the surface and hydrogen atoms in the gas phase, or hydrogen molecules on the surface and hydrogen molecules in the gas phase. Both equilibria can be described by first order kinetics, but for hydrogen atoms on the surface which leave or arrive at the surface as molecules it is not valid to use first order kinetics. The following equation gives the rate of desorption of a species from a surface for first order kinetics

$$\frac{dN}{dt} = Nv \exp\left(-E/kT\right), \tag{1}$$

where N is the surface density of atoms (molecules), E is the adsorption energy of the atoms (molecules) on the surface, T is the absolute temperature of the surface, v is a factor with dimensions of frequency and k is Boltzmann's constant. Usually v is assumed to be 10^{12} s^{-1}. Alternatively Equation (1) can be rearranged to define a mean residence or stay time

$$t = v^{-1} \exp\left(E/kT\right). \tag{2}$$

Both v and E depend upon temperature but only weakly, and so such variations will be ignored in this paper.

The rate at which atoms or molecules in a gas arrive at a surface can be written as

$$R = \tfrac{1}{4}n\langle v\rangle, \tag{3}$$

where n is the density of the species in the gas phase and $\langle v\rangle$ is the mean thermal speed of the atoms or molecules. The rate at which adsorption takes place is given by

$$R = \tfrac{1}{4}\gamma n\langle v\rangle, \tag{4}$$

where γ is the sticking coefficient of two incident species on the surface. In equilibrium Equations (1) and (4) give

$$n = \frac{4v}{\gamma\langle v\rangle} N \exp\left(-E/kT\right). \tag{5}$$

Experimental measurements of n as a function of T will then yield values of E. Another experimental method is to measure the mean residence time, t, as a function of temperature (cf. Lee and Stickney, 1972). In this case v as well as E can be determined directly.

Equation (5) also relates the surface density, N, to the gas density, n, and the surface temperature. One use of this is to determine the conditions at which certain surface coverages are reached, and for determining when condensation of a solid takes place. In this context it is important to take variations of E with coverage into account. For uniform surfaces at low coverages E is independent of coverage, but as one monolayer is approached E tends towards the sublimation energy of the adsorbing species. This value is attained for one monolayer in certain cases but ten or more monolayers may be required in others. N is of course the number of atoms (molecules) exposed on the surface and not the total number if more than one layer is formed.

The sticking coefficient γ is coverage dependent and temperature dependent. For example in the case of H_2 molecules on a copper surface it varies between about 0.3 for incident molecules at 500 K and 0.7 for incident molecules at 80 K. As molecular hydrogen is adsorbed on the surface the sticking coefficient increases and for tens of monolayers the sticking coefficient is greater than 0.9 for a gas temperature of 500 K, and approaches unity for 80 K molecules.

3. Association of Hydrogen Atoms on a Surface

Several authors (Gould and Salpeter, 1963; Stecher and Williams, 1966; Knaap *et al.*, 1966; Augason, 1970) have treated the case of association to form molecules of two hydrogen atoms physically adsorbed on a uniform grain surface. Hollenbach and Salpeter (1971) (hereafter referred to as H & S) have extended the treatment to surfaces which have sites with enhanced binding energies. In what follows, the results will be summarized and quoted from H & S. The mean time between successive adsorptions of gas species on a surface is

$$t_s = (\gamma n \langle v \rangle o)^{-1}, \tag{6}$$

where σ is the cross-section of the grains and the other symbols retain their previous meaning. When the mean time between adsorption of atoms is less than the mean stay time there are two atoms on a surface for a significant period of time. If the atoms have sufficient mobility – that is, they can move over the whole grain surface in a time short compared to that between adsorptions – then hydrogen molecules may be formed. On the basis of quantum mechanical arguments the mobility of hydrogen atoms physically adsorbed is expected to be high. Equations (2) and (6) can be combined to define a critical temperature below which hydrogen molecule formation is efficient

$$T_c = E \left\{ k \ln \left(\frac{v_H}{\gamma_H n_H \langle v_H \rangle \sigma} \right) \right\}^{-1}. \tag{7}$$

For $T < T_c$ the rates of molecule formation can approach

$$\tfrac{1}{2}(\gamma n \langle v \rangle \sigma) \quad \text{per grain.}$$

For a grain with enhanced binding sites with certain applied conditions H & S find that

$$T_c' = E' \left\{ k \ln \left(\frac{v_H N}{\gamma_H n_H \langle v_H \rangle \sigma} \right) \right\}^{-1}, \tag{8}$$

where N is the number of regular sites on the grain surface, E is the energy of the sites for H. In the above equations γ_H is strictly the recombination efficiency – i.e., the fraction of atoms sticking on the grain which leave it as molecules.

Molecules, either hydrogen or other interstellar gas cloud constituents can also be adsorbed by grains. The temperature at which a complete monolayer is just formed, assuming that the adsorption energy remains constant up to monolayer coverage is given by

$$T_m = E_m \left\{ k \ln \left(\frac{v_m}{\gamma_m n_m \langle v_m \rangle \sigma} \right) \right\}^{-1},$$ (9)

where E_m is the adsorption energy of the molecule on the grain surface. If this is the sublimation energy of the condensed molecules, and N the surface density, then T_m defines the temperature below which condensation takes place, under the physical conditions implied by the other parameters on the right-hand side of Equation (9).

Temperatures defined by Equations (7), (8) and (9) allow examination of conditions for hydrogen molecule formation. To do this, values for gas densities of hydrogen atoms and various molecules and their adsorption energies on grain surfaces are required.

4. Physical Adsorption Energies

Physical adsorption energies are difficult to calculate, especially for light gas atoms and molecules where the interaction is particularly small and the wave functions are large. Methods consist of calculating an adsorptive potential by summing the van der Waals interactions between an adsorbed molecule and the molecules of the solid on which adsorption takes place. The wave equation is then solved for the motion of the molecule in this potential to arrive at the zero point kinetic energy. The physical adsorption energy is then the nett energy. Several calculations have been made, e.g., by Ricca and Garrone (1970) or Hollenbach and Salpeter (1971), but the methods currently used are subject to possible error. Few experimental results have been available for comparison and some of them were not obtained under ideal conditions. There has thus been little chance for the development of successful techniques for estimating physical adsorption energies of light gas molecules. Our first measurements of desorption energy of molecular hydrogen H_2 on H_2O (cf. Lee, 1972) gave results of the order of the adsorptive potential calculated by Hollenbach and Salpeter (1971), being in fact about 70% greater than the Hollenbach and Salpeter value. It was therefore not possible to accept their results for atomic hydrogen physisorption energies. For the purpose of estimating the temperature at which hydrogen molecule formation might take place on grains, the atomic hydrogen physisorption energy was assumed to be 80% of that measured for hydrogen molecules. This is somewhat arbitrary though there is some justification since the adsorption potential should be approximately the same in both cases, and the zero point kinetic energy is greater for the smaller mass. Temperatures below which hydrogen monolayers can form were estimated from the molecular adsorption energies. Measurements have now been made for adsorption energies of H_2 on solid CO_2, 'dirty' copper and 'dirty' graphite in addition to the previous substrates H_2O and CO. The measurement method was the same as that in Lee (1972), the one based on Equation (5). The dirty copper was a bare machined copper surface baked to 700 K, at which temperature the residual gas pressure was 10^{-9} torr. Residual gas adsorption probably of CO_2 and H_2O on the copper must have taken place while the cryostat was cooled down to working tempera-ture. Thus the surface is described as dirty. To produce a graphite surface the copper was painted with a graphite suspension. The apparatus was reassembled, evacuated and baked out for more than 48 hr. The residual pressure was 2×10^{-8} torr; com-ponents of the residual gas were hydrocarbons, water, CO_2 and CO. Physical adsorp-tion energies for hydrogen on this surface are less than those for clean graphite.

In Table I the measured adsorption energies are expressed as E_2/k for hydrogen molecules and E/k for hydrogen atoms. This conversion to temperature units (Kelvins) facilitates the use of Equations (7), (8) and (9). For all substrates the desorption energy falls to the sublimation energy of solid H_2 over a number of monolayers, rather than a single monolayer in a manner similar to that illustrated for CO and H_2O substrates in Lee (1972).

TABLE I

Desorption energies and critical temperatures

Substrate	E_2/k	E/k	T_m K	T_c K
H_2O	860	690	17	22
CO	350	280	6	9
CO_2	800	640	16.5	21
Cu(1)	850	680	17	22
Cu(2)	740	590	14.5	19
Graphite	980	785	19	25

Physical conditions in clouds for calculating T_m and T_c:

$$n_H = 10^3 \text{ cm}^{-3}, \quad n_{H_2} = 10 \text{ cm}^{-3}, \quad T_{gas} = 100 \text{ K},$$
$$\sigma = 10^{-9} \text{ cm}^2, \quad \gamma_H = \gamma_{H_2} = 0.5.$$

5. Astrophysical Implications

Critical temperatures below which molecule formation is efficient, T_c, are listed in the last columns of Table I. The physical conditions are those used by H & S: $n_H = 10^3$ cm^{-3}, $n_{H_2} = 10$ cm^{-3}, $T_{gas} = 100$ K, $\sigma = 10^{-9}$ cm^2, $\gamma_H = \gamma_{H_2} = 0.5$ and the number of adsorption sites per grain, N, is 10^6. Except for the case of CO, T_c is no higher than the grain temperatures expected in cool clouds. Silicate and silicon carbide have been proposed as probable grain materials. They should have properties as physisorption substrates somewhere between those of H_2O and graphite and so for these substrates T_c should be in the higher range. Lee (1972) has shown that this gradual fall off in adsorption energy with hydrogen coverages makes it unlikely that the formation of hydrogen monolayers will inhibit H—H association on this surface. Thus for grains with mantles of H_2O or CO_2, molecule formations can take place if grains cool to between about 20 K and 16 K and probably below 16 K as well. Slightly higher temperature ranges should apply to bare grains. More recently Duley (1974) has argued that H_2O is insufficiently abundant for mantles of these molecules to be condensed but that CO mantles are far more likely. How will such mantles affect H_2 formation? Equation (9) with E_m equal to the sublimation energy of CO and N_m equal to its gas density will give us the temperature at which this gas can condense in solid form. Before the grains are so cool, given sufficient time, they can accrete one or more monolayers of CO since the adsorption energy on any likely grain surface will be

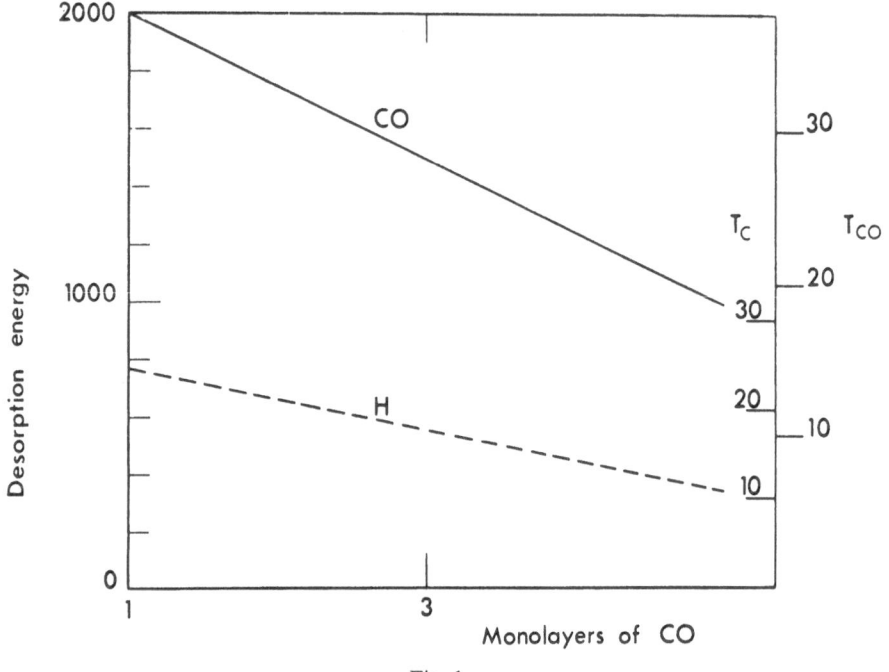

Fig. 1.

greater than its sublimation energy. To a first approximation the adsorption energy of H on a grain with m monolayers of CO adsorbed will have the same functional relation as the adsorption energy of CO on the same surface, as m increases from zero to the number of monolayers at which the sublimation energy of CO is attained. Figure 1 illustrates this idealized situation in which the adsorption energies are assumed to fall linearly with the logarithm of the number of monolayers. Figure 2 in Lee (1972) suggests this kind of dependence.

To calculate T_c, the temperature at which successive monolayers are complete, $N = 10^6$ and the CO abundance given by Duley (1974), $n_N/n_{CO} = 2 \times 10^3$, and values of the other parameters above are used. Figure 1 can then be used by referring to the scale on the right-hand side to show the variations of T_c and T_{CO} as the number of monolayers of absorbed CO increases. It is clear that sufficient CO monolayers will form to reduce the H adsorption energy and hence the H atom stay time, to such an extent that H—H association is no longer efficient if T_{CO} always exceeds T_c.

When T_{CO} exceeds T_c CO monolayers will form to reduce the H adsorption energy, and hence the H atom stay time, to such an extent that H—H association is no longer efficient. With the variations of adsorption energy with number of CO monolayers used here, this is true down to the temperature at which the sublimation energy is attained if T_{CO} exceeds T_c for the bare substrate.

Under the assumptions presented here, and with the value of the desorption energy of hydrogen on dirty graphite, this is so when the adsorption energy of CO on the bare substrate exceeds $E/k = 1320$ K. For heavier molecules the adsorption energy of one species on another is approximately equal to the geometric mean of the sublimation energies. For CO on CO_2 this gives $E/k = 1950$. It should be little different for all likely grain materials. This means that CO adsorption can indeed inhibit H—H association.

Obviously CO adsorption can take place only if CO is available. The way in which T_c and T_{CO} are calculated assumes that cooling takes place infinitely slowly. To illustrate this effect of finite gas density, consider the following. For 0.1 cm^{-3} molecules of CO, a relative abundance of 5×10^{-4}, and a grain abundance of 4×10^{-13} n_H with 10^6 sites per grain there are 1250 monolayers of CO per grain. Now the time for all molecules of a species in a gas to strike a grain is proportional to $M^{1/2}$, where M is the molecular weight of that species. If the sticking coefficient of CO is the same as the recombination coefficient for H, then at least 10 monolayers of CO can condense before 4% of the H atoms associate, and in this case the grain will effectively have a CO mantle throughout the period of H_2 formation. For such a surface, T_c is low, at 9 K. For a lower CO abundance the amount of H_2 found will be proportionally higher for a grain coverage of CO; a relative abundance of 10^{-5} or less will not have a great effect.

Even for high CO abundance there is a factor which should enable H_2O formation to take place on the surface for $T > 9$ K.

According to Duley (1974), O atoms in the gas will react with a CO mantle to give CO_2 molecules. These in turn will be adsorbed on the CO mantles to form enhanced binding sites at which H_2 formation can take place in the manner described by H & S. Equation (8) gives about 14 K for the temperature below which this process is efficient.

6. Conclusion

In conclusion, from experimental results on H_2 physisorption, it is apparent that of the substrates studied CO is the material of astrophysical interest which adsorbs H_2, and by inference H atoms least strongly. For a pure CO mantle, H_2 formation is not efficient for grain temperatures above 9 K, but for other grain surfaces the critical temperature is about 20 K. Mantles of CO can form for relative abundances of CO of a few times 10^{-4}, and will thus inhibit H_2 formation, but abundances 10^{-5} should have little effect. However, the reactions of O atoms with CO mantles to form CO_2 molecules enhance H_2 formation, since adsorbed CO_2 molecules act as high binding energy sites enabling efficient formation up to temperatures of 14 K. This temperature is the lowest critical temperature in the model considered. On the basis of their moderate polarizability methane and ammonia may behave in a similar way to CO.

No account has been taken of the effects of photons or energetic charged particles on gas phase molecules, adsorbed molecules or grain surfaces. These are all likely to play some part in molecule formation processes. The discussion does illustrate how atomic and molecular surface phenomena may interact in the process of hydrogen molecule formation.

References

Augason, G. C.: 1970, *Astrophys. J.* **162**, 463.

Duley, W. W.: 1974, *Astrophys. Space Sci.* **26**, 199.

Gould, R. J. and Salpeter, E. E.: 1963, *Astrophys. J.* **138**, 393.

Hollenbach, D. and Salpeter, E. E.: 1971, *J. Chem. Phys.* **53**, 59.

Knaap, H. F. P., van den Meijdenberg, C. J. N., Beenakker, J. J. M., and van de Hulst, H. C.: 1966, *Bull. Astron. Inst. Neth.* **18**, 256.

Lee, T. J.: 1972, *Nature Phys. Sci.* **237**, 99.

Lee, T. J. and Stickney, R. E.: 1972, *Surface Science* **32**, 100.

Ricca, F. and Garrone, E.: 1970, *Trans. Farad. Soc.* **166**, 959.

Stecher, T. P. and Williams, D. A.: 1966, *Astrophys. J.* **146**, 88.

GRAIN CHARGING IN H II REGIONS

A. F. M. MOORWOOD and B. FEUERBACHER

Astronomy Division, European Space Research Organisation, Noordwijk, Holland

Abstract. Equilibrium grain potentials have been calculated as a function of radial position and for a wide range of electron densities in H II regions ionized by stars of spectral type O5 and B0. Results are presented for both graphite (low yield photoemitter) and aluminium oxide (high yield photoemitter) for which laboratory photoemission data has been obtained. The results for aluminium oxide should approximate the behaviour expected of dielectric grains – e.g., silicates which may be present in H II regions. The importance of charging is discussed in relation to the growth and motion of grains in these regions.

1. Introduction

Evidence for the presence of dust grains within H II regions comes from visual observations of scattering (O'Dell *et al.*, 1966) and nebular emissions (Münch and Persson, 1971) and from a number of recent infrared observations (see for example the review of Wynn Williams and Becklin, 1974). In general the observations indicate that dust and ionized gas are fairly well mixed. Some nebulae are known to have their minimum density in the central regions and in the case of the Rosette nebula, Mathews (1967) has shown that the observed central 'hole' could have been produced by the action of radiation pressure on dust grains which are positively charged and effectively frozen into the plasma. In general it is to be expected that the lifetime of grains against expulsion due to radiation pressure will be increased if the grains are charged and that the magnitude and sign of the charge will also have important consequences for the growth and destruction of grains and the efficiency of molecule formation on grain surfaces.

The composition of the grains is not well established although it is generally believed that the spectral features observed at 10 μ and 18 μ are due to dielectric silicate grains (Gillett and Forrest, 1973; Aitken and Jones, 1973; Frogel and Persson, 1975).

Ney *et al.* (1972) have observed emission from circumstellar silicate shells with diameters ~ 0.06 pc around six early type stars in Orion but generally the silicate features appear in absorption and could be due to cooler silicate material surrounding the H II regions. This dust could be responsible also for the far infrared emission measured from H II regions (Harper and Low, 1971; Emerson *et al.*, 1973).

Whether the 'hot' dust which is mixed with the ionized gas is also silicate, some other material or a mixture of materials is still not known in most cases.

The grain charge calculations discussed here therefore have been performed both for dielectric and graphite grains which represent the extremes in charging behaviour, dielectrics being high yield photoemitters and graphite being a low yield material. Although there is no direct evidence for the existence of graphite within H II regions its presence can be argued for on the grounds that graphite in the general interstellar medium provides the best explanation for the 2200 Å hump in the interstellar extinction curve.

2. Grain potential

Equilibrium grain potentials have been calculated as a function of position within H II regions by establishing a current balance at the grain surface between photoelectron emission on the one hand and the net effect of proton and electron capture on the other. Details of the procedure used have already been published (Feuerbacher et al., 1973). Calculations have been performed for both graphite and aluminium oxide for which the required photoemission data exist in the form of photoelectric work functions, yields per incident photon and sets of electron energy distribution spectra for a range of ultraviolet photon energies (Feuerbacher and Fitton, 1972).

Graphite is a low yield photoemitter while aluminium oxide has a high yield, typical of dielectrics, and should approximate the behaviour of silicates for which no direct photoemission data exists.

The H II regions themselves have been considered to be 'classical' Strömgren spheres and calculations have been performed separately for ionizing stars of spectral type O5 and B0. The stellar parameters used were taken from Spitzer (1968) and are given in Table I. (More recently derived temperature and luminosity scales for OB stars are somewhat different but the general conclusions reached here are not affected.)

TABLE I

Stellar parameters

Spectral type	T_c (K)	$R/R_\odot$	N_c (s^{-1})	$r_s N_e^{2/3}$ (pc cm^2)
O5	5.6×10^4	7	3.1×10^{49}	100
B0	2.1×10^4	5.3	4.1×10^{46}	11

The variation of grain potential with distance from the ionizing star is shown in Figures 1 and 2 for a range of representative electron densities. Radial distance is given in units of the Strömgren radius as calculated from the data in Table I. Attenuation of the stellar ultraviolet flux due to dust absorption has not been taken into account in calculating these curves but as dust absorption also reduces the Strömgren radius their appearance is not substantially altered if the ultraviolet dust absorption

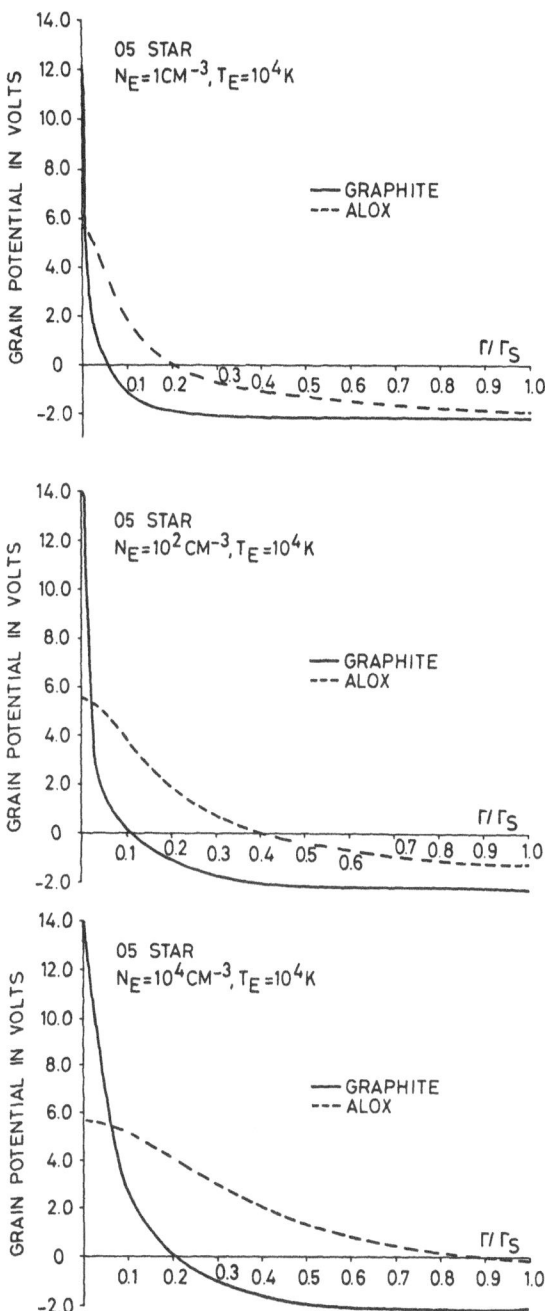

Fig. 1.   Grain potential plotted against distance (in units of the Strömgren radius) from an O5
star.

depth $\lesssim 1$. Recent infrared results indicate that this condition holds for a large number of H II regions and that the dust to gas ratio is less than the interstellar ratio
(Furniss *et al.*, 1974; Wright, 1973).

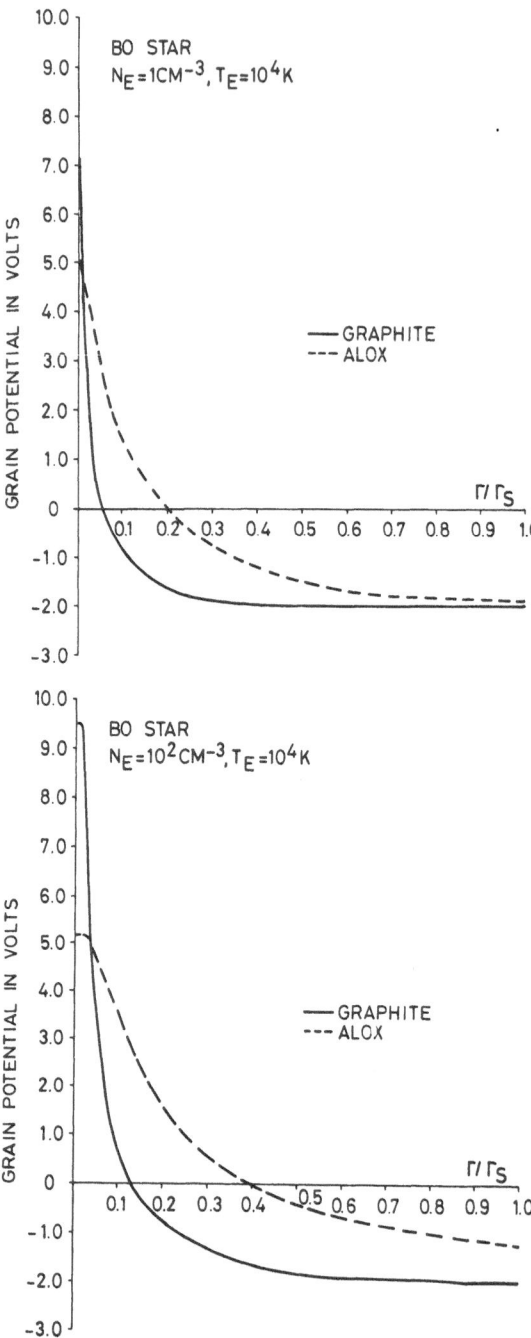

Fig. 2. Grain potential, plotted against distance (in units of the Strömgren radius) from a B0 star.

Qualitatively the charging behaviour is similar over the wide range of conditions considered. Grains are positively charged in the inner regions but become negatively charged beyond some distance from the star. Graphite grains typically become negatively charged within the inner 10%–20% of the Strömgren radius, reaching a limiting potential before the H II boundary, while dielectric grains become negatively charged only at larger distances from the star due to their higher photoyield. Accretion of ions by graphite grains therefore is much more likely than by dielectric grains throughout most of an H II region. As the two materials are oppositely charged over considerable regions there is also the possibility that collisions will lead to the formation of composite grains. The cross-section for this process under normal density conditions is low however and this possibility is not pursued here.

The curves presented can be considered to apply to 'normal' size grains having radii of the order of 6×10^{-6} cm. The situation is complicated however if we consider a size distribution of grains due to the fact that the photoyield per absorbed photon is dependent on grain size as pointed out by Watson (1973) and, in addition, the ratio of absorbed to incident photons is size dependent through $Q_{abs}(a/\lambda)$. An example of the variation to be expected for dielectric grains is shown in Figure 3

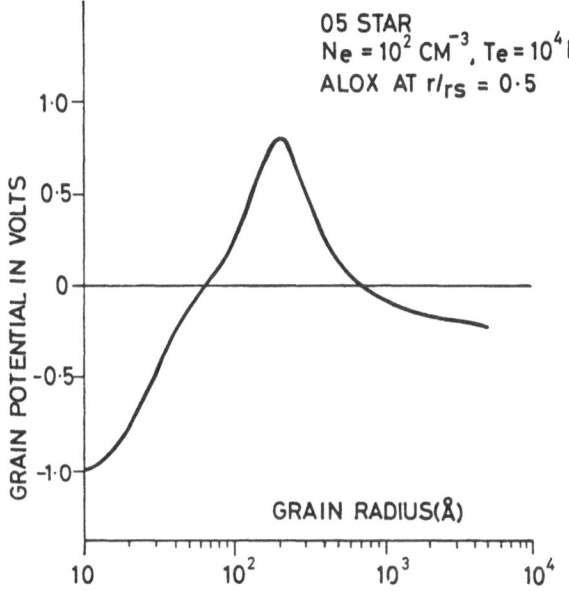

Fig. 3.   Example of the dependence of grain potential on size.

where potential has been plotted as a function of size for the case of an O5 star, $N_e = 100$ cm^{-3} and $r/r_s = 0.5$. Thus at a given position, different size grains of the same material can be oppositely charged. This may lead either to the separation of different size grains or conversely to the growth of medium size grains through accretion of small grains.

3. Motion of grains

The dominant forces acting on grains in H II regions are radiation pressure due to the central star and collision and electrostatic drag in the plasma. Grains move outwards relative to the gas; but, as shown by Ireland (1970) for the case of $N_e =$ 10 cm^{-3}, they are unlikely to escape from the H II region during the lifetime of the ionizing star. In what follows we consider a wider range of electron densities and investigate also whether grains of different composition within the same H II region are likely to separate as a result of differences in their charging behaviour.

In the simple calculations outlined here we ignore the problems involved during the formation of an H II region and in its subsequent expansion. The motion of grains before ionization of gas around a protostar occurs is of course a separate problem which does not include the effects of grain charging. This stage of evolution may be important, however, in determining the final dust to gas ratio in the H II region. As proposed by Davidson (1970) for example, grains may be forced outwards by radiation pressure to form a dust front which moves ahead of the ionization front. Such a picture is consistent with the infrared measurements already referred to which indicate the presence of dust surrounding H II regions and a lower than 'normal' dust to gas ratio inside the H II region. Assuming, however, that this or a similar process leads to the formation of an H II region surrounding a stable Main Sequence star in a relatively short time (say $< 10^4$ yr), we now consider the motion of grains which are left within the ionized gas.

We approximate the outward radiation force as

$$F_r = \frac{La^2 Q_{pr}}{4r^2 c} \text{ dyn},$$

where L is the total stellar luminosity (erg s^{-1}), a the grain radius (cm), r the distance from the star (cm) and $Q_{pr} \simeq 1$ is the efficiency factor for radiation pressure averaged over the stellar spectrum.

In addition to the radiation pressure, grains may also experience an outward force due to momentum transfer from ejected photoelectrons. This is due to the fact that when the grain size is comparable with or exceeds the effective wavelength for photoemission, more photoelectrons will be ejected from the illuminated hemisphere of the grain than from the hemisphere away from the star. The critical size is around a few hundred ångströms. In order to estimate the maximum effect we consider dielectric grains illuminated by an O5 star and assume that all photoelectrons are emitted from the illuminated hemisphere. The effective force on the grain is then given by

$$F_p \simeq 5.4 \times 10^{-20} \frac{\kappa \overline{Q_{abs}} a^2 N_c \overline{Y}}{4r^2} \int\limits_{E_{min}}^{E_{max}} E^{3/2} e^{-E} \, dE \text{ dyn},$$

where κ is a geometrical factor equal to about 0.5 if the photoelectrons obey a cosine distribution. N_c is the number of Lyman-continuum photons s^{-1} radiated by the star and $\overline{Y}$ is the mean photoyield which is essentially flat above 13 eV. The integral contains the $\sqrt{E}$ dependence (E in eV) of momentum on energy and a Maxwellian fit to the energy distribution of emitted photoelectrons. For positively charged grains the lower limit $E_{min} = V$, the grain potential, while for negatively charged grains we have set $E_{min} = 0$. Under these conditions the photoelectron pressure can exceed the radiation pressure by up to a factor of four. For graphite grains the photoelectron pressure is less than the radiation pressure and for B0 stars the effect is unimportant for both types of grains.

The electrostatic drag experienced is due mainly to distant rather than close encounters with protons and has been formulated by Spitzer (1956) as

$$F_e = \frac{M_H(\ln \Lambda)N_e Z_g^2 \omega}{11.7T_e^{3/2}} \text{ dyn},$$

where

$$\Lambda = \frac{3}{2Z_g e^3}\left(\frac{k^3 T_e}{\pi N_e}\right)^{1/2};$$

Z_g is the grain charge in units of the electron charge, ω the grain speed in cm s^{-1}, and T_e the electron temperature.

Rewriting this expression in terms of the grain potential V (volts), approximating $\ln \Lambda = 15$ and with $T_e = 10^4$ K we have

$$F_e = 10^{-16}a^2 V^2 N_e \omega \text{ dyn},$$

where a is the grain radius in cm.

Following Mathews (1967) we take the gas collision drag to be given by

$$F_c = \tfrac{8}{3}(2\pi k T_e M_H)^{1/2}N_e a^2\omega \text{ dyn} \simeq 10^{-17}N_e a^2\omega \text{ dyn}.$$

Under most of the conditions considered this is considerably smaller than the electrostatic drag.

The distance travelled by grains from the central star as a function of time has been computed under the assumption that the grains always move at the equilibrium speed determined by equating the radiation and drag forces – i.e.,

$$F_r = F_e + F_c.$$

If we consider all other uncertainties involved, this is a reasonable simplification to make as grains are initially accelerated to their equilibrium speed in a time which is very much shorter than the transit time. Note that the grain speed only depends on grain size through the size dependence of Q_{pr} and V. The effect of including the 'photoelectron pressure' is shown separately for the case of an O5 star and $N_e = 100$ cm^{-3}.

Grain distance in units of the Strömgren radius is shown as a function of time in Figure 4 for the B0 star. For $N_e = 10^2$ cm^{-3} the transit times for dielectric and graphite grains are $\simeq 2$–3×10^6 yr and $\simeq 10^7$ yr respectively which means that both types of grain can escape the H II region during the expected Main Sequence lifetime of 10^7–10^8 yr. Grains will first of all move out of the central region and there will be some separation of dielectric and graphite grains. Assuming that initially the grain

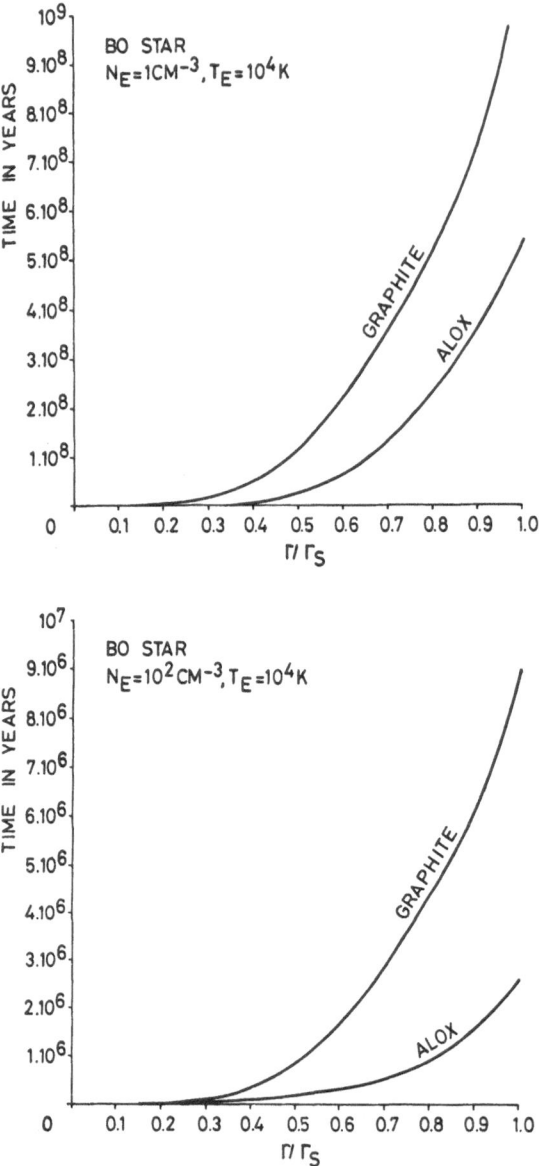

Fig. 4.   Time vs distance (in units of the Strömgren radius) for grains moving out from a B0 star.

distribution is uniform, the central 'hole' formed will be surrounded by a graphite shell and the outer regions will contain a mixture of grains until first the dielectric and eventually the graphite grains escape completely. (Of course the H II region itself may disperse in a time shorter than the M.S. lifetime of the central star.) For the case of $N_e = 1$ cm^{-3} the relative motion of the grains is similar but the timescale is longer. A central hole is still likely to form but grains in the outer regions are essentially frozen into the plasma and escape of grains is unlikely. Results for the O5 star are shown in Figures 5 and 6. If we assume a Main Sequence lifetime of

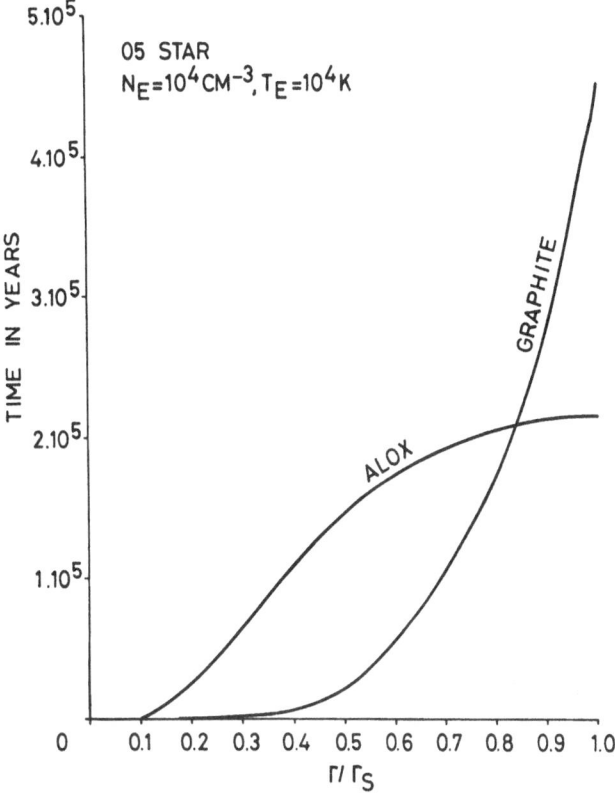

Fig. 5.   Time vs distance (in units of the Strömgren radius) for grains moving out from an O5 star
for an electron density of 10^4 cm^{-3}.

about 10^6 yr, the electron density required for escape of grains is $\sim 10^4$ cm^{-3}. In
this case, graphite grains move out of the central region rapidly while dielectric grains
escape easily from the outer regions. The effect of photoelectron pressure on the dielectric grains will further increase their probability of escape relative to the graphite
grains. The effect of the pressure due to photoemitted electrons is shown as a dashed
curve for the case where $N_e = 100$ cm^{-3}. As explained earlier this curve corresponds
to the maximum effect expected and the actual motion should lie somewhere between
the dashed and solid curves depending on the grain size. At this electron density, however, no significant motion of grains occurs outside the inner 20% of the Strömgren
radius during a realistic time period. For $N_e = 1$ cm^{-3} there is essentially no motion
of grains relative to the plasma.

4. Summary and Conclusions

Significant differences are to be expected in the variation of electric charge with position for dielectric and graphite grains in H II regions. Under a wide range of conditions it is possible for the two types of grain to be oppositely charged in the same

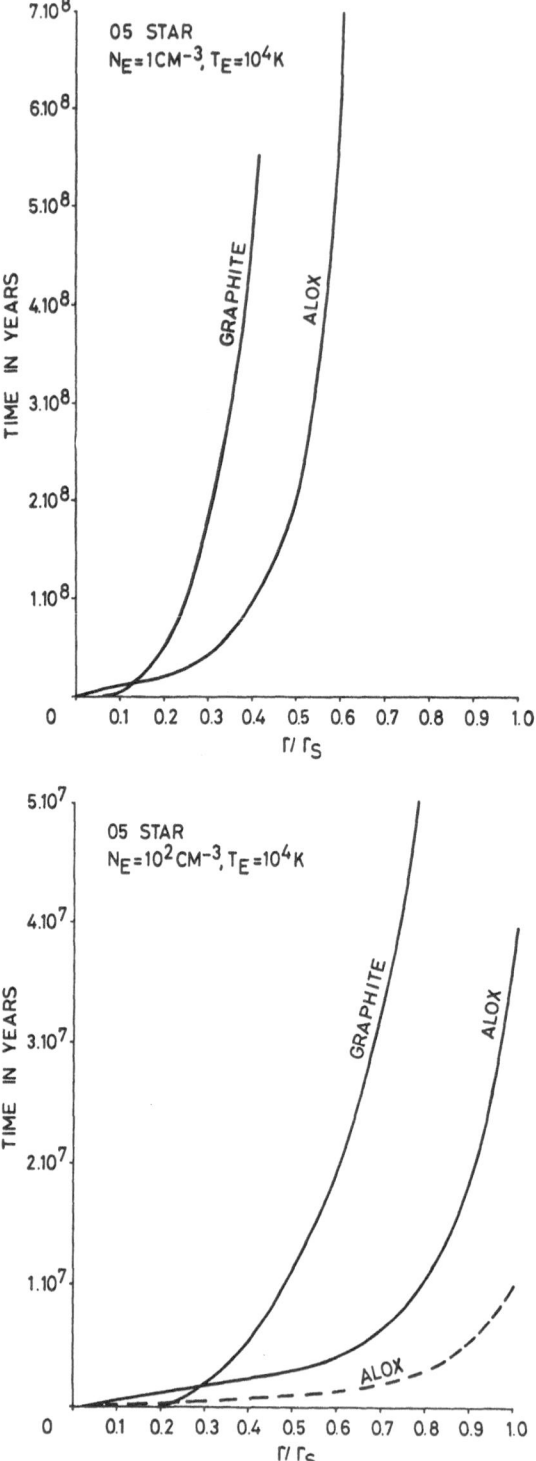

Fig. 6.   Time vs distance (in units of the Strömgren radius) for grains moving out from an O5 star
for electron densities of 1 and 10^2 cm^{-3}.

region of space. Electrostatic drag resulting from the grain charge increases the life-time of grains against expulsion due to radiation pressure from the central star but grains can escape from high density, compact H II regions providing such regions do not disperse in a time short compared with the Main Sequence lifetime of the central star. Probability for escape of dielectric grains is increased if the grains are large enough for the photoelectron pressure to be significant. Except in very low density regions it is likely that a central 'hole' is formed fairly rapidly and that there is some separation of graphite and dielectric grains in the central regions. The fact that silicate features in the infrared are generally seen in absorption supports the conclusion that silicate grains should be found mainly in the outer parts of H II regions or outside the H II regions completely. It is possible that graphite rather than silicate grains are responsible for the underlying infrared emission spectrum.

Acknowledgement

We appreciate the support given to this work by Dr B. Fitton.

References

Aitken, D. K. and Jones, B.: 1973, *Astrophys. J.* **184**, 127.

Davidson, K.: 1970, *Astrophys. Space Sci.* **6**, 422.

Emerson, J. P., Jennings, R. E., and Moorwood, A. F. M.: 1973, *Astrophys. J.* **184**, 401.

Feuerbacher, B. and Fitton, B.: 1972, *J. Appl. Phys.* **43**, 1563.

Feuerbacher, B., Willis, R. F., and Fitton, B.: 1973, *Astrophys. J.* **181**, 101.

Frogel, J. A. and Persson, S. E.: 1975, *Astrophys. J.* **195**, L15.

Furniss, I., Jennings, R. E., and Moorwood, A. F. M.: 1974, to appear in *Proceedings of the 8th ESLAB Symposium, H II Regions and the Galactic Centre.*

Gillett, F. C. and Forrest, W. J.: 1973, *Astrophys. J.* **179**, 483.

Harper, D. A. and Low, F. J.: 1971, *Astrophys. J.* **165**, L9.

Ireland, J. G.: 1970, *Astrophys. Space Sci.* **6**, 107.

Mathews, W. G.: 1967, *Astrophys. J.* **147**, 965.

Münch, G. and Persson, S. E.: 1971, *Astrophys. J.* **165**, 241.

Ney, E. P., Strecker, D. W., and Gehrz, R. D.: 1972, *Astrophys. J.* **180**, 809.

O'Dell, C. R., Hubbard, W. B., and Peimbert, M.: 1966, *Astrophys. J.* **143**, 743.

Spitzer, L.: 1956, *Physics of Fully Ionized Gases*, Interscience, New York.

Spitzer, L.: 1968, *Diffuse Matter in Space*, Interscience, New York.

Watson, W. D.: 1973, *J. Opt. Soc. Am.* **63**, 164.

Wright, E. L.: 1973, *Astrophys. J.* **185**, 569.

Wynn Williams, G. and Becklin, E.: 1974, *Publ. Astron. Soc. Pacific* **86**, 5.

OPTICAL PROPERTIES OF PARTICULATES

DONALD R. HUFFMAN

Physics Department, University of Arizona, Tucson, Ariz., U.S.A.

Abstract. Optical properties of small particles of olivine (less than 0.1 μ) have been studied in the ultraviolet as an example of an insulating solid. Very little structure survives in the ultraviolet extinction curves for such small particles. By contrast 'surface modes', observed for graphite small particles in the ultraviolet and for olivine particles in the infrared, produce dominant and persistent structure in extinction. The general trend of optical properties of graphite is surprisingly similar to the behavior required to explain all features of the interstellar extinction and albedo curves from near visible to 1000 Å. Measured extinction of small olivine particles in the infrared agrees with calculations based on newly measured optical constants, but dominant sharp structure in the 10 μ region still presents a bit of a problem in explaining 'silicate' features in astronomical data.

1. Introduction

Of the various features due to interstellar grains discussed in this Symposium, some general trends can be explained by a number of supposed grains, while certain discrete features seem to be more restrictive in limiting the proposed solid materials. The success of various grain models in fitting the general rise of extinction from near infrared to near ultraviolet, the far ultraviolet rise of extinction and the shape of the interstellar polarization curve, is an indication of the insensitivity of these effects to the grain composition. By contrast, the extinction features near 3 μ, 10 μ, and 2200 Å, and the rise of albedo in the far ultraviolet seem to be unique to only a few solids, and the diffuse bands in the visible are so restrictive that no solids known have convincingly explained any of these features.

The studies in our laboratory have been aimed at finding the characteristics of solids which can explain the restrictive features of interstellar grains. Part of our work has been aimed at measuring the optical constants from far infrared to far ultraviolet of likely solids for which such constants are either unavailable or questionable. New candidates for explaining the 2200 Å feature or the diffuse bands may arise when such optical constants are put into scattering calculations. However, such scattering calculations are only possible for very restricted geometries, and are done under the assumption that the measured properties of the bulk solids apply for small particles. The second part of our experimental program is designed to study whether these assumptions are indeed valid in all cases, by making actual collections of small particles, determining their size distributions by electron microscopy, and measuring the extinction and polarization at various wavelengths. This program is summarized below:

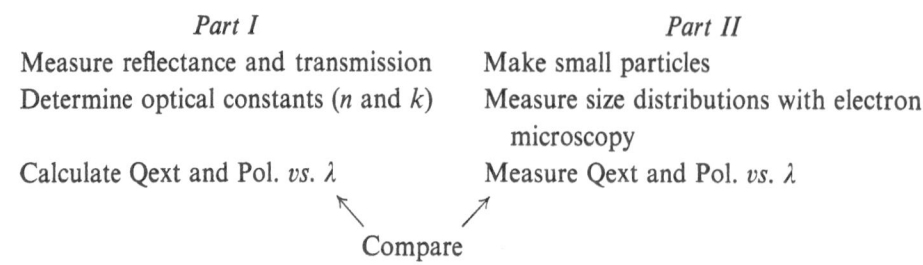

                    Part I   *Part II*
Measure reflectance and transmission   Make small particles
Determine optical constants (n and k)   Measure size distributions with electron
                                            microscopy
Calculate Qext and Pol. *vs.* λ Measure Qext and Pol. *vs.* λ

                              Compare

In this paper we will give results of this program for the following cases:
  (1) Olivine in the ultraviolet (4000 Å → 1000 Å)
  (2) Graphite in the ultraviolet (4000 Å → 1000 Å)
  (3) Olivine in the infrared (2 μ → 30 μ)
These examples are chosen to illustrate important classes of optical properties of small particles. Case (1) illustrates the behavior of typical electronic absorption in insulators, while cases (2) and (3) illustrate the collective modes in small particles, commonly referred to as 'surface plasmons' and 'surface phonons'.

2. Optical Properties of an Insulator-Olivine

Since small particle extinction spectra are presented in this paper for both ultraviolet and infrared, and in order to illustrate typical optical properties of insulators, Figure 1 shows the imaginary part k of the complex index of refraction $N = n + ik$ for olivine as a function of wavelength from 300 μ to 400 Å. The variation of the real part is not shown. It can be taken from the detail publications (Huffman and Stapp, 1973; Steyer and Huffman, 1974). Of the three possible polarization directions for olivine, results for only one direction are presented for simplicity. Characteristic features common to all non-metallic solids are the very intense lattice absorption bands in the infrared, a region of relative transparency ($k < 0.001$), and strong ultraviolet absorption with structure characteristic of the particular solid. In the case of olivine, relatively weak absorption features between 2 μ and 3600 Å are due to iron impurities. The k's of Figure 1 along with their associated real parts n, were put into the Mie calculations with varying size distributions to study features in extinction that might be caused either by the absorption edge near 3000 Å or by the strong structure toward shorter wavelengths. This essentially completes part I of our program for olivine in the ultraviolet.

To compare these results with actual measurements it is first necessary to prepare olivine as small particles. We have satisfactorily produced particles less than 0.1 μ by embedding pieces of natural crystal olivine into a depression in carbon-arc-lamp electrodes, and striking a D.C. arc of about 5 A between the electrodes in air. The resulting smoke, upon analysis by X-ray diffraction, electron diffraction and electron microscopy is found to consist of rather spherical particles of crystalline olivine. A

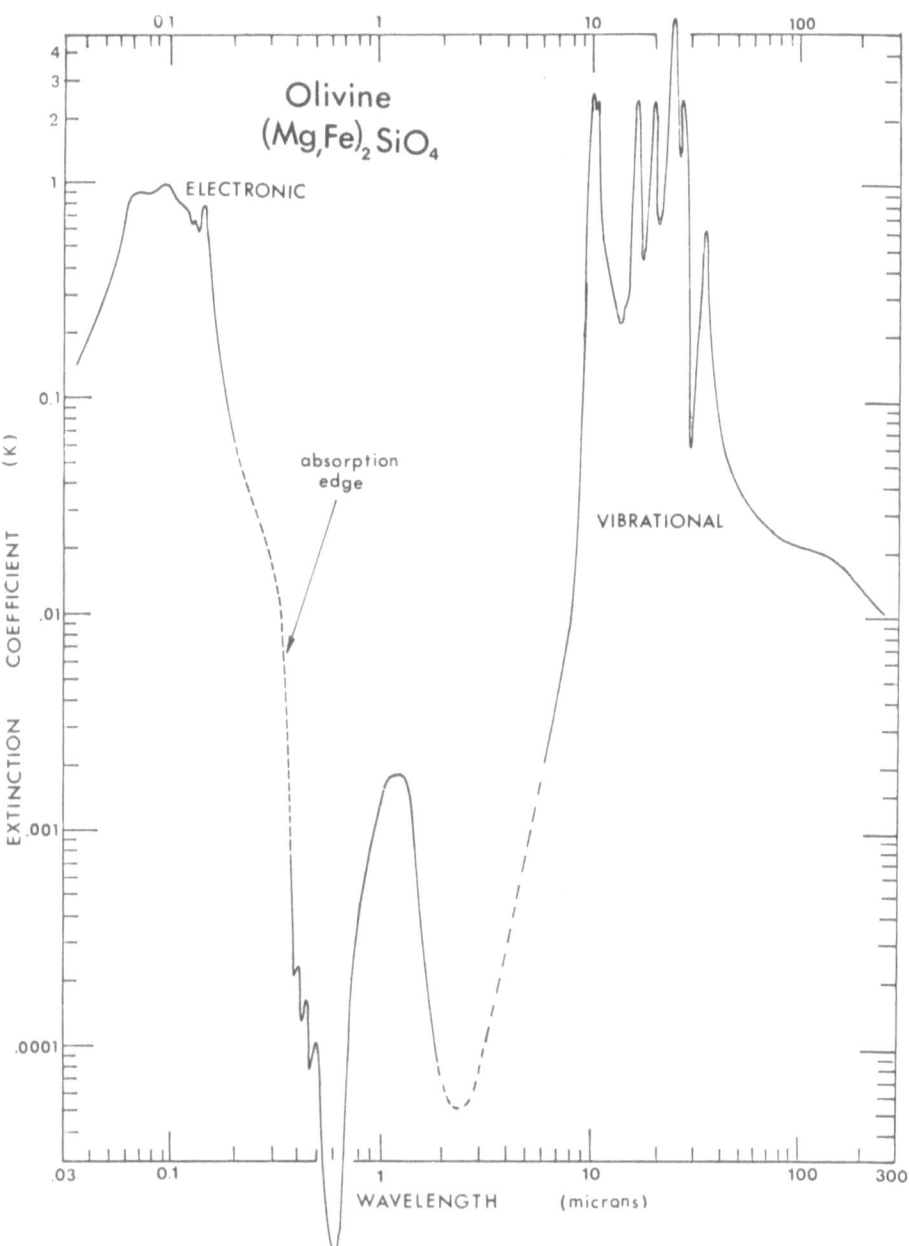

Fig. 1. The absorptive part (*k*) of the complex index of refraction $N=n+ik$ for a crystal of olivine. The determinations are for an arbitrary polarization direction, since this figure is only meant to show the general features of absorption in a typical silicate.

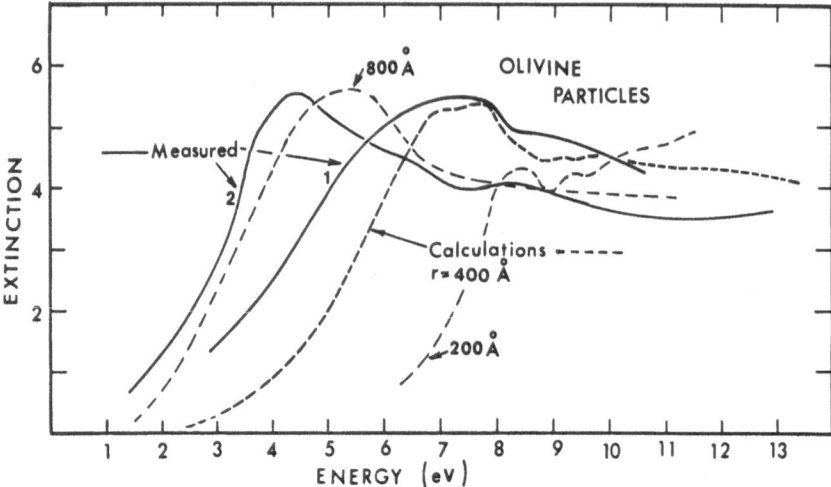

Fig. 2.   Measured and calculated extinction curves for small particles of olivine. The dashed curves are calculations for the mean radii indicated, assuming Gaussian size distributions with widths at half maxima equal to the mean radii. Measurements shown with solid lines are for two separate samples made under differing conditions.

thin coating of the smoke is trapped on glass slides and the extinction determined by transmission measurements up to the transmission cut-off of the glass slide. Reflectance measurements from the coated slide continue the data to higher incident energies. These extinction results can now be compared with extinction calculated from measured optical constants. These results are shown in Figure 2. The rather broad hump near 7 eV in both the measurements (1) and the 200 Å calculations is due to a combination of a rounded scattering feature and the absorption edge in the ultraviolet which sets in near 7.5 eV. This is the type of feature that has been discussed (Huffman and Stapp, 1971) as a possibility for producing the 2200 Å interstellar bump, in solids where the absorption edge is properly placed. However, such a feature appears to be neither terribly unique nor very persistent since it is fairly dependent on size. A careful look at Figure 2 also shows that the most prominent feature in the ultraviolet absorption curve of olivine (the peak near 0.14 μ in Figure 1) is easily washed out. Only in the calculations for 200 Å radius does this peak near 8.5 eV appear. In all larger sizes the calculations and the two sets of measurements show that this feature is essentially absent. Similar experiments on many other solids convince us that very strong and persistent features such as the 2200 Å interstellar band are rare. Almost all such experimental small particle systems show a smooth rise in extinction at longer wavelengths and a rather neutral and featureless curve at higher energies.

3. Surface Modes

Considerably more persistent and prominent are the extinction features expected in connection with collective modes of oscillation in small particles. Papers by Gilra (1972, 1973) have well emphasized to the astronomy community the importance of these so-called 'surface modes'. Although the most important 'surface mode' in small particles is actually a mode of uniform polarization, it is the first of a series of polarization modes that are confined more nearly to the surface of the solid in the higher modes (Ruppin and Englman, 1970). Because of the popular usage we will continue to use the term 'surface mode' with the quotation marks to remind the reader that this term is used with reservation.

While the references are to be consulted for details, we can briefly summarize some of the important points.

(1) There will be a peak in extinction for spheres near the point where the real part ε_1 of the complex dielectric function $(\varepsilon_1 + i\varepsilon_2)$ is given by

$$\varepsilon_1 = -2\varepsilon_m,$$

where ε_m is the dielectric constant of the surrounding medium $(\varepsilon_m \cong 1$ for air).

(2) Shapes other than spheres are expected to give extinction maxima in general regions of negative ε_1.

(3) The height of the extinction bump will be inversely related to the magnitude of the absorptive part of the dielectric function, ε_2.

There are two common situations in solids which give rise to strongly negative ε_1, and hence to the possibility for the kind of intense resonances we are discussing. These are (1) spectral regions where electrons behave as free electrons, such as the visible and near ultraviolet regions in many metals and some semiconductors, and (2) the spectral region for lattice vibrations in ionic solids (Reststrahl region). Theoretical and experimental examples of both of these types of 'surface modes' will now be presented, chosen from a group of solids that have been mentioned frequently as possible interstellar constituents.

3.1. GRAPHITE IN THE ULTRAVIOLET

Figure 3b is a plot of the dielectric functions for graphite measured by Taft and Phillipp (1965) for the electric vector of the light lying in the basal plane of the graphite. The point at which $\varepsilon_1 = -2$ is noted by an arrow. In Figure 3a, a series of Mie calculations for spheres of sizes indicated are shown. In all cases a Gaussian size distribution was used with a width at half maximum equal to the mean radius. One sees clearly the 'surface mode' or 'surface plasmon' which persists for various sizes near the $\varepsilon_1 = -2$ condition. This is, of course, the extinction feature which has been often associated with the 2200 Å interstellar feature. The peak also persists for varying shapes since the requirement that ε_1 be negative is fulfilled over a fairly narrow range of energies from about 4 eV to about 6.8 eV.

Such calculations can convincingly demonstrate an extinction near the 2200 Å interstellar feature, but additional extinction in the visible and especially in the ultraviolet shorter than 2000 Å wavelength have been found necessary to match the complete interstellar extinction curve (see for example Gilra, 1971). When measurements were made of extinction from actual smoke particles of graphite (Day and Huffman, 1973),

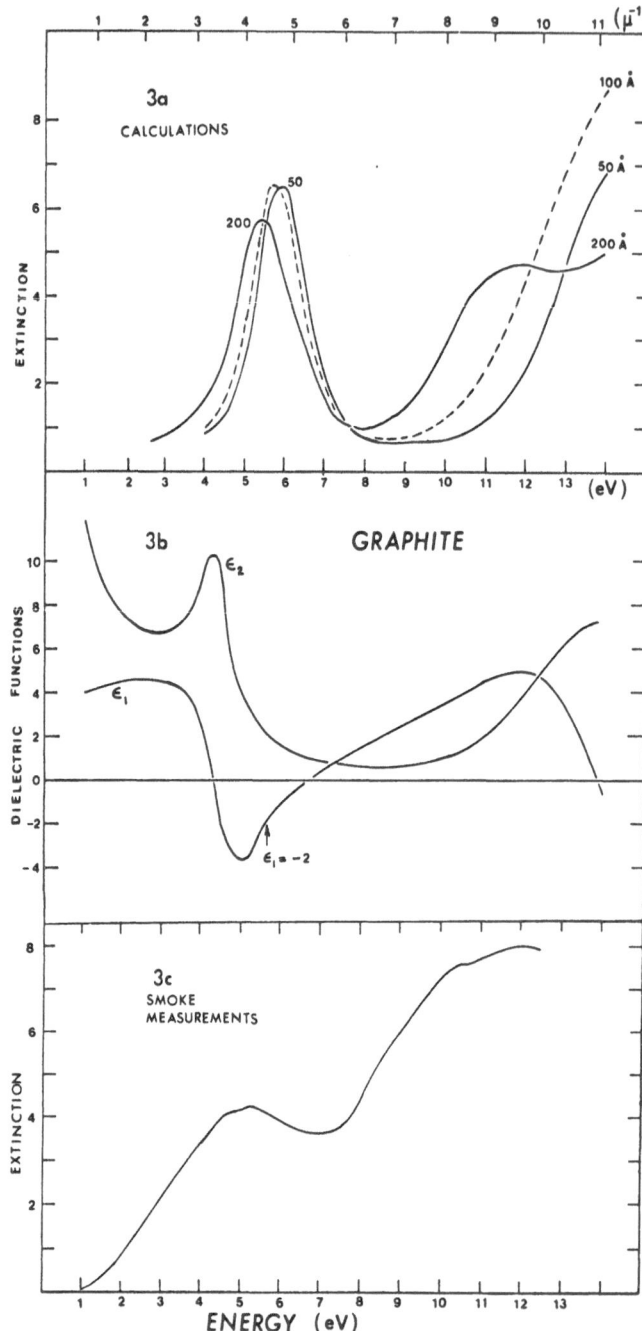

Fig. 3. The center figure shows the dielectric functions for graphite as measured by Taft and Phillipp (1965). Figure 3a shows results of Mie calculations using the optical properties of (b). Sizes shown are mean radii of Gaussian size distributions selected as in Figure 2. Figure 3c shows the results of extinction determined on graphite particles produced in an arc. The arrow in (b) calls attention to the 'surface mode condition', $\varepsilon_1 = -2$.

the extinction was found to increase even faster toward high energy than the inter-stellar extinction, producing an enormous discrepancy in the 10 eV region between calculations for any reasonable size distribution and the measurements. The main point of this work was to show that, in contrast to the results of Mie calculations, graphite could reasonably explain both the 2200 Å bump and the rapid increase of interstellar extinction toward higher energies, without the need for another dust component. Figure 3c shows the results of more recent measurements extending to higher energies for graphite particles similar to those of the previous work. The 'surface plasmon' resonance is clearly seen along with the rapid rise toward higher energy. A leveling off in the experimental curve for graphite particles near 10 eV does not agree with the recent interstellar extinction measurements (York *et al.*, 1973). We do not at this time consider this to be a serious drawback to the graphite hypothesis since we have had to be content with the rather special size distribution characteristic of our production method of graphite particles. Although calculated extinction is so badly in error compared to the measurements that the calculations do not provide very much firm guidance, they do suggest that a shoulder near 10 eV may be a rather size dependent feature. (In this connection, compare the calculations of Figure 3a for 100 Å and 200 Å mean size.)

Before leaving the discussion of graphite, we should consider the comparison of graphite calculations with the albedo of interstellar grains in the far ultraviolet as determined by Witt and Lilly (1972). Their extremely surprising results show the albedo of interstellar grains becoming very large at the highest energies, after undergoing a dip corresponding to the 2200 Å feature. The amazing import of these results is that the interstellar grain materials are becoming rather transparent (low absorption) near the highest energies. This very fact, if true, tends to rule out almost all solids yet proposed, since all non-metallic solids (semiconductors and insulators) have regions of quite high absorption for photon energies higher than the band gaps. This would seem to imply that at about 1500 Å we have not yet reached the onset of absorption across the band gap for interstellar grains, if they are non-metallic.

What is the situation of graphite in this regard? Mie calculations based on the optical constants of Figure 3b do not show a high and rising albedo in the 1500 Å region. However, the lack of agreement between experiment and theory in just this region makes us very distrustful of the calculations. If the very large extinction ob-served experimentally from graphite in this region is mostly due to scattering rather than absorption, a high albedo would be produced. In fact if one looks at the optical properties of graphite as surveyed in Figure 3b, one sees the onset of a low absorption region (low ε_2) between about 7 eV (1770 Å) and about 10 eV (1240 Å). Although ε_2 is not quite low enough to allow a large predominance of scattering over absorption in 200 Å particles for example, the general trend of the optical properties is surprisingly similar to what one would require to explain all features of the interstellar grains from near the visible to the far ultraviolet, including the 2200 Å band, the rapid and con-tinuing upturn of extinction extending to 1000 Å, and the surprisingly high and rising

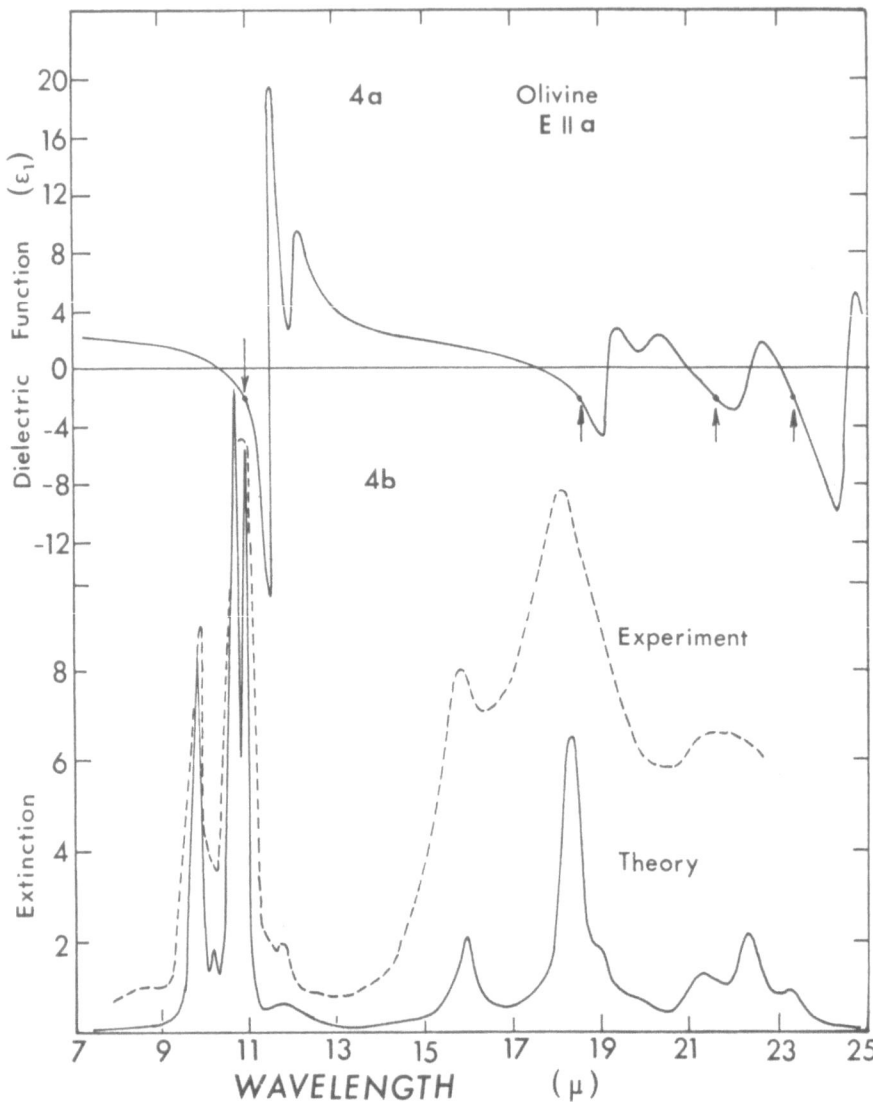

Fig. 4. The upper curve shows the real part ε_1 of the complex dielectric function $(\varepsilon_1 + i\varepsilon_2)$ measured for one of three major orientations of the light (E‖c) incident on a single crystal of olivine. The lower curve shows theoretical extinction calculations compared to extinction measured on small olivine particles ($< 0.1\ \mu$) produced in an arc. The mean size for the calculations was taken to approximate the size of the experimental particles.

albedo near 1500 Å. Could it be that the widely used optical constants for graphite are just slightly incorrect or inappropriate in the 2000 Å to 1000 Å region?

3.2. Olivine in the infrared

A pertinent example showing strong resonances from small particles in the infrared lattice absorption region is olivine. T. R. Steyer of our laboratory has just completed a determination of optical constants of olivine from $2\,\mu$ to $300\,\mu$ including the three polarization directions. Results for ε_1 for one of the three polarization directions are plotted in Figure 4a. Results for the other polarization directions have been omitted for purposes of clarity. As in Figure 3, the arrows denote the $\varepsilon_1 = -2$ condition in the region of which strong 'surface phonon' modes are expected. In Figure 4b are shown the results of calculations for small particles including all three sets of optical constants, and the results of measurements of small particle extinction. Details of these optical constants and these measurements will be presented elsewhere. Here we merely want to note the importance of such 'surface modes' by pointing out that dominant extinction occurs near $\varepsilon_1 = -2$, and is confined generally to the negative ε_1 region. It is worth observing that the negative ε_1 regions are quite narrow in the vicinity of $10\,\mu$, so that neither particle shape variations nor reasonable size variations are able to destroy the rather sharp structures present in extinction. We have found this to be generally true of all silicates studied thus far. Examination of rather high resolution observational data on extinction and infrared excess in the $10\,\mu$ region (Gillett and Forrest, 1973; Gammon *et al.*, 1972) has repeatedly failed to show structure such as these calculations and measurements show for silicate particles of the order of $0.1\,\mu$ in size. There therefore seems to be a bit of a problem remaining in this area. In an effort to resolve this we are looking into optical properties of amorphous silicates, effects of certain coatings, and the effects of temperature on the optical constants.

4. Conclusions

In this paper we have attempted to summarize some work on small particles of potential interest to astronomers and to demonstrate the following points:

(1) 'Surface modes' in small particles are persistent producers of structure in extinction in the ultraviolet ('surface plasmons') and in the infrared ('surface phonons').

(2) When accurate optical constants have been measured, optical properties of these modes can be reasonably well predicted.

(3) Graphite has been shown to be highly unusual in the ultraviolet region in ways which are suggestive of the strange behavior of interstellar grains.

Acknowledgements

This work has been partially supported by the Atmospheric Sciences Section, National Science Foundation.

References

Day, K. L. and Huffman, D. R.: 1973, *Nature* **243**, 50.
Gammon, R. H., Gaustad, J. E., and Treffers, R. R.: 1972, *Astrophys. J.* **175**, 687.

Gillett, F. C. and Forrest, W. J.: 1973, *Astrophys. J.* **179**, 483.

Gilra, D. P.: 1971, *Nature* **229**, 237.

Gilra, D. P.: 1972, *The Scientific Results from the Orbiting Astronomical Observatory OAO-2* (ed. by A. D. Code), NASA SP-310, p. 295.

Gilra, D. P.: 1973, in J. M. Greenberg and H. C. van de Hulst (eds.), 'Interstellar Dust and Related Topics', *IAU Symp.* **52**, 517.

Huffman, D. R. and Stapp, J. L.: 1971, *Nature Phys. Sci.* **229**, 45.

Huffman, D. R. and Stapp, J. L.: 1973, in J. M. Greenberg and H. C. van de Hulst (eds.), 'Interstellar Dust and Related Topics', *IAU Symp.* **52**, 297.

Ruppin, R. and Englman, R.: 1970, *Rept. Progr. Phys.* **33**, 149.

Steyer, T. R. and Huffman, D. R.: 1974, in preparation.

Taft, E. A. and Phillipp, H. R.: 1965, *Phys. Rev.* **138**, A197.

Witt, A. N. and Lillie, C. F.: 1972, *The Scientific Results from the Orbiting Astronomical Observatory OAO-2* (ed. by A. D. Code), NASA SP-310, p. 295.

York, D. G., Drake, J. F., Jenkins, E. B., Morton, D. C., Rogerson, J. B., and Spitzer, L.: 1973, *Astrophys. J.* **182**, L1.

POLARIZATION PROPERTIES OF SILICATE-LIKE GRAINS
IN CIRCUMSTELLAR ENVELOPES OF LATE-TYPE
STARS DUE TO TEMPERATURE VARIATIONS

J. SVATOŠ, M. ŠOLC, and V. VANÝSEK

Dept. of Astronomy and Astrophysics, Charles University, Prague, Czechoslovakia

Abstract. The influence of temperature changes in circumstellar silicate-like envelopes upon the polarization effects is investigated. It is shown that under the assumption that $\Delta T_g > 50°$ and conductivity of silicate grains is indirectly proportional to T_g this mechanism can be responsible for the observed dependence of intensity vs polarization in some late-type stars, e.g. V CVn. The same effects can be produced by dirty ices and graphite grains. It is suggested that irradiation by electrons and/or protons can affect the circumstellar envelopes in a similar way, especially those of early-type stars, and irradiation by neutrons can exert an influence on the envelopes of supernovae.

1. Introduction

It was shown in our previous paper (Svatoš *et al.*, 1975) that temperature variations of graphite grains can, in general, account for time-dependent changes of intrinsic polarization observed in some late-type variables, e.g. in V CVn. It is, however, well known that the formation of silicate-like grains in the atmospheres of these stars is more probable than that of the graphite ones. An attempt is, therefore, made to find whether the temperature changes of the silicate grains can account for the perfect correlation of intensity vs polarization, too. Since the real composition of the silicate grains is unknown a hypothetical model of possible silicate-like grains must be adopted. We therefore assume grains with the refractive index $m = 1.7 - 0.03\,i$ at room temperature and $\lambda = 5000$ Å. The temperature changes of the grains in a circumstellar envelope must be evaluated in order to obtain the corresponding values of refractive indices. Using these refractive indices polarization degrees in a disc-like circumstellar envelope at minimum and maximum of the light (temperature) are computed with the aid of the Mie theory.

2. Evaluation of the Temperature Changes and the Corresponding Refractive Indices

The temperature T_g of the grains can be computed by help of the well-known equation

$$\frac{R^2}{4r^2} \int_0^\infty Q_{abs}(a, \lambda) B(\lambda, T_s)\, d\lambda = \int_0^\infty Q_{abs}(a, \lambda) B(\lambda, Tg)\, d\lambda, \tag{1}$$

where $B(\lambda, T)$ = black-body intensity distribution; T_s = temperature of the stellar atmosphere; R = radius of the star; r = distance of the grains from the centre of the star; and a = radius of the grains.

Silicate grains radiate at far infrared $9.7 - 20\ \mu$ (Humphreys, 1974) and absorb in the visible and UV region. Since the imaginary part of the grains of interest is small, Q_{abs} at the left-hand side of Equation (1) can be replaced by the mean value of $\bar{Q} \to 1$. Due to $x = 2\pi a/\lambda \ll 1$ the right-hand side of Equation (1) can be expressed in terms of the mean value of the imaginary part of the refractive index $\bar{n}$ related to the far infrared (Kaplan and Pikelner, 1963). Thus Equation (1) after rearrangement takes the form

$$\left(\frac{R}{r}\right)^2 T_s^4 \cong 0.018\ T_g^5 \bar{n}. \tag{2}$$

Since in the region $9.7 - 20\ \mu$ $\bar{Q}_{abs} \cong 0.02$, $\bar{n}$ can be obtained according to van de Hulst's (1957) formula

$$Q_{abs} = -4x\ \mathrm{Im}\left(\frac{m^2 - 1}{m^2 + 2}\right). \tag{3}$$

The mean value 0.02 is also in good agreement with the course of the absorption efficiency factor for spherical quartz grains obtained by laboratory experiments (Martin, 1971) in the above mentioned region. From Equation (3)

$$\bar{m} \cong 1.8 - 0.1\ i \qquad (\text{i.e. } \bar{n} = 0.1)$$

is obtained.

For a semiregular M-type star (V CVn)

$$T_s^{\mathrm{eff}} = 2700\ \mathrm{K} \text{ at maximum of the light and}$$
$$T_s^{\mathrm{eff}} = 2400\ \mathrm{K} \text{ at minimum of the light.}$$

Using Equation (2) we find that

$$T_g \cong 320\ \mathrm{K} \text{ at maximum of the light and}$$
$$T_g \cong 285\ \mathrm{K} \text{ at minimum of the light}$$

for $r \cong 2.5\ R$. This distance is considered as a 'mean' distance of the adopted circumstellar disc-like envelope. Due to reasons discussed in more detail in our previous paper (Svatoš et al., 1975) $\Delta T_g \gg 50°$ so that we assume T_g to be at least about 450 K at maximum of the light. Since no exact relation of conductivity vs temperature is known, analogously with some other silicates (Pearson and Bardeen, 1949)

$$\sigma \sim T^{-3/2} \tag{4}$$

may be considered in the above-obtained temperature region.

The relations combining conductivity with the real and imaginary parts of the refractive index

$$n^2 = 0.5[(\varepsilon^2 + 4\lambda^2 c^{-2}\sigma^2)^{1/2} + \varepsilon], \tag{5}$$
$$k^2 = 0.5[(\varepsilon^2 + 4\lambda^2 c^{-2}\sigma^2)^{1/2} - \varepsilon],$$

together with relation (4) are used to obtain the corresponding refractive index. As already mentioned above $m_1 = 1.7 - 0.03\,i$ is assumed. This value corresponds to $T_g = 285$ K, i.e. at minimum of the light. Using (4) and (5) the computed value

$$m_2 \simeq 1.7 - 0.01\,i \quad \text{for} \quad T_g \simeq 450 \text{ K (i.e. at maximum of the light)}$$

adopting the dielectric constant $\varepsilon \simeq 3$ for both cases and the wavelength $\lambda = 0.5\,\mu$.

3. Model of the Circumstellar Envelope and the Results

A disc-like model of circumstellar nebula lying in the plane to the observer was used. The particle density was assumed to be constant so that the optical depth of the star was about 0.14. RTE along the line in the direction to the observer is

$$S_j(\xi + d\xi) = S_j(\xi)(1 - NC_e d\xi) + NC_e I_s r^{-2} e^{-\tau} i_j(\vartheta)d\xi, \tag{6}$$

where $S_j(\xi)$, $S_j(\xi+d\xi)$ = intensity of the entering and emerging beams, respectively, from an elementary volume of the width $d\xi$; N = number of particles in the elementary volume placed at $d\xi$; C_e = extinction cross-section; $i_j(\vartheta)$ = Mie's scattering functions ($j = 1, 2$); I_s = luminosity of the star; r = distance of the element $d\xi$ from the star; τ = optical depth of the distance r.

The solution to Equation (6) was obtained by computer using the Cauchy-Euler method step by step along the lines of sight to the observer at equidistant distances

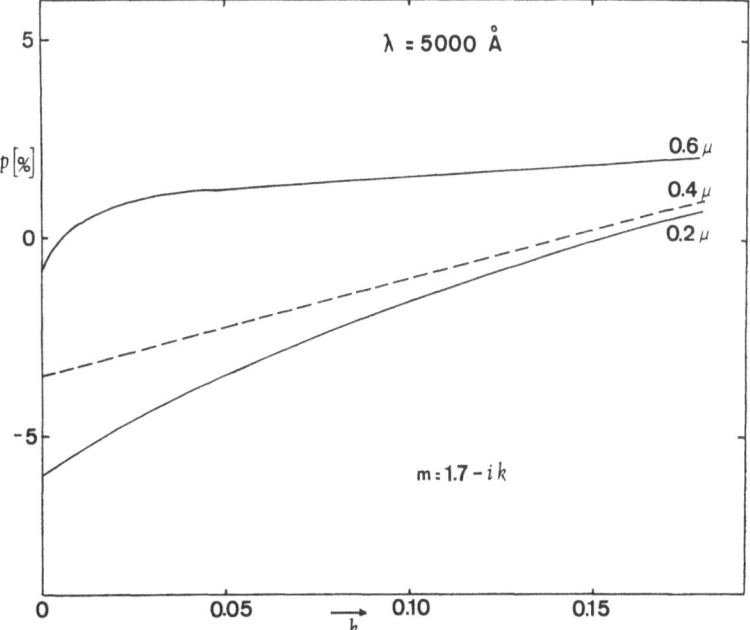

Fig. 1.   Percentual dependence of polarization on the imaginary part of refractive index of silicate grains in the circumstellar disc-model envelope for different sizes of grains.

TABLE I

Percentual polarization for different radii of grains in dependence on the refractive index

Refractive index		Particle radius in microns							
Real	Imaginary	0.2	0.3	0.4	0.6	0.7	0.8	0.9	1.0
1.7	−0.00	−5.95	−7.68	−3.45	−0.84	−4.51	−2.56	1.01	0.19
1.7	−0.01	−5.39	−8.34	−3.19	0.40	−3.91	−2.11	1.01	0.36
1.7	−0.05	−3.41	−7.90	−2.30	1.27	−0.88	0.35	1.50	0.85
1.7	−0.10	−1.50	−5.70	−0.81	1.65	0.89	1.34	1.67	1.27
1.7	−0.18	0.76	−2.26	1.03	2.18	1.92	1.86	1.87	1.63

from the star. If the whole object is so far away that the star + nebula are seen as a single point (the case of intrinsic polarization) the resulting polarization is given as a sum of the emerging beams – i.e.,

$$p = \frac{\Sigma S_1 - \Sigma S_2}{\Sigma S_1 + \Sigma S_2}. \tag{7}$$

The results obtained are presented in Table I and Figure 1.

4. Discussion

Our results show that:

(1) In general, the polarization degree increases with increasing imaginary part of the refractive index;

(2) This increase seems to be valid for silicates (this paper), graphites (Svatoš *et al.*, 1975) and for dirty ices as well, as can be seen from Figure 2. This figure shows the results obtained by Čermák (1972). His models of reflection nebulae are planparallel slabs with the star inside and the particle size distribution functions $N_a = = N_0 \exp\{(-5a/a_0)^3\}$ with the lower and upper size limits $x = 2\pi a/\lambda = 5$ and $x = 10$, respectively. The wavelength $\lambda = 0.55\,\mu$ is used.

(3) Geometry of the nebula is not essential. When using a disc model the forward scattering with a small degree of polarization dominates for the grains of interest and the resulting polarization is small, too. In the plan-parallel slab with the star inside the angles near 90° with higher polarization degree predominate and, consequently, the resulting polarization is greater. The course of the polarization mentioned at (1) is, of course, the same in the two geometrical models.

(4) Since the particles of all sizes of interest exhibit the same behaviour the distribution function plays no essential role.

(5) Under the assumption that in the circumstellar envelope

(a) $\Delta T_g > 50°$,

(b) the imaginary part of the refractive index of the grains is not too small, i.e. if $k > 0.001$,

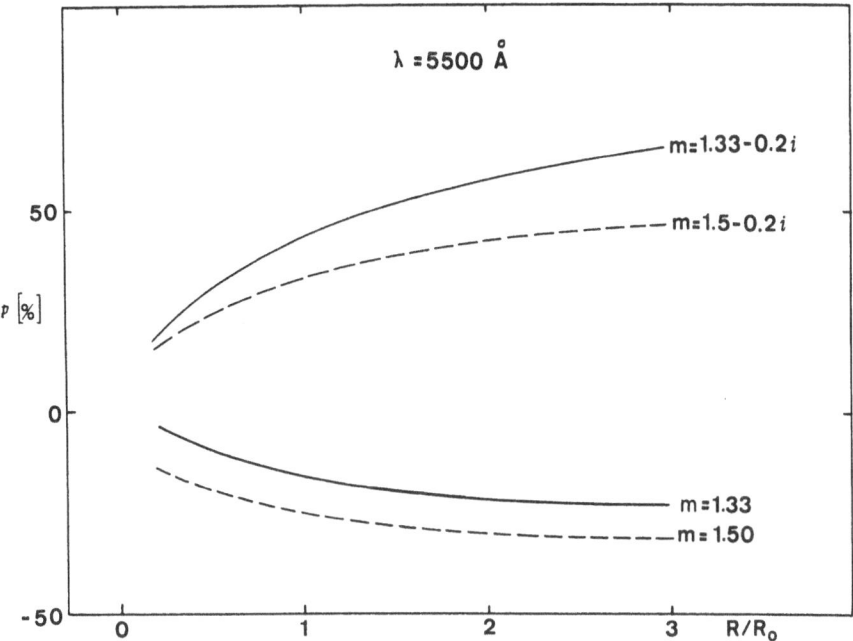

Fig. 2. The course of polarization in a slab-model reflection nebula in dependence on the distance from the illuminating star for different refractive indices. The optical depth of the nebula is equal to 0.1.

(c) $\sigma \sim T_g^{-1}$ in the temperature region of interest, then the silicate grains, especially those of greater sizes $a \sim 0.6 \, \mu$ (see Table I) can fully account for the perfect time-dependent correlation of intensity versus polarization, observed e.g. in V CVn.

5. Conclusion

It is evident that the investigation of temperature dependences is very important in astrophysical problems since these dependences can in a simple way account for some phenomena concerning intrinsic polarization.

It would be desirable to carry out laboratory measurements of temperature versus refractive index or conductivity of possible candidates responsible for intrinsic polarization.

Another process which could affect the polarization properties of silicate semiconductors is their irradiation by electrons and/or protons. As is known (Dunlap, 1957), the properties of silicon change strongly after irradiation by electrons or by nuclear particles in such a manner that the silicon resistance always increases independently of the original sample being p-type or n-type.

It may be expected that even in the case of semiregular variables this influence (irradiation by electrons) can be strong enough to contribute to the changes of optical properties. According to Ikaunieks (1970) the electron density in long-periodic Mira Ceti-type variables changes from 10^{10} cm^{-3} at maximum of intensity to 10^7 cm^{-3}

at minimum of intensity. This means that with increasing irradiation (temperature maximum) σ diminishes and, consequently, the polarization does so too, so that this influence could act simultaneously with the above-discussed mechanism.

The changes of optical properties caused by irradiation can play a dominant role in the envelopes of early-type stars and the changes due to irradiation by neutrons can be important for envelopes of the supernovae.

References

Čermák, V. M.: 1972, Thesis (unpublished), Charles University, Prague.

Dunlap, W. C., Jr.: 1959, *An Introduction to Semiconductors* (in Russian), IIL, Moscow, p. 293.

Hulst, H. C. van de: 1957, *Light Scattering by Small Particles*, John Wiley, New York.

Humphreys, R. M.: 1974, *Astrophys. J.* **188**, 75.

Ikaunieks, Ya. Ya.: 1970, *Pulsiruyushchiye zvyozdy*, Nauka, Moscow, p. 329.

Kaplan, S. A. and Pikelner, S. B.: 1963, *Mezhzvyozdnaya sreda*, Gosud. Izdat. Fiziko-Matem. Literatury, Moscow, p. 263.

Martin, P. G.: 1971, *Astrophys. Letters* **7**, 194.

Pearson, G. L. and Bardeen, J.: 1949, *Phys. Rev.* **75**, 865.

Svatoš, J., Šolc, M., and Vanýsek, V.: 1975, *Publ. Astron. Inst. Charles Univ.*, Prague, in press.

UV RADIATION FIELDS IN DARK CLOUDS

A. P. WHITWORTH**

Dept. of Applied Mathematics and Astronomy, University College, Cardiff, Wales

Abstract. How dark is it inside a dark cloud? If – as is currently believed – interstellar extinction at UV wavelengths is mainly due to scattering with a strongly forward throwing phase-function, the interior of a dark cloud may be much better illuminated at UV wavelengths than its measured extinction would suggest. We consider the penetration of radiation into a dark cloud against scattering and absorption by grains; and we define a new group property for interstellar grains, the exclusion optical depth τ_d. τ_d is a measure of the ability of the grains to exclude radiation from the interior of an externally illuminated cloud. Radiation – as measured by the radiation energy density – penetrates the cloud approximately as if against pure absorption only, with effective optical depth τ_d. Thus τ_d is a conceptually and numerically useful quantity when estimating the role of UV radiation in the thermal and chemical balance within a dark cloud. Computations are made of the radiation fields in (1200, 4500) Å, at the centres of dark clouds with measured visual extinctions. It is found that even in very dark clouds, the radiation energy density in (1200, 1800) Å may be significant, due to the high grain albedo at these short wavelengths.

1. Introduction

Dark clouds – and in general regions of heavy obscuration – are currently the objects of intensive investigation. On the observational side, they have been found to contain a wide range of molecular, IR continuum, thermal radio continuum and radio recombination line sources. The intimate relationships between these and other observations are not yet well understood, but interpretations usually involve the assumption that dark clouds are regions of imminent or on-going star formation. In this scheme the early evolution of the dark cloud is determined by a magneto- and/or gravothermal energy imbalance, which ultimately forces some parts of the gas to form local subcondensations or protostars: the later evolution involves the contraction of these protostars to form stars, and their subsequent interaction with the remaining gas of the dark cloud. During the early evolution of the dark cloud (i.e., until the gas is heated substantially by the release of its own gravitational and nuclear energy in forming a star, or is ionized by radiation from a newly formed star in the vicinity), the physical state of the gas is observable mainly through its molecular lines. However, the interpretation of these observations, either in terms of chemical balance, or in terms of excitation conditions, is difficult. One critical uncertainty is the UV radiation field within a dark cloud. Firstly, we do not know what flux of radiation is incident on the cloud. Secondly, we do not know how efficiently the flux penetrates the interior of the cloud. It is this second problem which we consider here.

On the assumption that the transfer of radiation at most wavelengths is dominated by its interaction with dust grains, we proceed in Section 2 to outline the grain parameters involved in the equation of transfer. In Section 3 we define two simple radiative transfer models; and in Section 4 we derive, in integral form, the equations of transfer for radiation being scattered and absorbed, without any (frequency co-herent) re-emission. These equations are then translated to a simple finite difference form, in order to facilitate the treatment of multiple scattering (Section 5). The numer-ical results are presented in Section 6. It is found that radiation penetrates into the dark cloud approximately as if against pure absorption only, with effective optical depth τ_d, which we call the exclusion optical depth. τ_d is a function of the orthodox grain parameters τ_e (extinction optical depth), γ (grain albedo), and g (mean scatter-ing-angle cosine). In Section 7 we discuss the conceptual and numerical usefulness of the quantity τ_d, in the context of the thermal and chemical balance within a dark cloud. Finally, we estimate the radiation fields in $(1200, 4500)$ Å at the centres of dark clouds with measured visual extinctions.

2. Grain Properties

The interaction between radiation and spherical homogeneous grains is described by three quantities: the absorption cross-section σ_a; the scattering cross-section σ_s; and the scattering phase-function $P_s(\theta)2\pi \sin (\theta) d\theta$, which gives the probability that radiation is scattered through an angle in $(\theta, \theta+d\theta)$. In astronomy, σ_a and σ_s are commonly replaced by the extinction cross-section $\sigma_e = \sigma_a + \sigma_s$, and the albedo $\gamma = \sigma_s/\sigma_e$. Likewise, $P_s(\theta)$ is often characterized by a parameter

$$g = \overline{\cos (\theta)} = \frac{1}{\sigma_s} \int\limits_0^\pi P_s(\theta)2\pi \cos (\theta) \sin (\theta) d\theta, \qquad (1)$$

where the integration is over all solid angles. g is the mean scattering-angle cosine. (Thus a large positive g implies strong forward scattering; whilst $g=0$ suggests iso-tropic scattering, and a large negative g implies strong backward scattering.) With each cross-section is associated an efficiency q (as, for example, the extinction effi-ciency $q_e = \sigma_e/\sigma_g$, where σ_g is the geometric cross-section of the grain); and an optical depth τ (e.g. the extinction optical depth $\tau_e = N_g\sigma_e$, where N_g is the column density of grains).

The extinction cross-section σ_e measures the ability of the grain to remove radia-tion from a parallel beam by absorption *and* scattering. If we map a dark cloud by performing star counts, we only detect direct (i.e. parallel) stellar flux, and thus we obtain the extinction optical depth τ_e through the cloud. Similarly, if we measure the reddening of individual stars, we obtain the wavelength-differential of the extinction optical depth through the cloud $\Delta\tau_e/\Delta\lambda$.

However, τ_e is not in itself a reliable measure of how dark it is within the cloud. There is a growing body of evidence which indicates that a large part of the interstellar extinction at UV wavelengths derives from scattering, $0.5 \lesssim \gamma \lesssim 1.0$; and that the scattering phase-function is strongly forward throwing over a large part of the UV wavelength range, $0.25 \lesssim g \lesssim 1.0$ (see, for example, Witt and Lillie, 1973). If this is true, then diffuse stellar flux penetrates *into* a dark cloud much more efficiently than direct stellar flux penetrates *through* the cloud. Thus the interior of the cloud may be quite well illuminated. For consider the flux $F=1$ associated with a single photon incident normally on the cloud boundary. Following an absorption event, $F=0$. But if the photon is scattered through an angle θ, then $F=\cos(\theta)$. Thus if θ is small ($g \rightarrow 1$), the flux is only slightly attenuated by the scattering event. Multiple scattering will randomize the photon direction ($\bar{F}=0$) after

$$\mathcal{N} \sim \left(\frac{\pi}{2 \cos^{-1}(g)}\right)^2 \tag{2}$$

scattering events (thus after one scattering if the phase-function is isotropic, $g=0$). However, even then the flux is not necessarily zero, since there is a probability of $\frac{1}{2}$ that the photon is still moving forward into the cloud. Nevertheless, dark clouds do appear as noticeably dark patches on the sky, even if individual stars are not resolved. Therefore we do still require that diffuse stellar flux is in some sense attenuated as it penetrates the cloud.

It should be noted that all the above grain properties (σ_a, σ_s, $P_s(\theta)$, etc) depend on the wavelength of the radiation involved. This dependence has been suppressed here for the following reason. In general, the transfer of continuum radiation against grain opacity involves two interactions: scattering events, which are essentially frequency coherent; and the absorption/emission balance of the grain, which is not frequency coherent, but rather tends to degrade the photon energies. More specifically, most grains absorb a large fraction of their thermal energy from dilute optical and UV radiation fields; whilst this absorption is offset by thermal emission at much longer IR wavelengths. To a good approximation, the absorption/emission balance determines the temperature of the grain in most astronomical situations. If we accept current ideas about the constitution of grains, we find that grains cannot emit any appreciable UV radiation, because if they do, they are so hot that they will rapidly vaporize, and will not be easy to replace. Grains in dark clouds are expected to be much cooler than this, and to be emitting thermal radiation at far infrared wavelengths. Thus, in evaluating the UV radiation field inside a dark cloud, we can ignore thermal emission by the grains, and limit our considerations to the externally incident photons. These are preserved during scattering events, and are destroyed by absorption: they are never replaced in any significant numbers by thermal emission. It is for this reason that we need only to know the grain properties at the wavelength under consideration.

By contrast, if thermal emission were important, as it is at longer wavelengths, we should need to know the grain properties over a wide range of wavelengths. We should then have to solve a system of coupled equations describing the simultaneous transfer of radiation at all wavelengths and the thermal balance of the grains at each position in the cloud (e.g. Werner and Salpeter, 1969; Whitworth, 1972). This more comprehensive problem is in fact no more complex algebraically than the one treated here; but it immediately involves us in a detailed discussion of the wavelength dependence of the grain properties which we wish to avoid at this stage (see, for example, Wickramasinghe and Nandy, 1972). We are concerned here with the description of a purely radiative transfer phenomenon. We shall see later how this is directly relevant to the evaluation of UV radiation fields in dark clouds. In the meantime, we can meaningfully continue to suppress the wavelength dependence of the grain properties: this will serve to underline the generality of the formulism.

In addition, we note that the optical properties of a spherical grain (σ_a, σ_s, $P_s(\theta)$, etc.) are functions of the grain size. In the case of a grain whose structure is anisotropic, the optical properties are also functions of the orientation of the grain with respect to the incoming photon, and of the polarization of the incoming photon. We shall assume that the distribution of grain sizes, shapes, structures, constituents and orientations in a dark cloud can be adequately represented by the optical properties of a single average spherical grain (thus σ_a, σ_s, $P_s(\theta)$, etc.), and we shall use these properties in our subsequent calculations. (This approximation is only strictly acceptable if any anisotropic grains are not strongly aligned, and/or if the illumination of the cloud is not very anisotropic. In reality, there may be strong grain alignment due to a frozen-in magnetic field (Carrasco et al., 1973); and the illumination may be highly anisotropic due to newly formed bright stars in the immediate vicinity.)

3. Radiative Transfer Model

Firstly, we model the cloud by a semi-infinite, plane-parallel slab. Positions within the cloud are measured in terms of the optical depth τ, normal to the 'near' cloud boundary at $\tau=0$. The 'far' cloud boundary is at $\tau=\tau_0$. Outside the cloud, i.e. outside the range $0 \leqslant \tau \leqslant \tau_0$, the medium is completely transparent.

Secondly, for the radiation field incident on the cloud boundary, we consider two extreme cases. Case a: *diffuse illumination*. The radiation field at the cloud boundary is semi-isotropic. In terms of the intensity $I(\tau, \mu)$, we write

$$I_0(\tau = 0, \mu > 0) = I_* = \frac{U_* c}{4\pi} = I_0(\tau = \tau_0, \mu < 0). \tag{3}$$

$\mu = \cos(\phi)$, and ϕ is the angle between the intensity and the τ-axis. U_* is to be identified with the radiation density in an unobscured region far outside the cloud.

Case b: *direct illumination*. A parallel beam of radiation is incident normally on both cloud boundaries:

$$I_0(\tau = 0, \mu > 0) = \frac{U_*\delta(1 - |\mu|)c}{4\pi} = I_0(\tau = \tau_0, \mu < 0). \qquad (4)$$

Thirdly, we approximate the scattering phase-function $P_s(\theta)$ by a probability g that the scattered photon is scattered straight forward (which is equivalent to its not being scattered at all), and a probability $(1-g)$ that the scattered photon is scattered isotropically. g is defined by Equation (1). Evidently this approximation suppresses some details of the phase-function. The phase-function of an individual grain is generally a rather complex function of θ, and possibly of other angles (see, for example, Wickramasinghe, 1973). However, when we allow for a realistic distribution of grain sizes, shapes, structures, constituents and orientations in the interstellar medium, it is likely that very little remains except a tendency for the grains to scatter radiation forward or backward, more or less strongly. In the first approximation, g is a convenient measure of this tendency (cf. Aannestad and Purcell, 1973). With the scattering phase-function approximated in this way, we define two new variables: the effective extinction cross-section,

$$\sigma = \sigma_e(1 - g\gamma); \qquad (5)$$

and the effective albedo,

$$\alpha = \frac{\gamma(1 - g)}{(1 - g\gamma)}. \qquad (6)$$

The radiative transfer equations are now formulated in terms of α and the effective extinction optical depth

$$\tau = N_g\sigma. \qquad (7)$$

4. Equations of Transfer: Integral Formulation

We denote by a subscript 'i' the density of radiation which has been scattered i times. The penetration of radiation into the cloud is measured by the total radiation density $U_t(\tau)$, normalized to the density of radiation U_* far outside the cloud as

$$u(\tau) = \frac{U_t(\tau)}{U_*}. \qquad (8)$$

In principle, $u(\tau)$ can be derived from the following three equations

$$u_0(\tau) = \begin{cases} \frac{1}{2}(E_2(\tau) + E_2(\tau_0 - \tau)), & \text{case a}; & (9a) \\ \frac{1}{2}(\exp(-\tau) + \exp(\tau - \tau_0)), & \text{case b}; & (9b) \end{cases}$$

$$u_i(\tau) = \frac{\alpha}{2} \int_0^{\tau_0} u_{i-1}(\tau')E_1(|\tau - \tau'|)\, d\tau', \qquad i \geqslant 1; \qquad (10)$$

$$u(\tau) = \sum_{i=0}^{\infty} (u_i(\tau)) = \frac{U_i(\tau)}{U_*}. \tag{11}$$

The functions $E_1(x)$, $E_2(x)$, etc. are exponential integrals defined by

$$E_n(x) = \int_1^{\infty} \frac{\exp(-xy)}{y^n} \, dy, \qquad n \geqslant 1. \tag{12}$$

Their analytic and numerical properties are discussed in Abramowitz and Stegun (1964, p. 227f.).

5. Equations of Transfer: Finite Difference Formulation

The iterative routine contained in Equation (10) must be treated numerically. We divide the cloud into $2\mathcal{N}$ plane-parallel slab elements of equal optical thickness $\delta\tau = \tau_0/2\mathcal{N}$. The j-th element is contained in $(\tau(j-1), \tau(j))$, where $\tau(j) = j\tau_0/2\mathcal{N}$. The normalized radiation densities $u_i(\tau)$ are averaged over τ within each element, at each iteration (i.e. at each scattering order); and are then treated as being uniform across the element in computing the next iteration. We denote these average values by a bar as $\bar{u}_i(j)$. The equations of transfer (9 to 11) reduce to the form

$$\bar{u}_0(j) = \frac{\begin{aligned}(E_3(\tau(j-1)) - E_3(\tau(j)) + E_3(\tau(2\mathcal{N}-j)) - \\ - E_3(\tau(2\mathcal{N}+1-j)))\end{aligned}}{(2\delta\tau)}, \qquad \text{case a;} \tag{13a}$$

$$\bar{u}_0(j) = \frac{\begin{aligned}(\exp(-\tau(j-1)) - \exp(-\tau(j)) + \exp(-\tau(2\mathcal{N}-j)) - \\ - \exp(-\tau(2\mathcal{N}+1-j)))\end{aligned}}{(2\delta\tau)},$$
$$\text{case b;} \tag{13b}$$

$$\bar{u}_i(j) = \frac{\alpha}{2\delta\tau} \sum_{k=1}^{2\mathcal{N}} (\bar{u}_{i-1}(k) \, d(k,j)); \tag{14}$$

$$d(k,j) = E_3(\tau(j-k-1)) + E_3(\tau(j-k+1)) - $$
$$- 2E_3(\tau(j-k)), \qquad k < j; \tag{15a}$$
$$= 2\delta\tau - 1 + 2E_3(\delta\tau), \qquad k = j; \tag{15b}$$
$$= d(j,k), \qquad k > j. \tag{15c}$$

$$\bar{u}(j) = \sum_{i=0}^{\mathcal{M}_c} (\bar{u}_i(j)). \tag{16}$$

The numerical procedure contained in Equations (13a) through (16) divides into two parts. Firstly, we must evaluate the zero order solution (Equation (13a) or (13b)), and the elements of the coupling matrix $d(j, k)$ (Equations (15a, b, c)). The computing time required for this is governed by the time required to evaluate the exponential integrals with sufficient accuracy. Secondly, we enter the iterative routine described by Equation (14), and use this to generate the contribution $\bar{u}_i(j)$ of each scattering order to the total normalized radiation density. These contributions are then summed according to Equation (16), to obtain the total mean normalized radiation density $\bar{u}(j)$ in each element. This summation is terminated for $\mathcal{M}_c$ terms (i.e. $\mathcal{M}_c$ orders of scattering) as soon as adequate convergence is obtained. Since the iterative routine is extremely simple, it is not prohibitive to follow a large number of scatterings when this is necessary (i.e. for large τ_{e0}, and/or large γ, and/or small g).

6. Numerical Results

We have solved the equations of transfer in finite difference form for a grid of: (i) grain albedos, $\gamma = 0.10, 0.30, 0.50, 0.70, 0.90, 0.95$; (ii) mean scattering-angle cosines, $g = 0.00, 0.10, 0.25, 0.50, 0.75, 0.90, 0.95$; and (iii) cloud sizes, as measured by the total extinction optical depth through the cloud, $\tau_{e0} = 1, 2, 4, 8, 16, 32$. We have considered the two geometries defined in Section 3, namely a plane-parallel slab illuminated (a) diffusely by a semi-isotropic radiation field, and (b) directly by a parallel beam at normal incidence. Eighty optical depth elements were used for each case. The results are determined in terms of the dimensionless variable $\bar{u}(j)$, which is the mean radiation energy density in element j, normalized to a value unity far outside the cloud. It is not feasible or useful to present the results in full. Rather, we shall simply note and discuss the most important features.

In Table I we give values of u_c, the normalized radiation density at the cloud centre, on a grid of (γ, g, τ_{e0}) values for illumination case a. The first column gives the value of γ, and the second column gives the value of g. The remaining six columns give values of u_c in floating point notation. (In this and all subsequent tables, an entry of the form $1.23(-45)$ is equivalent to 1.23×10^{-45}.) Each of the last six columns is headed by the appropriate value of the total extinction optical depth τ_{e0}. Table II is the same as Table I, but for illumination case b. Evidently, if γ and/or g is large (strong scattering/forward throwing), u_c may be substantially larger than if the extinction were due purely to absorption ($\gamma = 0$). In fact, for illumination case b (Table II) and sufficiently large γ, u_c may actually exceed unity at the centres of thin clouds. This is a real effect due to the anisotropic illumination in case b, and is not a violation of basic physical principles.

The form of $u(\tau)$ in $(0, \tau_0)$ has in all cases a shape which approximates to that which would obtain if the grains did not scatter, but only absorbed radiation, with an effective optical depth τ_d:

$$u(\tau) = \tfrac{1}{2}(E_2(\tau_d) + E_2(\tau_{d0} - \tau_d)), \quad \text{case a;} \tag{17a}$$

$$u(\tau) = \tfrac{1}{2}(\exp(-\tau_d) + \exp(\tau_d - \tau_{d0})), \quad \text{case b;} \tag{17b}$$

$$\tau_d = \tau_e G_d. \tag{18}$$

A. P. WHITWORTH

TABLE I

The normalized central radiation density u_c, for various total extinction optical depths τ_{e0}, and illumination case a

γ	g	u_c					
		$\tau_{e0} = 1$	$\tau_{e0} = 2$	$\tau_{e0} = 4$	$\tau_{e0} = 8$	$\tau_{e0} = 16$	$\tau_{e0} = 32$
0.10	0.00	3.53(−1)	1.66(−1)	4.37(−2)	3.93(−3)	4.49(−5)	8.98(−9)
0.10	0.10	3.53(−1)	1.66(−1)	4.42(−2)	4.04(−3)	4.77(−5)	1.03(−8)
0.10	0.25	3.54(−1)	1.67(−1)	4.49(−2)	4.20(−3)	5.23(−5)	1.25(−8)
0.10	0.50	3.55(−1)	1.69(−1)	4.62(−2)	4.50(−3)	6.10(−5)	1.75(−8)
0.10	0.75	3.55(−1)	1.71(−1)	4.75(−2)	4.81(−3)	7.11(−5)	2.46(−8)
0.10	0.90	3.56(−1)	1.72(−1)	4.83(−2)	5.01(−3)	7.80(−5)	3.01(−8)
0.30	0.00	4.18(−1)	2.12(−1)	6.22(−2)	6.46(−3)	8.97(−5)	2.33(−8)
0.30	0.10	4.19(−1)	2.15(−1)	6.41(−2)	6.95(−3)	1.06(−4)	3.37(−8)
0.30	0.25	4.20(−1)	2.18(−1)	6.70(−2)	7.75(−3)	1.36(−4)	5.87(−8)
0.30	0.50	4.23(−1)	2.24(−1)	7.22(−2)	9.33(−3)	2.09(−4)	1.51(−7)
0.30	0.75	4.25(−1)	2.29(−1)	7.78(−2)	1.13(−2)	3.22(−4)	3.95(−7)
0.30	0.90	4.26(−1)	2.33(−1)	8.14(−2)	1.26(−2)	4.20(−4)	7.09(−7)
0.50	0.00	5.07(−1)	2.86(−1)	9.71(−2)	1.25(−2)	2.43(−4)	1.08(−7)
0.50	0.10	5.09(−1)	2.90(−1)	1.01(−1)	1.39(−2)	3.09(−4)	1.83(−7)
0.50	0.25	5.11(−1)	2.97(−1)	1.08(−1)	1.63(−2)	4.46(−4)	4.11(−7)
0.50	0.50	5.15(−1)	3.07(−1)	1.20(−1)	2.14(−2)	8.38(−4)	1.68(−6)
0.50	0.75	5.17(−1)	3.17(−1)	1.34(−1)	2.82(−2)	1.62(−3)	7.32(−6)
0.50	0.90	5.18(−1)	3.23(−1)	1.42(−1)	3.35(−2)	2.43(−3)	1.83(−5)
0.70	0.00	6.38(−1)	4.17(−1)	1.77(−1)	3.23(−2)	1.15(−3)	1.55(−6)
0.70	0.10	6.40(−1)	4.23(−1)	1.85(−1)	3.62(−2)	1.49(−3)	2.71(−6)
0.70	0.25	6.43(−1)	4.32(−1)	1.98(−1)	4.32(−2)	2.23(−3)	6.52(−6)
0.70	0.50	6.46(−1)	4.47(−1)	2.22(−1)	5.85(−2)	4.55(−3)	3.17(−5)
0.70	0.75	6.46(−1)	4.60(−1)	2.49(−1)	8.02(−2)	9.82(−3)	1.82(−4)
0.70	0.90	6.44(−1)	4.67(−1)	2.65(−1)	9.75(−2)	1.60(−2)	5.68(−4)
0.90	0.00	8.45(−1)	6.99(−1)	4.46(−1)	1.62(−1)	2.00(−2)	3.07(−4)
0.90	0.10	8.46(−1)	7.05(−1)	4.60(−1)	1.76(−1)	2.41(−2)	4.55(−4)
0.90	0.25	8.48(−1)	7.13(−1)	4.81(−1)	2.00(−1)	3.24(−2)	8.50(−4)
0.90	0.50	8.49(−1)	7.26(−1)	5.19(−1)	2.50(−1)	5.56(−2)	2.75(−3)
0.90	0.75	8.46(−1)	7.35(−1)	5.56(−1)	3.16(−1)	1.02(−1)	1.12(−2)
0.90	0.90	8.39(−1)	7.33(−1)	5.73(−1)	3.62(−1)	1.53(−1)	3.04(−2)
0.95	0.00	9.16(−1)	8.26(−1)	6.32(−1)	3.23(−1)	7.28(−2)	3.58(−3)
0.95	0.10	9.17(−1)	8.30(−1)	6.44(−1)	3.43(−1)	8.36(−2)	4.77(−3)
0.95	0.25	9.18(−1)	8.36(−1)	6.64(−1)	3.75(−1)	1.04(−1)	7.57(−3)
0.95	0.50	9.19(−1)	8.44(−1)	6.96(−1)	4.39(−1)	1.55(−1)	1.81(−2)
0.95	0.75	9.16(−1)	8.49(−1)	7.25(−1)	5.14(−1)	2.44(−1)	5.21(−2)
0.95	0.90	9.10(−1)	8.45(−1)	7.35(−1)	5.61(−1)	3.28(−1)	1.13(−1)

TABLE II

The normalized central radiation density u_c, for various total extinction optical depths τ_{e0}, and illumination case b

γ	g	u_c					
		$\tau_{e0} = 1$	$\tau_{e0} = 2$	$\tau_{e0} = 4$	$\tau_{e0} = 8$	$\tau_{e0} = 16$	$\tau_{e0} = 32$
0.10	0.00	6.51(−1)	4.05(−1)	1.53(−1)	2.13(−2)	4.05(−4)	1.42(−7)
0.10	0.10	6.50(−1)	4.05(−1)	1.54(−1)	2.19(−2)	4.31(−4)	1.63(−7)
0.10	0.25	6.48(−1)	4.05(−1)	1.56(−1)	2.27(−2)	4.72(−4)	2.00(−7)
0.10	0.50	6.45(−1)	4.06(−1)	1.59(−1)	2.42(−2)	5.50(−4)	2.82(−7)
0.10	0.75	6.41(−1)	4.06(−1)	1.62(−1)	2.57(−2)	6.41(−4)	3.97(−7)
0.10	0.90	6.39(−1)	4.06(−1)	1.64(−1)	2.67(−2)	7.03(−4)	4.88(−7)
0.30	0.00	7.62(−1)	5.03(−1)	2.04(−1)	3.09(−2)	6.56(−4)	2.65(−7)
0.30	0.10	7.57(−1)	5.03(−1)	2.08(−1)	3.32(−2)	7.79(−4)	3.91(−7)
0.30	0.25	7.50(−1)	5.04(−1)	2.15(−1)	3.68(−2)	1.01(−3)	7.03(−7)
0.30	0.50	7.36(−1)	5.03(−1)	2.26(−1)	4.36(−2)	1.56(−3)	1.88(−6)
0.30	0.75	7.21(−1)	5.01(−1)	2.36(−1)	5.16(−2)	2.40(−3)	5.07(−6)
0.30	0.90	7.12(−1)	4.99(−1)	2.43(−1)	5.70(−2)	3.11(−3)	9.21(−6)
0.50	0.00	9.14(−1)	6.55(−1)	2.94(−1)	5.15(−2)	1.35(−3)	7.80(−7)
0.50	0.10	9.03(−1)	6.55(−1)	3.03(−1)	5.70(−2)	1.75(−3)	1.38(−6)
0.50	0.25	8.87(−1)	6.53(−1)	3.17(−1)	6.64(−2)	2.58(−3)	3.33(−6)
0.50	0.50	8.57(−1)	6.45(−1)	3.38(−1)	8.52(−2)	4.96(−3)	1.50(−5)
0.50	0.75	8.21(−1)	6.31(−1)	3.56(−1)	1.08(−1)	9.56(−3)	7.03(−5)
0.50	0.90	7.97(−1)	6.17(−1)	3.64(−1)	1.24(−1)	1.41(−2)	1.80(−4)
0.70	0.00	1.13(0)	9.20(−1)	4.90(−1)	1.11(−1)	4.62(−3)	6.72(−6)
0.70	0.10	1.11(0)	9.16(−1)	5.06(−1)	1.24(−1)	6.13(−3)	1.23(−5)
0.70	0.25	1.08(0)	9.05(−1)	5.28(−1)	1.47(−1)	9.49(−3)	3.23(−5)
0.70	0.50	1.02(0)	8.74(−1)	5.58(−1)	1.94(−1)	2.02(−2)	1.83(−4)
0.70	0.75	9.52(−1)	8.22(−1)	5.70(−1)	2.50(−1)	4.35(−2)	1.19(−3)
0.70	0.90	9.00(−1)	7.78(−1)	5.63(−1)	2.83(−1)	6.83(−2)	3.80(−3)
0.90	0.00	1.48(0)	1.48(0)	1.12(0)	4.51(−1)	5.78(−2)	8.91(−4)
0.90	0.10	1.45(0)	1.46(0)	1.13(0)	4.90(−1)	7.06(−2)	1.34(−3)
0.90	0.25	1.39(0)	1.41(0)	1.15(0)	5.54(−1)	9.69(−2)	2.59(−3)
0.90	0.50	1.28(0)	1.31(0)	1.15(0)	6.70(−1)	1.71(−1)	9.11(−3)
0.90	0.75	1.15(0)	1.16(0)	1.07(0)	7.62(−1)	3.10(−1)	4.15(−2)
0.90	0.90	1.04(0)	1.03(0)	9.51(−1)	7.52(−1)	4.17(−1)	1.12(−1)
0.95	0.00	1.60(0)	1.73(0)	1.54(0)	8.52(−1)	1.96(−1)	9.62(−3)
0.95	0.10	1.56(0)	1.70(0)	1.55(0)	9.02(−1)	2.26(−1)	1.29(−2)
0.95	0.25	1.49(0)	1.63(0)	1.54(0)	9.81(−1)	2.85(−1)	2.08(−2)
0.95	0.50	1.37(0)	1.50(0)	1.49(0)	1.11(0)	4.31(−1)	5.19(−2)
0.95	0.75	1.21(0)	1.29(0)	1.33(0)	1.15(0)	6.57(−1)	1.60(−1)
0.95	0.90	1.09(0)	1.12(0)	1.13(0)	1.04(0)	7.67(−1)	3.38(−1)

TABLE III

Values of $G_d = \tau_d/\tau_e$, for various total extinction optical depths τ_{e0}, and illumination case a

γ	g	G_d					
		$\tau_{e0} = 1$	$\tau_{e0} = 2$	$\tau_{e0} = 4$	$\tau_{e0} = 8$	$\tau_{e0} = 16$	$\tau_{e0} = 32$
0.10	0.00	0.9110	0.9263	0.9420	0.9566	0.9692	0.9793
0.10	0.10	0.9099	0.9236	0.9378	0.9510	0.9623	0.9714
0.10	0.25	0.9081	0.9196	0.9315	0.9425	0.9520	0.9596
0.10	0.50	0.9054	0.9130	0.9210	0.9283	0.9347	0.9398
0.10	0.75	0.9026	0.9065	0.9105	0.9142	0.9173	0.9199
0.10	0.90	0.9010	0.9026	0.9042	0.9057	0.9070	0.9080
0.30	0.00	0.7228	0.7642	0.8094	0.8528	0.8913	0.9230
0.30	0.10	0.7200	0.7575	0.7985	0.8377	0.8725	0.9013
0.30	0.25	0.7161	0.7476	0.7820	0.8150	0.8443	0.8685
0.30	0.50	0.7100	0.7313	0.7546	0.7770	0.7968	0.8132
0.30	0.75	0.7044	0.7154	0.7272	0.7386	0.7487	0.7570
0.30	0.90	0.7017	0.7060	0.7109	0.7155	0.7195	0.7229
0.50	0.00	0.5199	0.5783	0.6476	0.7169	0.7799	0.8329
0.50	0.10	0.5166	0.5697	0.6327	0.6958	0.7534	0.8020
0.50	0.25	0.5122	0.5569	0.6103	0.6639	0.7129	0.7547
0.50	0.50	0.5057	0.5364	0.5730	0.6100	0.6439	0.6729
0.50	0.75	0.5014	0.5172	0.5361	0.5553	0.5728	0.5879
0.50	0.90	0.5000	0.5066	0.5143	0.5222	0.5293	0.5355
0.70	0.00	0.3038	0.3622	0.4420	0.5287	0.6096	0.6775
0.70	0.10	0.3012	0.3543	0.4271	0.5067	0.5816	0.6452
0.70	0.25	0.2977	0.3428	0.4047	0.4733	0.5383	0.5946
0.70	0.50	0.2936	0.3248	0.3679	0.4163	0.4629	0.5042
0.70	0.75	0.2930	0.3096	0.3323	0.3581	0.3833	0.4059
0.70	0.90	0.2957	0.3028	0.3123	0.3231	0.3337	0.3432
0.90	0.00	0.0865	0.1135	0.1632	0.2353	0.3116	0.3770
0.90	0.10	0.0855	0.1100	0.1552	0.2215	0.2931	0.3554
0.90	0.25	0.0844	0.1051	0.1434	0.2006	0.2642	0.3212
0.90	0.50	0.0835	0.0980	0.1245	0.1651	0.2129	0.2580
0.90	0.75	0.0855	0.0935	0.1078	0.1298	0.1572	0.1849
0.90	0.90	0.0907	0.0943	0.1006	0.1102	0.1224	0.1352
0.95	0.00	0.0380	0.0508	0.0781	0.1266	0.1879	0.2440
0.95	0.10	0.0375	0.0492	0.0738	0.1181	0.1754	0.2289
0.95	0.25	0.0370	0.0468	0.0675	0.1052	0.1559	0.2050
0.95	0.50	0.0367	0.0435	0.0577	0.0838	0.1214	0.1608
0.95	0.75	0.0380	0.0418	0.0494	0.0633	0.0845	0.1094
0.95	0.90	0.0416	0.0432	0.0466	0.0528	0.0623	0.0741

TABLE IV

Values of $G_d = \tau_d/\tau_e$, for various total extinction optical depths τ_{e0}, and illumination case b

γ	g	G_d					
		$\tau_{e0} = 1$	$\tau_{e0} = 2$	$\tau_{e0} = 4$	$\tau_{e0} = 8$	$\tau_{e0} = 16$	$\tau_{e0} = 32$
0.10	0.00	0.8577	0.9046	0.9390	0.9620	0.9767	0.9859
0.10	0.10	0.8616	0.9039	0.9350	0.9557	0.9690	0.9773
0.10	0.25	0.8676	0.9030	0.9290	0.9464	0.9575	0.9644
0.10	0.50	0.8779	0.9017	0.9192	0.9308	0.9383	0.9430
0.10	0.75	0.8887	0.9007	0.9085	0.9154	0.9192	0.9215
0.10	0.90	0.8953	0.9003	0.9038	0.9061	0.9077	0.9086
0.30	0.00	0.5439	0.6874	0.7958	0.8690	0.9164	0.9468
0.30	0.10	0.5564	0.6864	0.7850	0.8517	0.8949	0.9225
0.30	0.25	0.5762	0.6858	0.7692	0.8258	0.8624	0.8858
0.30	0.50	0.6125	0.6871	0.7441	0.7831	0.8083	0.8243
0.30	0.75	0.6537	0.6917	0.7209	0.7411	0.7541	0.7623
0.30	0.90	0.6808	0.6962	0.7081	0.7163	0.7216	0.7249
0.50	0.00	0.1809	0.4225	0.6116	0.7416	0.8257	0.8793
0.50	0.10	0.2029	0.4232	0.5965	0.7161	0.7937	0.8434
0.50	0.25	0.2392	0.4264	0.5747	0.6780	0.7453	0.7885
0.50	0.50	0.3095	0.4387	0.5424	0.6156	0.6635	0.6943
0.50	0.75	0.3946	0.4618	0.5165	0.5556	0.5813	0.5978
0.50	0.90	0.4551	0.4825	0.5052	0.5215	0.5323	0.5392
0.70	0.00	−0.2503	0.0831	0.3568	0.5501	0.6723	0.7446
0.70	0.10	−0.2171	0.0882	0.3411	0.5216	0.6369	0.7066
0.70	0.25	−0.1615	0.1003	0.3196	0.4789	0.5823	0.6464
0.70	0.50	−0.0484	0.1349	0.2921	0.4097	0.4880	0.5381
0.70	0.75	0.0973	0.1955	0.2809	0.3468	0.3918	0.4210
0.70	0.90	0.2097	0.2511	0.2872	0.3157	0.3355	0.3484
0.90	0.00	−0.7825	−0.3920	−0.0554	0.1989	0.3563	0.4390
0.90	0.10	−0.7363	−0.3766	−0.0631	0.1783	0.3313	0.4135
0.90	0.25	−0.6579	−0.3467	−0.0710	0.1477	0.2918	0.3723
0.90	0.50	−0.4955	−0.2726	−0.0696	0.1000	0.2205	0.2937
0.90	0.75	−0.2716	−0.1480	−0.0332	0.0679	0.1464	0.1989
0.90	0.90	−0.0803	−0.0257	0.0251	0.0713	0.1093	0.1366
0.95	0.00	−0.9379	−0.5481	−0.2159	0.0400	0.2038	0.2903
0.95	0.10	−0.8879	−0.5283	−0.2180	0.0257	0.1857	0.2718
0.95	0.25	−0.8033	−0.4914	−0.2174	0.0048	0.1571	0.2420
0.95	0.50	−0.6279	−0.4032	−0.1997	−0.0252	0.1052	0.1849
0.95	0.75	−0.3834	−0.2582	−0.1414	−0.0355	0.0524	0.1144
0.95	0.90	−0.1697	−0.1137	−0.0606	−0.0110	0.0331	0.0678

We shall call τ_d the exclusion optical depth, and G_d the exclusion factor. A small value of G_d ($G_d \ll 1$) implies that the grains exclude radiation from the cloud interior much less efficiently than their extinction optical depth τ_e would suggest. Conversely, if G_d approximates to unity ($G_d \sim 1$), then the extinction is mainly due to absorption, and the extinction optical depth τ_e is a good measure of the ability of the grains to exclude radiation from the cloud interior.

We have estimated τ_{d0} and G_d on the grid of (γ, g, τ_{e0}) values by matching the central normalized radiation density u_c – as derived above using the accurate numerical Equations (13a) through (16) and presented in Tables I and II – to the value given by the approximate Equations (17a, b). When this procedure is used to establish the values of τ_{d0} and G_d, the greatest divergence between the accurate numerical solutions (Equations (13a) through (16)) and the approximate solutions (Equations (17a, b)) occurs at the cloud boundary. In the worst case, this divergence may amount to a factor of two, but in general it is much smaller. Clearly the error might be reduced somewhat by matching the solutions at some intermediate optical depth; or alternatively by a least squares fitting procedure. This was deemed unnecessary, in the sense that the central radiation density is probably the most relevant quantity. For in

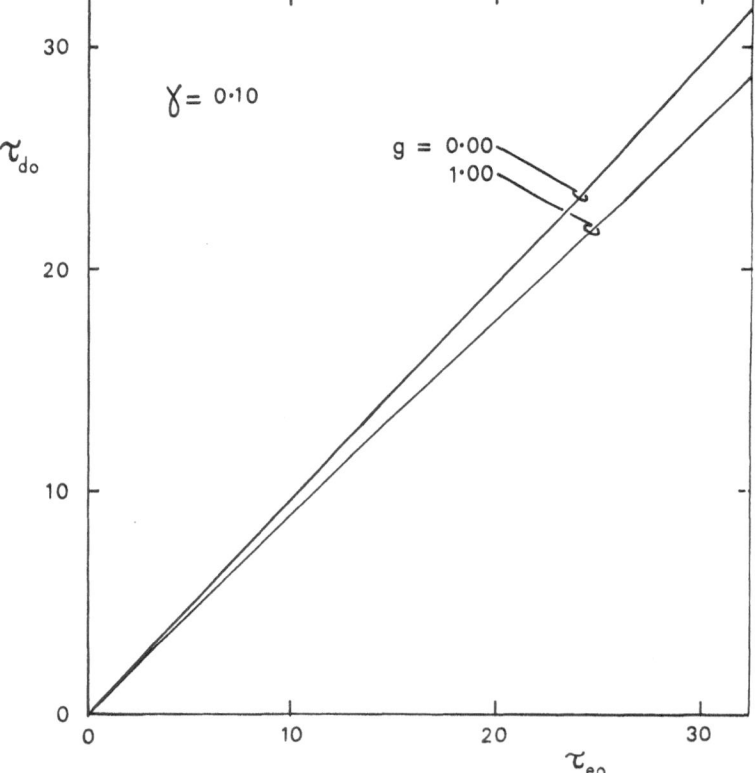

Fig. 1.   The variation of τ_{d0} with τ_{e0} for both illumination cases (a and b); $\gamma=0.10$; and $g=0.00$, 1.00; the results are effectively identical for the two different illumination cases.

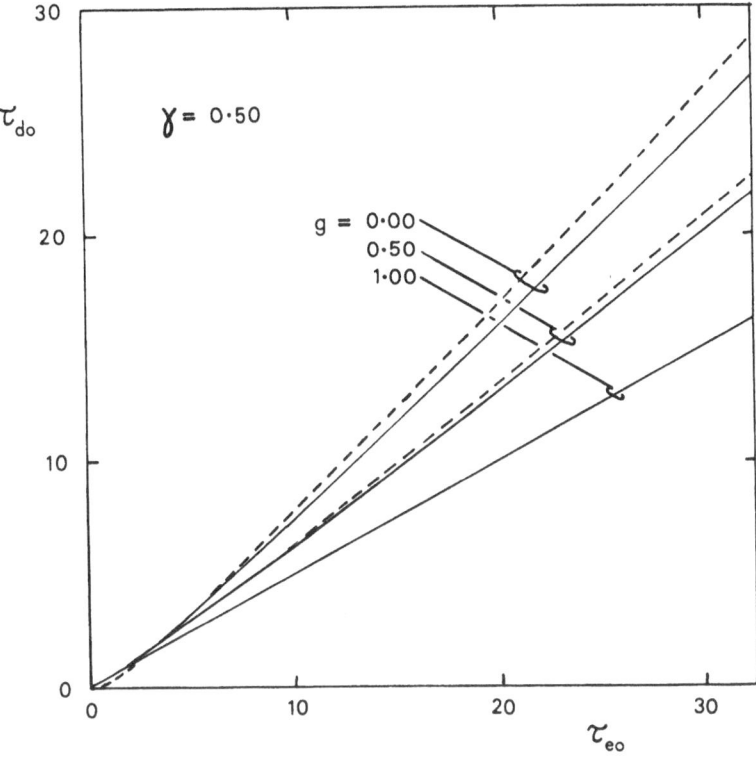

Fig. 2a.   As Figure 1, but for $\gamma=0.50$; and $g=0.00, 0.50, 1.00$: the full curves correspond to illumination case a, and the dashed curves to illumination case b.

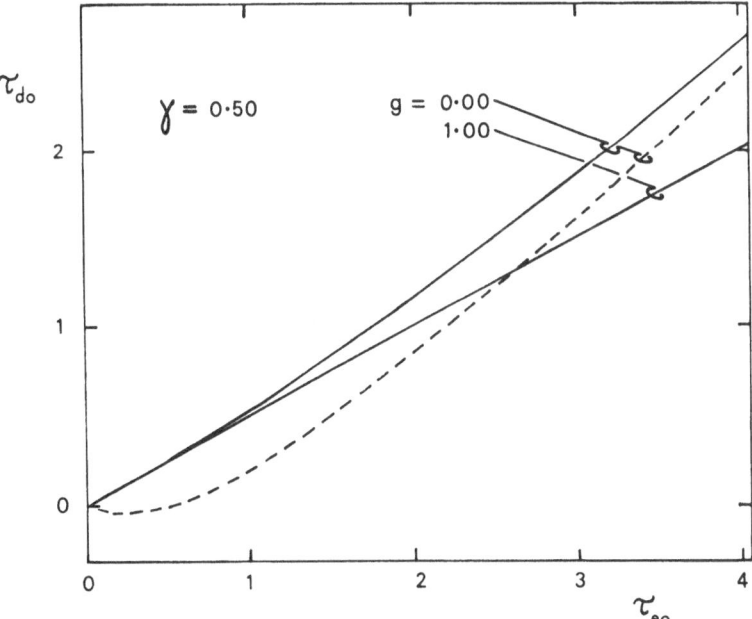

Fig. 2b.   An enlargement of the part of Figure 2a near the origin.

regions of star formation, the evolution is most strongly influenced by the conditions
in the dark, dense cores of protostellar or protocluster fragments.

   Table III gives values of G_d the exclusion factor, on a grid of (γ, g, τ_{eo}) values for
illumination case a. The first and second columns give γ and g. The remaining six
columns contain the values of G_d, each column being headed by the appropriate
value of τ_{eo}. Table IV is the same as Table III, but for illumination case b. The
negative values of G_d in Table IV correspond to negative values of τ_{do}, which in
turn correspond to the values of u_c exceeding unity which we discussed above.

   In Figures 1 through 3b, we have plotted τ_{do} against τ_{eo}. In all these figures, solid
curves correspond to illumination case a: the results for illumination case b are
marked by dashed curves, only where they differ appreciably from those for illumination case a. Figure 1 shows the results for small albedo: $(\gamma, g) = (0.10, 0.00)$ and
$(0.10, 1.00)$. For small albedo, the exclusion optical depth τ_{do} is effectively independent
of the illumination geometry, and varies approximately linearly with extinction

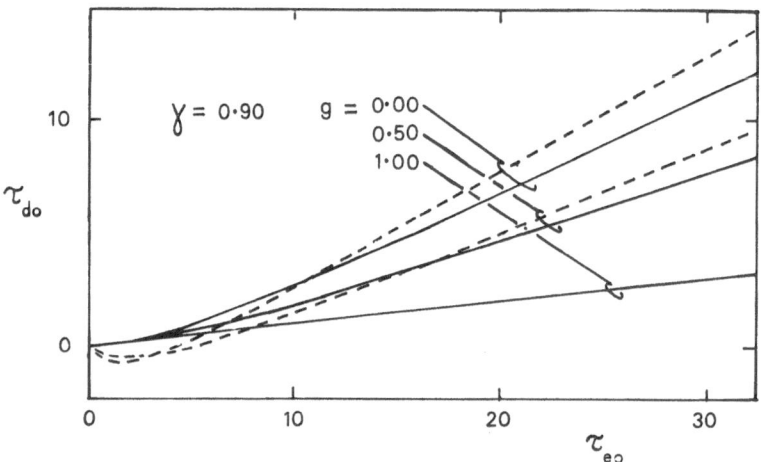

Fig. 3a.   As Figure 2a, but for $\gamma = 0.90$.

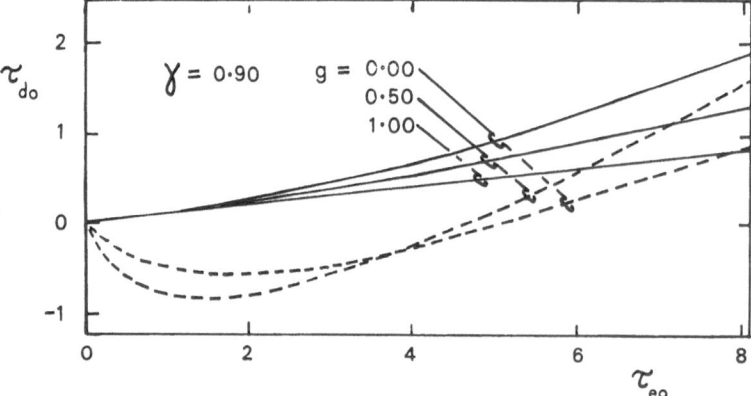

Fig. 3b.   An enlargement of the part of Figure 3a near the origin.

optical depth τ_{e0}. Figure 2a shows the results for intermediate albedo: $(\gamma, g) =$ (0.50, 0.00), (0.50, 0.50) and (0.50, 1.00). Figure 2b is an enlargement of Figure 2a near the origin. Figures 3a, b are the same as Figures 2a, b, except that here the albedo is large, $\gamma = 0.90$. We see that as γ is increased, the relationship between τ_{d0} and τ_{e0} comes to depend increasingly on the illumination geometry; is increasingly non-linear; and admits increasingly larger negative τ_{d0} values for illumination case b (corresponding to amplification of the radiation field in an anisotropically illuminated cloud).

Given the forms of the curves illustrated in Figures 1 through 3b, and the spacing of the tabulated points, it is clear that G_d values can easily be obtained for arbitrary (γ, g, τ_{e0}) by simple extrapolation using the grid of values in Tables III and IV. Equation (18) can then be used to evaluate the exclusion optical depth τ_{d0}.

7. Concluding Remarks

We submit that the concept of an exclusion optical depth is meaningful and numerically convenient in evaluating the role of UV radiation in dark clouds. With simple exponential integral or exponential forms for the attenuation of the radiation energy density (i.e. Equations (17a, b)), a phenomenon involving these radiations can be clearly formulated, without becoming lost in complicated auxiliary equations of radiative transfer. There are many phenomena involving UV radiations which may be important in the interiors of dark clouds. Firstly the radiations may be powerful heating agents (primarily for the grains). Secondly, they may be powerful ionizing agents, or rather more generally, agents of disruption.

UV and visual radiations are the main heating agents for grains, and thus play a dominant role in determining grain temperatures. Even in the dense dark regions where UV radiations do not penetrate, their influence is still felt by the grains, in the sense that the thermal IR radiation which does penetrate is mainly excited by UV radiations in less well shielded parts of the cloud. By influencing grain temperatures – either directly or indirectly – UV radiations can also strongly influence the state of the gas: firstly, because at sufficiently high densities and/or large optical depths, the gas and radiation field tend to thermal equilibrium with the grains; and secondly, because rates for molecule formation by surface recombination on grains are strongly dependent on the grain temperature (e.g. Hollenbach and Salpeter, 1970).

By interacting with a grain, an individual photon may cause the ejection of an electron, thus giving the grain a positive charge (Watson, 1972); or it may remove a newly formed molecule from the grain surface (Watson and Salpeter, 1972a). By interacting with a grain, a UV photon may also alter the IR radiative properties of the grain, and hence affect its thermal balance (Hoyle and Wickramasinghe, 1967); and it may affect the suitability of the surface for molecule formation by creating sites of peculiar binding (Hollenbach and Salpeter, 1970). In interacting with the gas,

TABLE V

UV radiation fields at the centres of dark clouds with a given extinction optical depth in the visual, $\tau_{eo}(V)$

λ (Å)	λ^{-1} (μ^{-1})	U_* (erg cm⁻³ Å⁻¹)	q_e	γ	g	U_c (erg cm⁻³ Å⁻¹) $\tau_{eo}(V)=1$	$\tau_{eo}(V)=2$	$\tau_{eo}(V)=4$	$\tau_{eo}(V)=8$	$T_{eo}(V)=16$	$\tau_{eo}(V)=32$
1500	6.67	1.05(−16)	7.22	0.60	0.00	3.05(−17)	9.78(−18)	1.09(−18)	1.52(−20)	3.28(−24)	1.81(−31)
				0.60	0.25	3.20(−17)	1.13(−17)	1.54(−18)	3.33(−20)	1.79(−23)	6.12(−30)
				1.00	0.00	1.05(−16)	1.05(−16)	1.05(−16)	1.05(−16)	1.05(−16)	1.05(−16)
				1.00	0.25	1.05(−16)	1.05(−16)	1.05(−16)	1.05(−16)	1.05(−16)	1.05(−16)
1680	5.95	7.15(−17)	6.92	0.50	0.00	1.80(−17)	5.44(−18)	5.63(−19)	7.05(−21)	1.30(−24)	5.39(−32)
				0.50	0.25	1.88(−17)	6.15(−18)	7.61(−19)	1.40(−20)	5.88(−24)	1.30(−30)
				0.50	0.50	1.96(−17)	6.97(−18)	1.04(−18)	2.88(−20)	2.89(−23)	3.93(−29)
				0.90	0.00	4.76(−17)	2.85(−17)	9.10(−18)	8.78(−19)	8.38(−21)	9.40(−25)
				0.90	0.25	4.89(−17)	3.12(−17)	1.16(−17)	1.52(−18)	2.62(−20)	8.68(−24)
				0.90	0.50	5.00(−17)	3.42(−17)	1.50(−17)	2.80(−18)	9.81(−20)	1.26(−22)
1920	5.21	3.88(−17)	7.31	0.40	0.50	8.32(−18)	2.52(−18)	2.82(−19)	4.64(−21)	1.74(−24)	3.60(−31)
				0.40	0.75	8.66(−18)	2.83(−18)	3.78(−19)	9.10(−21)	7.74(−24)	8.80(−30)
				0.60	0.50	1.22(−17)	4.68(−18)	7.80(−19)	2.59(−20)	3.50(−23)	7.99(−29)
				0.60	0.75	1.28(−17)	5.42(−18)	1.14(−18)	6.32(−20)	2.58(−22)	6.03(−27)
2040	4.90	3.11(−17)	8.00	0.05	0.75	3.31(−18)	6.42(−19)	3.31(−20)	1.33(−22)	3.54(−27)	4.52(−36)
				0.25	0.75	4.69(−18)	1.16(−18)	9.29(−20)	8.56(−22)	1.14(−25)	3.40(−33)
2380	4.20	2.85(−17)	7.30	0.10	0.75	3.83(−18)	8.85(−19)	6.34(−20)	4.78(−22)	4.40(−26)	6.55(−34)
				0.30	0.75	5.34(−18)	1.55(−18)	1.67(−19)	2.73(−21)	1.10(−24)	2.94(−31)
2460	4.07	3.06(−17)	6.86	0.25	0.75	5.72(−18)	1.67(−18)	1.84(−19)	3.12(−21)	1.38(−24)	4.49(−31)
				0.45	0.75	8.04(−18)	2.95(−18)	4.87(−19)	1.75(−20)	3.21(−23)	1.65(−28)
2940	3.40	3.02(−17)	5.50	0.35	0.75	8.49(−18)	3.34(−18)	6.34(−19)	3.04(−20)	1.00(−22)	1.72(−27)
				0.55	0.75	1.15(−17)	5.56(−18)	1.50(−18)	1.35(−19)	1.46(−21)	2.44(−25)
2980	3.36	3.02(−17)	5.43	0.35	0.75	8.60(−18)	3.42(−18)	6.60(−19)	3.27(−20)	1.16(−22)	2.27(−27)
				0.55	0.75	1.16(−17)	5.66(−18)	1.55(−18)	1.43(−19)	1.64(−21)	3.03(−25)
3330	3.00	3.11(−17)	5.00	0.45	0.75	1.10(−17)	5.09(−18)	1.29(−18)	1.05(−19)	9.61(−22)	1.19(−25)
				0.65	0.75	1.49(−17)	8.46(−18)	3.02(−18)	4.53(−19)	1.28(−20)	1.36(−23)
4250	2.35	6.75(−17)	4.20	0.38	0.75	2.51(−17)	1.21(−17)	3.37(−18)	3.30(−19)	4.39(−21)	1.17(−24)
				0.58	0.75	3.24(−17)	1.86(−17)	6.86(−18)	1.12(−18)	3.79(−20)	5.95(−23)

an individual photon may cause ionization, thereby increasing the electron density, and influencing the general ionization balance (e.g. Brown, 1972). An ion formed in this way may play an important role in the chemistry of the gas, either by undergoing a two-body reaction (in the gas phase) to form a molecule, or – if it is a molecular ion – by recombining dissociatively (Solomon and Klemperer, 1972; Watson and Salpeter, 1972b; cf. Herbst and Klemperer, 1973). By increasing the density of charged particles, UV radiation may also influence the dynamic evolution of the medium, by enhancing the frictional coupling between gas particles and magnetic flux (Mestel and Spitzer, 1956). (Either electrons or charged grains may be effective in this respect.) Finally, the grain charge can affect the growth of the grain itself, and hence the efficiency of depletion by accretion: it may also influence the alignment of the grain.

Evidently, it is important to know the UV radiation field within a dark cloud. To this end, we have taken the data of Witt and Johnson (1973) on the radiation energy density in the solar vicinity (assumed to be representative of the general interstellar medium), and the data of Witt and Lillie (1973) on the scattering properties of interstellar dust grains; and we have used them to compute the energy density in (1200, 4500) Å at the centre of a dark cloud. Witt and Lillie (1973) give errors for the albedo, and we have therefore done the computations for a maximum albedo, and for a minimum albedo. Similarly, at short wavelengths, Witt and Lillie (1973) give only an approximate upper limit to g, and we have therefore treated a range of g values consistent with this limit. We have used the illumination geometry of case a (diffuse illumination), and have labelled the clouds according to their total extinction optical depth in the visual $\tau_{e0}(V)$. The results are presented in Table V and Figures 4a, b, c, d, e. In Table V, the first and second columns give the wavelength λ in ångstroms, and the inverse wavelength λ^{-1} in inverse microns; the third column gives the unshielded radiation density U_* (Witt and Johnson, 1973) and this is plotted in Figure 4a; the fourth column gives the extinction efficiency q_e (Bless and Savage, 1972); the fifth and sixth columns give the albedo γ, and the mean scattering-angle cosine g (Witt and Lillie, 1973). The remaining six columns give the central radiation density u_c, each column being headed by the appropriate value of the total extinction optical depth in the visual $\tau_{e0}(V)$. In Figures 4b, c, d, e we have plotted the results for $\tau_{e0}(V)=4, 8, 16, 32$. In each case we give the maximum and the minimum radiation field allowed by the uncertainties in the albedo and the mean scattering-angle cosine: the area in between the maximum and the minimum is shaded. We see that in the interior of a dark cloud ($\tau_{e0}(V) \gtrsim 4$), the radiation field around the 2200 Å 'hump' is very weak, since the extinction here derives mainly from absorption. However, at shorter and longer wavelengths, the radiation field is much less strongly attenuated. This is because at shorter wavelengths ($\lambda < 2200$ Å) the albedo is very high ($\gamma \to 1$, i.e. the extinction derives mainly from scattering), but the scattering phase-function $P_s(\theta)$ is only weakly forward-throwing, if at all. At longer wavelengths ($\lambda > 2200$ Å), the albedo is moderately high ($\gamma \sim 0.5$, i.e. the extinction is approximately half

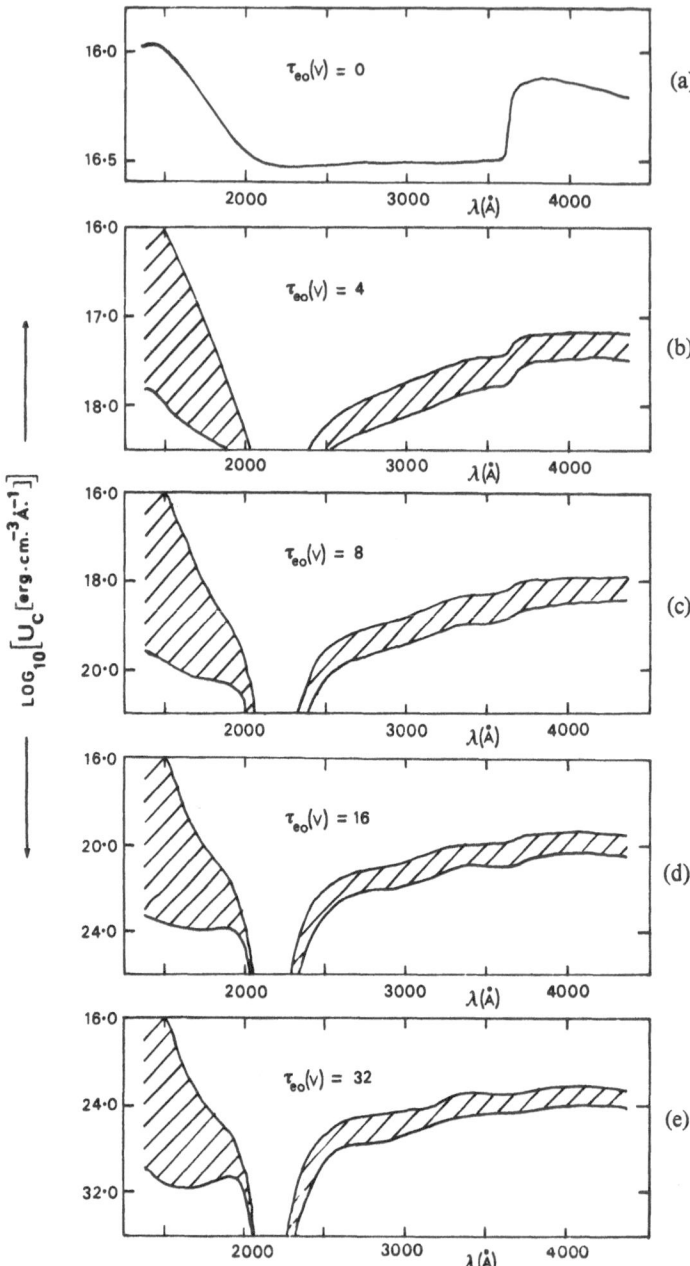

Fig. 4a. The unobscured radiation density U_* (erg cm^{-3} Å^{-1}) in the solar vicinity (from Witt and Lillie, 1973).

Fig. 4b. The radiation density U_c (erg cm^{-3} Å^{-1}) at the centre of a dark cloud having total extinction optical depth in the visual $\tau_{eo}(V)=4$. The upper and lower curves bounding the shaded area are the maximum and minimum densities allowed by the uncertainties in the grain properties.

Fig. 4c. As Figure 4b, but for $\tau_{eo}(V)=8$.

Fig. 4d. As Figure 4b, but for $\tau_{eo}(V)=16$.

Fig. 4e. As Figure 4b, but for $\tau_{eo}(V)=32$.

absorption and half scattering), and the scattering phase-function $P_s(\theta)$ is strongly forward-throwing ($g \sim 0.75$). Thus, with the exception of a narrow band around 2200 Å, UV radiation apparently penetrates rather efficiently into dark clouds, and must therefore play a number of important roles in the physics of the interiors of such clouds, as enumerated above. Whilst it is beyond the scope of this paper to evaluate these roles, we should note that the short wavelength photons ($\lambda < 2200$ Å) are the most energetic, and thus in general potentially the most influential. It is therefore unfortunate (although not surprising) that the results are most uncertain for these short wavelengths, and it is to be hoped that these uncertainties can be reduced in the near future. We should also note that most of the important phenomena involving UV photons which may take place in the interiors of dark clouds are not resonant. Thus they will be significant even if UV radiation penetrates the cloud efficiently only over a rather narrow band of wavelengths: the interacting species are, in the main, not very particular about the exact wavelength of the photon they receive, provided that this photon has sufficient energy.

References

Aannestad, P. A. and Purcell, E. M.: 1973, *Ann. Rev. Astron. Astrophys.* **11**, 309.

Abramowitz, M. and Stegun, I. A.: 1964, *Handbook of Mathematical Functions*, Dover, New York.

Bless, R. C. and Savage, B. D.: 1972, *Astrophys. J.* **171**, 293.

Brown, R. L.: 1972, *Astrophys. J.* **173**, 593.

Carrasco, L., Strom, S. E., and Strom, K. M.: 1973, *Astrophys. J.* **182**, 95.

Herbst, E. and Klemperer, W.: 1973, *Astrophys. J.* **185**, 505.

Hollenbach, D. and Salpeter, E. E.: 1970, *J. Chem. Phys.* **53**, 79.

Hoyle, F. and Wickramasinghe, N. C.: 1967, *Nature* **214**, 969.

Mestel, L. and Spitzer, L., Jr.: 1956, *Monthly Notices Roy. Astron. Soc.* **116**, 503.

Solomon, P. M. and Klemperer, W.: 1972, *Astrophys. J.* **178**, 389.

Watson, W. D.: 1972, *Astrophys. J.* **176**, 103.

Watson, W. D. and Salpeter, E. E.: 1972a, *Astrophys. J.* **174**, 321.

Watson, W. D. and Salpeter, E. E.: 1972b, *Astrophys. J.* **175**, 659.

Werner, M. W. and Salpeter, E. E.: 1969, *Monthly Notices Roy. Astron. Soc.* **145**, 249.

Whitworth, A. P.: 1972, unpublished Ph.D. dissertation, Manchester.

Wickramasinghe, N. C.: 1973, *Light Scattering Functions for Small Particles*, Hilger, London.

Wickramasinghe, N. C. and Nandy, K.: 1972, *Rep. Prog. Phys.* **35**, 157.

Witt, A. N. and Johnson, M. W.: 1973, *Astrophys. J.* **181**, 363.

Witt, A. N. and Lillie, C. F.: 1973, *Astron. Astrophys.* **25**, 397.

THE PLAUSIBILITY OF SILICATE-CORE ICE-MANTLE GRAINS

M. J. DEMPSEY and N. C. WICKRAMASINGHE

Dept. of Applied Mathematics and Astronomy, University College, Cardiff, Wales

Abstract. Extinction curves for silicate-core ice-mantle grains are computed and compared with the infrared spectral data on the BN object in the Orion nebula. A ratio of outer mantle to core radius of 1.3 which best fits this data suggests that silicate-core ice-mantle grains are unlikely to contribute a major part to the total visual extinction coefficient of interstellar material.

There is considerable evidence to support the hypothesis that silicate grains exist in the vicinity of certain cool stars (Woolf, 1973). It is also widely conjectured that silicate particles are responsible for a major part of the observed interstellar extinction and polarization at optical wavelengths. The evidence for this latter thesis is, however, not convincing. In fact, several counter-indications exist based on applications of the Kramers-Krönig relations to the interstellar extinction curve (Chiao *et al.*, 1973; Caroff *et al.*, 1973). If normal solar system abundances are appropriate for interstellar material, the available abundances of Mg, Si fall short of what is required to produce the observed visual extinction of ~ 2 mag. kpc^{-1} on the basis of silicate grains by a factor which may be as high as 10. Based on arguments of this type one might conclude that the only elements present in sufficient quantity to contribute to the main mass density of interstellar dust are C, O and possibly N. Since H_2O ice is a possibility which has been discussed for some time, it is of interest to investigate the properties of ice-coated silicate grains in relation to available astronomical data.

Silicates possess a characteristic absorption feature in the 8–12 μ waveband, whilst H_2O ice has an absorption band at $\lambda \simeq 3.1 \mu$. Attempts to detect this band in several moderately reddened stars produced negative results, and this has led to the conclusion that less than $\sim 10\%$ by mass of interstellar grains is composed of H_2O ice (Knacke *et al.*, 1969; Danielson *et al.*, 1965). Recently the spectra of a few extremely reddened stars such as NML Cyg and the BN object in the Orion nebula have revealed 3.1 μ absorption bands. The spectrum of the Becklin Neugebauer (BN) object in the Orion nebula is interesting since it has a visual extinction of ~ 70 mag. and spectral features at both 3.1 μ and over the 8–12 μ waveband (Gillett and Forrest, 1973) (Figure 1). If these features are due respectively to H_2O ice and silicates in the form of silicate-core ice-mantle grains, a comparison of observational data with this model becomes possible.

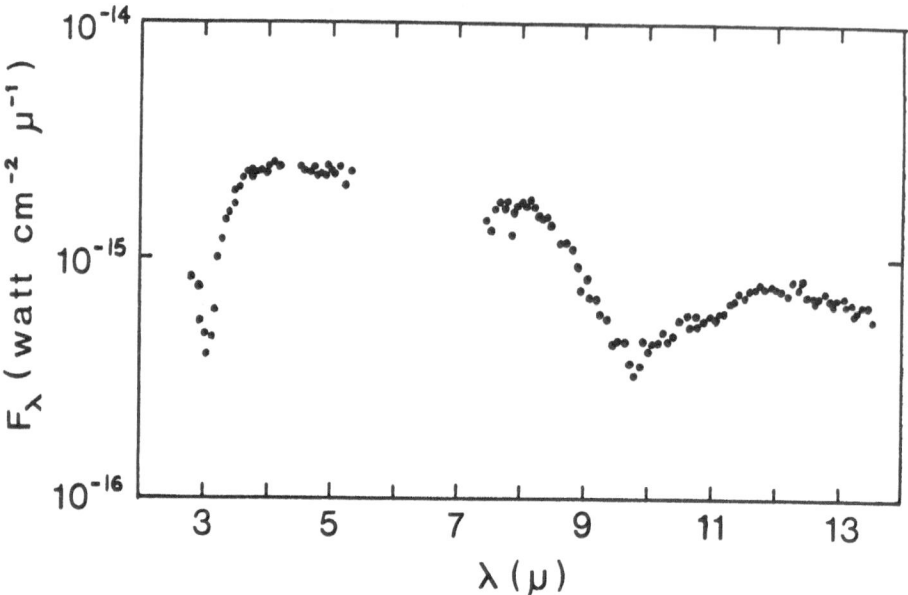

Fig. 1.   Infrared spectrum of the BN object in the Orion nebula showing bands at $\sim 10\,\mu$ and $\sim 3\,\mu$
(Gillett and Forrest, 1973).

For spherical concentric core-mantle grains the extinction cross-section in the long-wavelength limit $(2\pi r/\lambda \ll 1)$ is given by

$$C_{\text{ext}}(\lambda) = -\frac{8\pi^2}{\lambda}\,\text{Im}\,[\alpha],$$

$$\alpha = r^3\,\frac{\alpha^3(m_2^2 - 1)(m_1^2 + 2m_2^2) + (2m_2^2 + 1)(m_1^2 - m_2^2)}{\alpha^3(m_2^2 + 2)(m_1^2 + 2m_2^2) + (2m_2^2 - 2)(m_1^2 - m_2^2)},$$

(1)

where λ is the wavelength, r is the outer mantle radius, α is the ratio of outer to inner radii; m_1, m_2 are complex refractive indices of the core and mantle respectively. The quantities m_1, m_2 are in general functions of the wavelength λ.

For m_1 we adopt the optical constants of minerals measured by Pollack et al. (1973) and for m_2 those of H_2O ice measured by Irvine and Pollack (1968). The normalized extinction cross-sections defined by

$$Q_{\text{NOR}}(\lambda) = 0.01\,\frac{C_{\text{ext}}(\lambda)}{C_{\text{ext}}(\lambda_0)},$$

(2)

with $\lambda_0 = 2\,\mu$, were computed for various values of λ, α and for several types of silicate material. Our results for obsidian-core ice-mantle grains are plotted in Figure 2. It is clear that the observed approximate equality of optical depths at the centres of the $3\,\mu$ and $10\,\mu$ bands, as seen in Figure 1 for the BN object in Orion, implies a value of α close to 1.3. Higher values of α would result in a strong preponderance of the 'ice' band over the 'silicate' band which is contrary to observations.

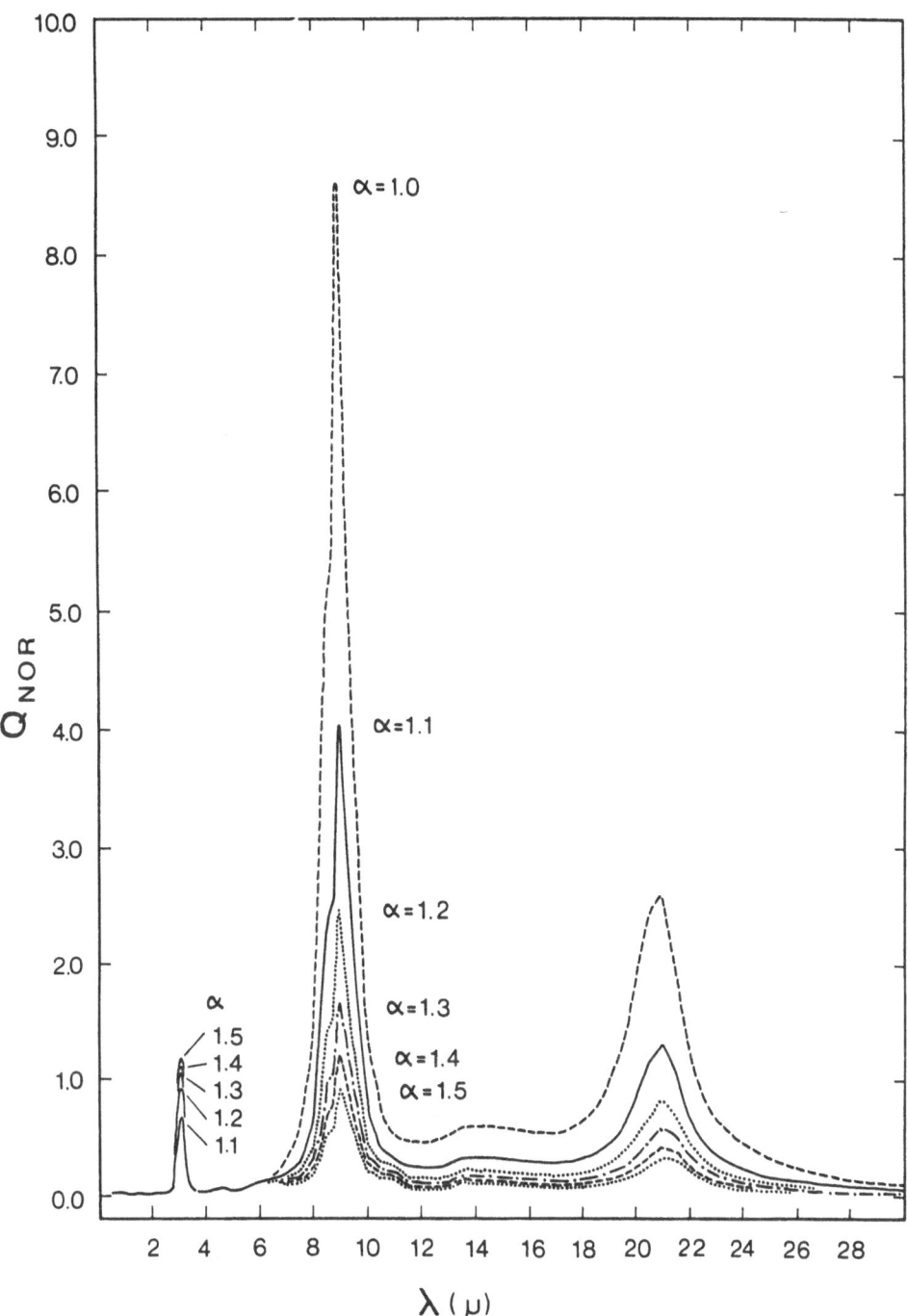

Fig. 2. Normalized extinction curve for obsidian-core H_2O ice-mantle grains computed from Equations (1) and (2), and experimental data of Pollack *et al.* (1973) and Irvine and Pollack (1968). The several curves are for various values of α, the ratio of outer mantle to core radius.

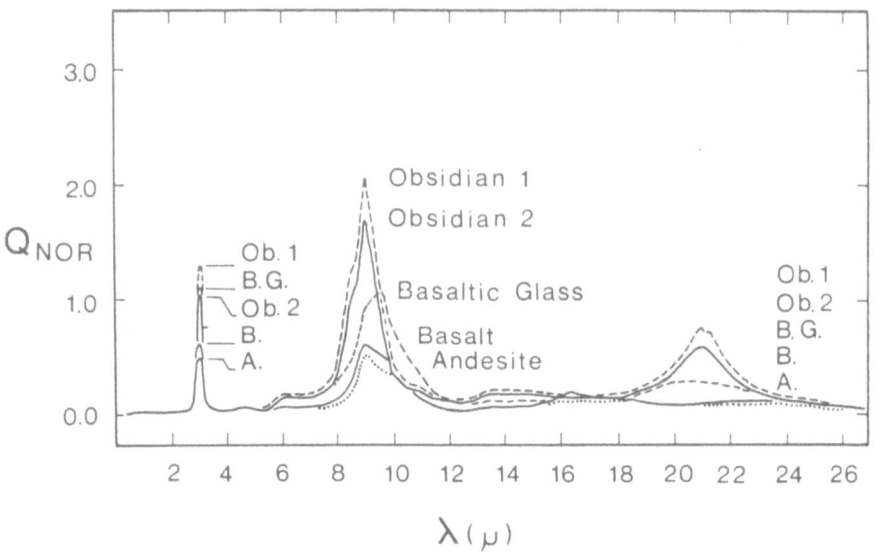

Fig. 3.   Normalized extinction curves for silicate-core H_2O ice-mantle grains with a ratio of outer mantle to core radius $\alpha = 1.3$. The several curves are for different types of silicate core materials for which refractive index data are given by Pollack *et al.* (1973).

These conclusions are not strongly dependent upon the type of silicate core considered. Figure 3 shows plots of $Q_{NOR}(\lambda)$ for several types of silicate core for which data is given by Pollack *et al.* (1973), for the case $\alpha = 1.3$. We note that an approximate equality of Q_{NOR} at the centres of the two bands is preserved in all cases.

For a core composed of obsidian, the mass ratio of ice to silicate in the form of core-mantle grains is

$$f \simeq \frac{\alpha^3 - 1}{s} \simeq 0.2, \tag{3}$$

with α set equal to 1.3, and $s \simeq 3.5 \, \text{g cm}^{-3}$ taken as the density of obsidian. The amount of ice which could be condensed on silicate grains, according to the present argument, is grossly inadequate to make any significant contribution to the visual extinction coefficient of the interstellar medium.

The optical depth of the 3.1 μ band in the spectrum of the BN object is unlikely to be anomalously low relative to estimates of the total visual extinction. Assuming a visual extinction of ~ 70 mag., as inferred from the near infrared spectral distribution of this source, the 3.1 μ band to visual extinction ratio is $\sim 1:60$. This is consistent with the limit set by the non-detection of the 3.1 μ band in a star such as VI Cyg No. 12 (Woolf, 1973).

We conclude that it is unlikely that ice mantles on silicate grains could significantly enhance the mean optical extinction coefficient of the interstellar medium which is obtained on the basis of uncoated silicate grains. If it is accepted that cosmic abundance constraints exclude silicate grains as the major contributor to optical interstellar extinction, other grain materials such as polyoxymethylene or polyformaldehyde (Wickramasinghe, 1974; Novak and Whalley, 1959) which provide extinction features at both 10 and 3 μ must be considered seriously.

References

Caroff, L. J., Petrosian, V., Salpeter, E. E., Wagoner, R. V., and Werner, M. W.: 1973, *Monthly Notices Roy. Astron. Soc.* **164**, 295.

Chiao, R. Y., Feldman, M. J., and Parrish, P. T.: 1973, in J. M. Greenberg and H. C. van de Hulst (eds.), 'Interstellar Dust and Related Topics', *IAU Symp.* **52**, 59.

Danielson, R. E., Woolf, N. J., and Gaustad, J. E.: 1965, *Astrophys. J.* **141**, 116.

Gillett, F. C. and Forrest, W. J.: 1973, *Astrophys. J.* **179**, 483.

Irvine, W. M. and Pollack, J. B.: 1968, *Icarus* **8**, 324.

Knacke, R. F., Cudaback, D. D., and Gaustad, J. E.: 1969, *Astrophys. J.* **158**, 151.

Novak, A. and Whalley, E.: 1959, *Trans. Far. Soc.* **55**, 1484.

Pollack, J. B., Toon, O. B., and Khare, B. N.: 1973, *Icarus* **19**, 372.

Wickramasinghe, N. C.: 1975, *Monthly Notices Roy. Astron. Soc.* **170**, 11P.

Woolf, N. J.: 1973, in J. M. Greenberg and H. C. van de Hulst (eds.), 'Interstellar Dust and Related Topics', *IAU Symp.* **52**, 485.

HOW TO MAKE METAL-POOR STARS, REDDEN OB ASSOCIATIONS AND GROW MANTLES ON GRAINS

M. G. EDMUNDS and N. C. WICKRAMASINGHE

Dept. of Applied Mathematics and Astronomy, University College, Cardiff, Wales, Great Britain

Abstract. Three consequences of the existence of grains with metal-rich ice mantles are considered: (i) The production of metal-poor stars by expulsion of protostellar grains by radiation pressure during star formation. (ii) The effects of these expelled grains in reddening massive stars in an OB association. (iii) The production of the icy mantles on grains in OB associations.

1. Introduction

Observations of the chemical composition of the interstellar medium by satellite 'Copernicus' (Morton *et al.*, 1973) have revealed abundances of heavy elements in the interstellar gas which are much lower than those observed in young stars. It has been suggested (Field, 1974) that the metals have been absorbed onto grains as icy mantles, and that the overall abundances of the gas/grain system are the same as for young stars. We consider here three problems associated with such metal rich grains: (i) the effect of transport of grains on stars which form out of the interstellar medium; (ii) the effects of transport of grains in OB associations; (iii) the growth of icy mantles in OB associations.

2. The Formation of Metal-Poor Stars

These calculations have been described in more detail in Edmunds and Wickramasinghe (1974). As a model of star formation we use the results of Larson and Starrfield (1971). In their model the protostellar cloud rapidly develops a stellar core of about half the total mass of the cloud, and then accretion continues for a time of the order of a free-fall time until the star has attained its final mass. During the accretion stage the density and temperature distributions are well approximated by simple formulae, and we consider the motion of grains in the envelope under radiation pressure from the central stellar condensation. Grains well away from the core will see re-radiated emission from the inner regions peaking typically at about 6 μ. The efficiency of radiation pressure on grains with mantles containing metallic impurities is given approximately by

$$Q \simeq \frac{0.5}{\lambda},\tag{1}$$

where λ is the wavelength in microns. For protostellar masses greater than $\sim 2\,M_\odot$, the radiation pressure on the grains will be stronger than the gravitational pull towards the accreting core, and we may ask what conditions will be sufficient to prevent the grains collapsing into the star. The terminal velocity of a grain through the gas at a distance r is given by

$$u \simeq \left(\frac{QL}{4\pi r^2 c\varrho(r)}\right)^{1/2},\tag{2}$$

where L is the luminosity of the central source and $\varrho(r)$ is the ambient gas density. If the grains are not to be dragged into the star, this velocity must exceed the infall velocity of the gas – i.e., we require that

$$u \geqslant \left(\frac{2GM}{r}\right)^{1/2},\tag{3}$$

where $2M$ is the total protostellar cloud mass. Using the standard mass-luminosity relation of Allen (1973) we find that grains could be inhibited from collapsing if the protostar has a total mass greater than $21\,M_\odot$. We now enquire how much of the star will be depleted of its grains.

We can define various critical radii in the collapsing envelope, and these are illustrated in Figure 1 for a cloud mass of $21\,M_\odot$. These are the initial radius of the

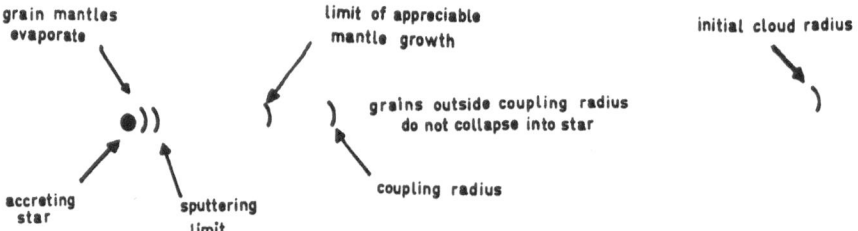

Fig. 1.   Schematic diagram of accretion stage of formation of $21\,M_\odot$ star.

cloud; the radius inside which grain growth is appreciable during a free-fall time, assuming a simple particle-growth law

$$\frac{da}{dt} = \frac{\alpha f n_H}{s}\left(\frac{kTm}{2\pi}\right)^{1/2},\tag{4}$$

where a is the grain radius, $\alpha=$ sticking probability $=1$, $f=$ fraction of heavy metals $= 10^{-3}$, $s=$ mean density of grain material $=1$ gm cm^{-3}, $m=$ mean molecular weight, $k=$ Boltzmann's constant, $T=$ temperature; the radius at which the grain temperature will be high enough to evaporate the mantle; the radius at which sputtering of the mantles could occur; and particularly the 'coupling radius' inside which the densities are high enough for forces acting on the grains to be transmitted to the gas on a timescale shorter than the free-fall time – in which case the gas and dust act as a coupled system, and a detailed hydrodynamic calculation would be required. It is found that nearly all of the mass of the envelope will be swept clear of grains, although the size of the grains will not be greatly enhanced during the process. The calculations of Larson and Starrfield (1971) suggest that the size of the Strömgren sphere will be small compared to the size of the accreting envelope except for very massive central core stars with masses over $25 M_\odot$ and perhaps not lower than $60 M_\odot$, and as our radiative transport mechanism will act for *core* masses greater than $10.5 M_\odot$ (i.e. resultant stellar masses greater than $21 M_\odot$), then there will exist a mass range in which the radiative separation of grains from the gas can occur without H II ionization halting the accretion or melting the mantles.

Hence we find that stars of mass $21 M_\odot$ or greater (i.e. O7, $M_v \sim -4$ or earlier) should have a total metal abundance a factor of two lower than the interstellar medium out of which they formed. The consequences of this are:

(1) The photospheric metal abundances of stars of $21 M_\odot$ or greater might be expected to be lower than for less massive stars of the same age. The actual photospheric abundances will depend on the extent of convection during and after the accretion of the outer envelope, mixing up the metals from inside the stellar core. If little convection occurs then the photospheric abundances could be much lower than a factor of two.

(2) For stars of less than $21 M_\odot$, the grains and gas are effectively coupled and the resultant total metalicity of the star will be similar to the matter out of which the stars formed. A more detailed calculation is required to determine whether the relative motion of grains through the collapsing envelope might give some enhancement of the final photospheric abundances.

(3) The metal formation yield of a massive supernova is sensitive to the initial metal content of the star, and hence if the central regions of the massive stars (where the explosive synthesis occurs) are on average low in metals by convective mixing of the material with the depleted out envelope, then the chemical evolution of the Galaxy may be somewhat affected.

(4) Massive stars will expel a considerable mass of grains during their formation.

3. OB Associations

From Blaauw (1964) we deduce the following model for a typical OB association; linear expansion from an initial dense state to dimensions of about 60 pc in 10^7 yr, with a stellar mass of $2000 M_\odot$ and perhaps fifteen O–B2 stars. Large masses of H I are found surrounding associations (Menon, 1962; Raimond, 1965; Simonson, 1968) and star formation appears to be only about 2% efficient, so we deduce a gas mass of perhaps $2 \times 10^5 M_\odot$ for the association.

Reddish (1967) has discussed the observed reddening of stars due to dust in associations, and confirms the earlier suggestion of Blanco and Williams (1959) that a correlation of interstellar reddening with the absolute luminosity of the OB stars occurs in young associations. The data are severely limited by observational cut-off, but for example in the association Cygnus II (Reddish *et al.*, 1966; Table 6) we note that the data are consistent with heavy reddening for *all* stars brighter than $M_v \sim -4$, and a spread of values due to rather patchy distribution of dust for lower luminosity stars. This agrees very well with our estimate for the mass at which grains are prevented from falling into the star and would form some sort of circumstellar shell. The masses of dust required to explain the reddening are also in good agreement with our model. The luminosity-reddening correlation is only observed in associations with stellar ages younger than 2×10^6 yr (Reddish, 1967), and this is also well explained on our model as follows. From Equation (2) we can write for the motion of a grain

$$\frac{dr}{dt} \simeq \left(\frac{QL}{4\pi r^2 \varrho}\right)^{1/2}. \tag{5}$$

We now assume that the star has settled down to Main Sequence life, that $Q \sim 1$ as the flux will now be strong at visible wavelengths, and integrating (5) for the motion of the expelled grain outside the star

$$r \simeq \tau^{1/2}\left(\frac{L}{\pi \varrho c}\right)^{1/4}, \tag{6}$$

where τ is the travel time. From our model for an OB association, stars of 21 $M_\odot$ will have formed after perhaps 3×10^6 yr and adding to this an apparent stellar age of the association of 2×10^6 yr we obtain a typical interstellar density of 5×10^{-22} gm cm^{-3} in the association. The distance travelled by a grain outside a 21 $M_\odot$ star at these densities in an association stellar age of 2×10^6 yr is given by (6) as $r \sim 8.7 \times 10^{18}$ cm. This is of the order of the interstellar distance at that stage in the expansion of the association. Thus we have a natural explanation of the disappearance of the luminosity-reddening correlation. After 2×10^6 yr the dust from the massive stars will be distributed effectively around several stars of different masses, and the luminosity correlation will have been washed out. The appearance of a Strömgren sphere around a massive star may melt the grain mantle as it is pushed out, but as accretion of gas onto the star will have stopped this should not affect the previous discussion. We finally note that the mass of metals pushed out by the massive stars is likely to be only a small perturbation on the overall gas mass of the association, and the metal abundance of subsequently forming stars is unlikely to be affected unless the formation occurs very near to and just after the formation of a massive star.

4. Mantle Growth

The calculations of the radius inside which appreciable grain growth occurs in a massive protostar indicate that this radius is inside the grain/gas coupling radius, and hence that the grains which are expelled by radiation pressure are unlikely to have grown much mantle during their expulsion. However, if gas densities as high as

$10^3 \, \text{cm}^{-3}$ exist during an OB association lifetime (as they will in our model for association ages less than $3 \times 10^6 \, \text{yr}$) then the mantle growth time given by Equation (4) for temperatures of 100 K is only $1.3 \times 10^6 \, \text{yr}$ and hence shorter than the time the association may spend at such high densities. Thus there seems little difficulty in growing mantles for the grains, and hence depleting the interstellar gas of its metals as the gas passes through OB associations. Even if the average density in the association is lower than $10^3 \, \text{cm}^{-3}$, there probably exist condensations of much higher densities where more rapid mantle growth would occur. However, if we use the local space density of OB associations (Blaauw, 1964) of one per $6 \times 10^7 \, \text{pc}^3$ and an interstellar gas density of $0.018 \, M_\odot \, \text{pc}^{-3}$ (Allen, 1973) then perhaps one atom in five of the interstellar gas is in an OB association. If we make the reasonable assumption that OB associations are formed in the galactic spiral density wave shock (inter-shock period about $2.8 \times 10^8 \, \text{yr}$ at the Sun), then we deduce that the typical interstellar metal atom will pass through an OB association, and hence be absorbed onto a grain only once every $10^9 \, \text{yr}$. Thus the typical cycling time of metal atoms into grain mantles is about $10^9 \, \text{yr}$, and unless high density condensations occur in other places in the interstellar medium, this cycle time may be too long to completely explain the observed depletion of interstellar metals, since the metals in the form of ices can be very easily evaporated from the grain surfaces. If the space density of OB associations in spiral arms is higher than the local value, then the cycling time could perhaps be reduced to just the intershock time.

5. Summary

We have shown that (i) if grains possess icy metallic mantles then stars of masses greater than $21 \, M_\odot$ may be poorer in metals than the medium out of which they formed. This implies that the photospheric abundances and nucleosynthetic yield of these stars may be different than for less massive stars. A detailed study of the grain/gas dynamics and convection of less massive stars might be worthwhile to investigate the setting up of abundance gradients within the stars. (ii) The correlation of absolute magnitude and reddening in young OB associations may be consistent with the model of expulsion of grains from massive protostars. (iii) Grain mantles can grow in OB associations, but the cycle time for growth may be rather slow.

References

Allen, C. W.: 1973, *Astrophysical Quantities*, 3rd Edn., Athlone Press, London.
Blaauw, A.: 1964, *Ann. Rev. Astron. Astrophys.* **2**, 226.
Blanco, V. M. and Williams, A. D.: 1959, *Astrophys. J.* **130**, 482.
Edmunds, M. G. and Wickramasinghe, N. C.: 1974, *Astrophys. Space Sci.* **30**, L9.
Field, G. B.: 1974, *Astrophys. J.* **187**, 453.
Larson, R. B. and Starrfield, S.: 1971, *Astron. Astrophys.* **13**, 190.
Menon, T. K.: 1962, *Astrophys. J.* **135**, 394.

Morton, D. C., Drake, J. F., Jenkins, E. B., Rogerson, J. B., Spitzer, L., and York, D. G.: 1973, *Astrophys. J. Letters* **181**, L103.

Raimond, E.: 1965, *Bull. Astron. Inst. Neth.* **18**, 191.

Reddish, V. C.: 1967, *Monthly Notices Roy. Astron. Soc.* **135**, 251.

Reddish, V. C., Lawrence, L. C., and Pratt, N. M.: 1966, *Publ. Roy. Obs. Edinburgh* **5**, No. 8.

Simonson, III., S. Ch.: 1968, *Astrophys. J.* **154**, 923.

RADICAL FORMATION, CHEMICAL PROCESSING,
AND EXPLOSION OF INTERSTELLAR GRAINS*

J. MAYO GREENBERG†‡

*State University of New York and Dudley Observatory, Albany,
New York, N.Y., U.S.A.*

Abstract. The ultraviolet radiation in interstellar space is shown to create a sufficient steady state density of free radicals in the grain mantle material consisting of oxygen, carbon, nitrogen, and hydrogen to satisfy the critical condition for initiation of chain reactions. The criterion for minimum critical particle size for maintaining the chain reaction is of the order of the larger grain sizes in a distribution satisfying the average extinction and polarization measures. The triggering of the explosion of interstellar grains leading to the ejection of complex interstellar molecules is shown to be most probable where the grains are largest and where radiation is suddenly introduced; i.e. in regions of new star formation. Similar conditions prevail at the boundaries between very dark clouds and H II regions. When the energy released by the chemical activity of the free radicals is inadequate to explode the grain, the resulting mantle material must consist of extremely large organic molecules which are much more resistant to the hostile environment of H II regions than the classical dirty ice mantles made up of water, methane, and ammonia.

1. Introduction

The chemical composition of interstellar dust is constrained by the availability of the elements to consist to a major extent of a composition of oxygen, carbon, and nitrogen (Greenberg and Hong, 1974a). Nevertheless, the observations of the infrared spectral absorption of dust at 3.07 μ imply that only a small part of the grains consist of ices of H_2O, CH_4, and NH_3 in the classical sense (Van de Hulst, 1947). In order to account for this discrepancy, Greenberg and Wang (1972) proposed that irradiation of the grains by ultraviolet photons could modify their molecular composition so that oxygen need not predominantly be in the form of H_2O. The consequences of ultraviolet radiation go beyond merely changing the molecular composition of the grains. Because the grains are extremely cold, they are ideally suited for storing the frozen free radicals which are created by ultraviolet photolysis. The chemical energy stored within the grains is then capable of producing, with adequate probability, violent reactions within the grain material. The resulting total disruption of the grain material is a likely source of the more complex molecules in interstellar space. Some of the possible implications of trapped radicals in interstellar matter were summarized in a review article by Donn (1960) but little quantitative work on the subject has been done. Perhaps this was because of the uncertainties in basic grain composition. Now that the evidence of O, C, N grain material has been reinforced, there is stronger justification for detailed

* Work supported in part by grants from the Research Foundation of the State University of New York and by grant #NGR-33-011-043 from the National Aeronautics and Space Administration.
† This paper was begun when the author was a visiting professor at the University of Tokyo under the auspices of the Japan Society for the Promotion of Science.
‡ Present address, Huygens Laboratory, University of Leiden, The Netherlands.

studies of which this is still a preliminary account. Laboratory experiments on the basic photochemical processes are currently planned or in progress.

2. The Grain Model

The classical part of the extinction and polarization are matched by a distribution of elongated particles with silicate cores upon which are accreted mantles of oxygen, carbon, and nitrogen in combination with hydrogen. A typical core radius of about $a_c = 0.06\,\mu$ and a mean mantle radius (representing a size distribution) of about $\bar{a}_m = 0.12\,\mu$ appear to give a reasonable representation of the average extinction and polarization curves (Greenberg and Hong, 1974a). In following the evolution of dust grains we must allow the mean mantle radius to vary from zero up to about $0.21\,\mu$, the latter value being the maximum (effective) size to which grains may grow by accretion from the interstellar medium.

Assuming that the general interstellar extinction outside of dense clouds and H II regions has the apparently ubiquitous rise in the far ultraviolet as given by OAO-2 (Bless and Savage, 1974) and *Copernicus* (York *et al.*, 1973), we must include among the interstellar dust population a large number of very small particles whose size is in the 30 Å to 50 Å range (Greenberg and Hong, 1974b). It has been shown (Greenberg, 1968) that such small particles are subject to sizable temperature fluctuations resulting from either ultraviolet absorption or surface formation of molecules (Greenberg and Hong, 1974b). Consequently, these particles tend to remain uncoated even in dense clouds and therefore the 'classical'-sized grains contain the bulk of the 'ices' in all regions of space.

We thus center our attention on the classical particles above in our further consideration of the free radical formation by ultraviolet photons. However, before continuing we wish to emphasize that the molecular formation *on* the very small particles must contribute a substantial fraction of the interstellar molecules and we shall make a rough estimate later of their contribution based on the *observational* fact that the bare particles remain bare.

3. H₂O Ice in Grains

The negative results of the first attempts at observing the $3.07\,\mu$ ice absorption band (Danielson *et al.*, 1965; Knacke *et al.*, 1969) give upper limits to the amount of H_2O ice in the interstellar dust of the order of 10% of that required to produce the measured extinction. It was noted, however (Greenberg *et al.*, 1972), that a likely explanation for this apparent inconsistency with the 'dirty ice' grain model was that the effect of ultraviolet photolysis of the grain material would be to reduce the amount of oxygen in the form of H_2O and that, if the oxygen were mainly in the form of the free radical OH, its absorption band would be shifted to $2.7\,\mu$ which is the same (almost) as that of water *vapor* and that it would therefore be entirely masked by absorption in the Earth's atmosphere.

Within the last few years two positive identifications of the $3.07\,\mu$ ice band have been made (Gillett and Forrest, 1973; Merrill and Soifer, 1974). In both of these observations, there was simultaneously observed the $9.7\,\mu$ absorption attributed to silicates. According to Greenberg (1973b) the relative strengths of these $3.07\,\mu$ and $9.7\,\mu$ bands imply a relative volume of H_2O ice to silicate of

$$\left[\frac{V_{H_2O}}{V_{sil}}\right]_{obs} = 0.21.$$

The average grain model with $0.12\,\mu$ mantle and $0.06\,\mu$ core implies a mantle to core volume ratio of 7. Assuming the mantle to have accreted O, C, and N in their relative cosmic abundance of $6.8:3.7:1.2$, the relative volumes should be

$$\left[\frac{V_{H_2O}}{V_{sil}}\right]_{model} = 4 \; \textit{if} \text{ the oxygen is in the } H_2O \text{ form}.$$

We infer that at most $0.21/4 = 5\%$ of the oxygen in the mantle occurs as H_2O. If we had assumed the very small bare particles to be entirely silicates and included them in our calculations, this ratio would be increased by a factor of about 1.5 to be $\sim 8\%$. This result is clearly consistent with the upper limits determined from the negative ice observations. We shall show later that as a result of ultraviolet photolysis there is good reason to expect the oxygen to be in a variety of molecular and free radical combinations rather than in the saturated molecule H_2O. Before considering the grains themselves we shall summarize the basic processes which can lead to grain disruption.

4. Conditions for Frozen Free Radical Chain Reactions

There exists a rich literature on the production and trapping of radicals in various mixtures. Work on substances with the complexity of interstellar ices is, however, limited to consideration of molecule production (Greenberg et al., 1972). The two examples we have chosen are for relatively simple systems.

It had been observed by Fontana (1958, 1959) that explosive reactions could occur in mixtures of nitrogen atoms and molecules at liquid helium temperature. The method used to store the free radicals was to deposit the mixture on a 4.2 K surface. At concentrations of the order of 0.2% N atoms in the beam, the time between explosions was of the order of a minute, while with beams containing 2% N atoms there was no noticeable explosion implying that the atoms recombined essentially instantaneously. Thus with lower concentrations of atoms in the beam, it takes a longer time between explosions because a larger critical size of the material must be built up in order for the mixture to be unstable. Another clue to the basic processes involved may be deduced from the thermal studies of nitrogen radical mixtures (Fontana, 1958). According to these studies there was no observable recombination of nitrogen atoms in a nitrogen molecule lattice when the temperature was kept below 10 K. Between 10 K and 36 K the atoms would recombine. These experiments suggested that the atoms are in traps in the solid which have a range of binding energies between 10 K and 36 K.

Wall *et al.* (1959) showed that in the radiolysis of methane at 4.2 K, H atoms and CH_3 radicals could each build up to 0.2% of the CH_4 with an efficiency of about 10% per photon of about 10 eV. It was also shown that by heating up the sample to liquid hydrogen temperatures (20.4 K) the radicals decayed rapidly. Later, Ausloos *et al.* (1965) using the xenon and krypton resonance lines at 20 K and 77 K on CH_4, CD_4 mixtures found that methylene and methyl radicals are produced and that the methylene radicals inserted in methane form ethane. Higher hydrocarbons up to ethylene, propane, propylene, isobutane, *n*-butane, isopentane, and *n*-pentane were observed to result from the photolysis.

While it is clear that solid complex molecular systems like those composing the mantles of interstellar grains are substantially different from the nitrogen atom-nitrogen molecule system, pure hydrocarbons, or methane-water systems (Stief *et al.*, 1965), nevertheless the basic concept of the trapping, diffusion, and reaction of the free radicals must be similar. Undoubtedly the diffusion rates of the radicals in a complex organic matrix will be quite a bit lower. Furthermore, the recombination processes may lead to the production of chains and polymerization (Morawetz, 1960) both of which are almost unique to the chemistry of hydrocarbon or organic systems.

The reactions of free radicals in a matrix appear to occur with zero activation energy (Pimentel, 1960) so that contact of one free radical with another will immediately lead to combination. Based on this fact and the experimental observations of Fontana, Jackson (1959a, b) developed a preliminary treatment of free radical mixtures by methods analogous to those used in nuclear reactors. We summarize here the main points in the theoretical development adapted by us to provide a basis for application to the astronomical problem of irradiation of interstellar dust.

Consider a matrix of material with n free radicals trapped per unit volume. Let M be the number of saturated molecules per unit volume. We assume that the initial temperature, T_d, of the lattice (dust) is less than the energy at which significant diffusion takes place. The binding energy of the radicals corresponds to a temperature T_f.

When two free radicals combine they release an amount of energy W which heats up the material surrounding the reaction site. Within a certain radius a_f the temperature may rise to $T > T_f$ in which case the frozen radicals in this volume are free and may collide with each other to react. If the probability is unity that one or more reactions are generated from the initial reaction, a chain reaction ensues.

The number of radicals freed following a release of energy W is nV_f. These radicals may either fall into another trap or recombine. We assume there will be a recombination if the radical comes to rest at a trap site occupied by a free radical or at a site which is a nearest neighbor to the trap site.

The number of nearest neighbor trap sites is equal to the number of nearest neighbor sites in the lattice (a term used for convenience). We call this number ξ. Thus the probability for a recombination is given by

$$P \simeq (nV_f)[(\xi + 1)n/M] \, ; \tag{1}$$

and the critical number density of free radicals is obtained when $P \geqslant 1$ or,

$$n_c = \left[\frac{M}{(\xi + 1)V_f} \right]^{1/2} . \tag{2}$$

We shall approximate the value of V_f by letting all the energy W released in a reaction be contained in that volume. We may deduce (Greenberg and Hong, 1974a) that in a volume V_f defined in this way, and at very low temperatures

$$T_f = T_d \left[1 + \frac{5\bar{\mu}M_H\theta^3 W}{3\pi^4 V_f sKT_d^4} \right]^{1/4} , \tag{3}$$

where $\bar{\mu}$ = average molecular weight of the matrix, θ = Debye temperature, s = specific density, k = Boltzmann constant. Using $s = M\bar{\mu}m_H$ and combining Equations (2) and (3) we obtain the result

$$\frac{n_c}{M} = \left[\frac{(T_f - T_d)\pi^4 k}{\frac{5}{3}(\xi + 1)[\theta/(T_f - T_d)]^3 W} \right]^{1/2} . \tag{4}$$

5. Free Radicals in Interstellar Grains

A. CRITICAL NUMBER DENSITY

We assume that most of the free radicals in a grain are created by absorption of ultraviolet photons from the interstellar medium. Free radicals like CO could also be directly accreted out of the gas. The first question to be answered is whether the rate of free radical production is such that the time required to produce the critical number density is substantially smaller than the characteristic life cycle of 10^7 yr for an interstellar cloud.

The number of photons with wavelength less than some threshold λ_t (for producing a free radical) impinging on a grain of radius, a, per unit time is

$$\frac{\mathrm{d}N_{UV}}{\mathrm{d}t} = (\pi a^2/h) \int_{912 \, \text{Å}}^{\lambda_t} u_\lambda \lambda \, \mathrm{d}\lambda, \tag{5}$$

where u_λ is the energy density.

If we approximate the Habing (1968) radiation field by letting $u_\lambda = 40 \times 10^{-18}$ erg cm^{-3} Å^{-1} throughout the ultraviolet, we may simply integrate Equation (5) to obtain

$$\frac{\mathrm{d}N_{UV}}{\mathrm{d}t} = 30\pi a^2(\lambda_t^2 - 912^2), \tag{6}$$

where λ_t is in Ångströms. Thus for $\lambda_t \simeq 2000$ Å the rate of UV photons impinging on a 0.12 μ size grain is

$$\frac{dN_{UV}}{dt} = 5 \times 10^{-2}\,s^{-1}. \tag{7}$$

Let the net efficiency for free radical production be ε, then the time required to achieve the critical number density is

$$\tau_c = n_c V_d / \varepsilon (dN_{UV}/dt) = (n_c/M)(M/\varepsilon)(dN_{UV}/dt), \tag{8}$$

where V_d=the volume of the grain. The value of (n_c/M) for the interstellar grain material is estimated from Equation (4) by letting $T_f = 50$ K, $\theta = 400$ K, $\xi + 1 = 10$, $T_d = 10$ K, $W = 2$ eV. With these values we get

$$\frac{n_c}{M} \simeq 0.01. \tag{9}$$

The value of ε for interstellar grain material is impossible to estimate with any precision. The 10% efficiency of free radical production in methane by 10 eV photons is probably much greater than the value of ε we should use. However, even allowing for an underestimate of ε by as much as a factor of 10^2, i.e. letting $\varepsilon = 10^{-3}$, we obtain

$$\tau_c \leqslant 10 \left[\bar{\mu} m_H \frac{dN_{UV}}{dt} \right]^{-1} \tfrac{4}{3}\pi a^3 = 5 \times 10^{10}\,s = 1.7 \times 10^3\,\text{yr}, \tag{10}$$

where we have approximated the mean molecular weight by $\bar{\mu} = 17$. We see that the critical free radical creation time is well within the limits prescribed by the grain lifetime.

B. CRITICAL SIZE

The fact that the critical density for chain reactions is achieved by no means guarantees that a chain reaction will take place. It is a necessary but not a sufficient condition. The value of a_f from $a_f \approx (V_f)^{1/3}$ is

$$a_f = (V_f)^{1/3} = \left[\frac{5}{3} \frac{[\theta/(T_f - T_d)]^3 W}{(T_f - T_d)\pi^4 k} \right]^{1/3} = 3 \times 10^{-7}\,\text{cm}, \tag{11}$$

which is considerably smaller than the grain size. The critical size for chain reactions is greater than this but apparently still perhaps of the order of the average grain size.

The critical size derived by Jackson is proportional to $(D\tau_{MR})^{1/2}$, where D is the diffusion constant and τ_{MR} is the lifetime for mobile radicals. As applied to the nitrogen system this leads to the order of 10^4 molecule layers for $n_c = 0.1\%$ or 10^3 layers for $n_c = 1\%$. The latter combination implies, for average molecular size of 2 Å, a minimum grain diameter of about 2×10^{-5} cm. Although this is within the range of average interstellar grain sizes we hasten to point out that there are enough uncertainties in its derivation to require much further experimental study before it may be considered conclusive. Among other things we have entirely ignored surface effects. However, in view of the fact that the probability for grain explosions must actually be small in order not to deplete the interstellar grain population, we should probably equally question a result which implied a high explosion probability during the lifetime of a grain.

C. ENERGY RELEASE

A final criterion for grain explosion is whether the energy released in the chain reaction is adequate to evaporate or at least break up the grain into smaller elements. It is conceivable that with the onset of rapid heating, a shock is propagated through the grain which may disrupt the grain without bringing about a complete separation of all the molecules. Although the critical number density of free radicals may be only $\sim 1\%$, it is known that amorphous, or glassy matrices are capable of supporting higher concentrations. Consequently, the energy stored in the grain may be larger than that given by the recombination of the critical concentration of free radicals. We shall assume, in our calculation of energy release, that all the free radicals recombine.

Let the energy released per radical-radical recombination be E. Then the total energy released per molecule in the material is $H > \frac{1}{2}(n_c/M)E$, where the $>$ sign is for the case where $n > n_c$. For $(n_c/M) = 0.01$ and $E = 5$ eV, $H > 0.025$ eV. The binding energy for an organic crystal is about 2 K cal/mole (Kittel, 1956) which is equivalent to about 0.1 eV/molecule. On estimating the break-up energy of the grain this quantity should be reduced by a substantial factor to account for the fact that the total number of molecules is generally reduced in the process of the chain reaction. If the average new molecule is five times larger than the original small ones (H_2O etc.) the grain will evaporate. This is a complicated process to treat theoretically, but experimentally it has been shown that under some conditions *very* large molecules can be created in solid water, methane, ammonia mixtures (Greenberg *et al.*, 1972) so that there is a strong presumption that it must take place in an interstellar grain.

D. TRIGGERING

Once the number density of frozen radicals and the critical size are reached, the probability for triggering a chain reaction is expected to become appreciable. In the case of the interstellar grains we must follow their history as well as the static requirements. During the course of the growth of a grain in an intermediate density cloud, the grain is subjected to the normal interstellar ultraviolet flux and therefore acquires free radicals in its mantle at the same time as it grows. During this stage, the grain temperature is of the order of 10 K. The number of grains with mantles greater than the critical size will increase during cloud contraction. As the cloud density increases, the ultraviolet flux decreases and the rate of formation of free radicals decreases. However, this rate of formation appears to remain substantial relative to the growth rate. Let the collision rate of heavy atoms dN_{HA}/dt be $\sim 10^{-3}$ that of hydrogen, then

$$\frac{dN_{HA}}{dt} = 4.5 \times 10^{-8} n_{\mathrm{H}} \text{ s}^{-1} \tag{12}$$

for a size 0.12 μ.

For comparison with Equation (12), the ultraviolet photon collision rate inside a cloud is that given in Equation (7) reduced by some attenuation factor. We estimate the effective ultraviolet optical depth as being about three times the visual extinction. This is adequate for an order of magnitude calculation (Greenberg, 1971). The visual extinction is obtained from the correlation factor $n_{\mathrm{H}} R / E(B - V) = 4.5 \times 10^{21}$ cm^{-2} (Morton, 1975), where R is the distance inside the cloud. The attenuation factor is then $e^{-\tau_{\mathrm{UV}}}$ where $\tau_{\mathrm{UV}} \sim 2 n_{\mathrm{H}} R \times 10^{21}$. The heavy atom and ultraviolet collision rates are equal when

$$e^{-2n_{\mathrm{H}} R \times 10^{21}} = 0.9 \times 10^{-6} n_{\mathrm{H}}. \tag{13}$$

For $R = 1$ pc the solution to Equation (13) is $n_{\mathrm{H}} = 10^3$. For $R = 0.1$ pc, one obtains $n_{\mathrm{H}} = 0.8 \times 10^4$. Thus, it is only in the very latest stages of cloud contraction that the photolysis may be negligible.

We thus expect that the combination of grain size and free radical concentration will be optimum for producing chain reactions at the latest stages of cloud contraction while the low temperatures and ultraviolet deficiency act in the direction of inhibiting the triggering. However, it is just such conditions which prevail at the time of star formation. As the star begins to turn on, the heating of the environment and the onset of ultraviolet flux enhance the probability for both triggering reactions in a grain and increasing the diffusion rate of the free radicals. Consequently, the production of copious quantities of complex molecules should accompany the process of star formation. The region around NGC 2264 is an excellent example of the existence of a wide range of (relatively simple) complex molecules accompanying star formation (Minn and Greenberg, 1975). Bearing in mind that the probability for grain explosions need not be large even when the chain reactions are frequent, most of the grains which survive after the star is formed and which are then found in the H II regions will have mantles whose composition consists substantially of large complex molecules.

An indirect measure of the grain explosion rate is obtained by equating the rate of destruction of interstellar molecules by photodissociation with the rate of production of molecules by grain explosion. Greenberg (1973a) estimates this in terms of a grain destruction lifetime of 4×10^9 yr which, when compared with a grain lifetime of 10^7 yr, implies an explosion probability of only 0.002 – not enough to modify the dust population significantly.

6. Concluding Remarks

We have shown that the explosion of interstellar grains is a likely source of complex interstellar molecules. The chemical reactions preceding the explosion lead to the formation of molecules which are substantially larger than the ones which have so far been detected. It is presumed that where the disruption of the grain is a dominant or important source of molecules the smaller ones are the result of photodissociation processes leading to a cascade downward from the larger ones.

The probability of finding grains with sizes larger than the critical size required for maintaining a chain reaction increases with the age and density of the dust cloud. Since new star formation occurs in regions of dense clouds and since the probability for triggering chain reactions in the grain increases as the radiant energy is turned on we expect to find the most copious interstellar molecules in regions where new stars are being born. The grain explosion mechanism is generally most probable where larger than average grains become subject to extra radiation as at the boundaries between dense clouds and H II regions.

If the formation of comets occurs by agglomeration of the dust grains and if the region where the comets form is sufficiently cold, as beyond the orbit of Jupiter, we may expect that a significant number of free radicals are stored in the cometary material. The comets appear to come from a region between 30 000 AU and 100 000 AU from the Sun (Oort, 1963) so that they are at a sufficiently low temperature during the major portion of their existence to store frozen radicals almost indefinitely until they come close to the Sun. If cometary outbursts can be attributed to the chemical energy released by the free radicals retained from the primeval dust grains then we must conclude that comets are likely to consist of complex mixtures of molecules with frozen radicals. The warmer the region in which the comets have formed, the more likely will have been the probability of generating chain reactions in the primeval grains and the less abundant will be the free radicals in the cometary composition which would then consist to a large extent of much larger complex organic molecules with little or no free radicals.

References

Ausloos, P., Rebbert, R. G., and Lias, S. G.: 1965, *J. Chem. Phys.* **142**, 540.
Bless, R. C. and Savage, B. D.: 1972, *Astrophys. J.* **171**, 273.
Danielson, R. E., Woolf, N. J., and Gaustad, J. E.: 1965, *Astrophys. J.* **141**, 110.
Donn, B.: 1960, in A. M. Bass and H. P. Broida (eds.), *Formation and Trapping of Free Radicals*, Academic Press, New York and London, Chapter 11, p. 347.
Fontana, B. J.: 1958, *J. Appl. Phys.* **29**, 1668.
Fontana, B. J.: 1959, *J. Chem. Phys.* **31**, 148.
Gillett, F. C. and Forrest, W. J.: 1973, *Astrophys. J.* **179**, 483.
Greenberg, J. M.: 1968, *Stars and Stellar Systems* **3**, 22.
Greenberg, J. M.: 1973a, in M. A. Gordon and L. E. Snyder (eds.), *Molecules in the Galactic Environment*, J. Wiley & Sons, New York, p. 94.
Greenberg, J. M.: 1973b, in J. M. Greenberg and H. C. van de Hulst (eds.), 'Interstellar Dust and Related Topics', *IAU Symp.* **52**, 1.
Greenberg, J. M. and Hong, S. S.: 1974a, in F. Kerr and S. C. Simonson (eds.), 'Galactic Astronomy', *IAU Symp.* **60**, 147.
Greenberg, J. M. and Hong, S. S.: 1974b, in A. F. M. Moorwood (ed.), 'H II Regions and the Galactic center', 8th ESLAB Symp., ESRO SP-105, Noordwijk, p. 153.
Greenberg, J. M. and Wang, R. T.: 1972, *Mem. Soc. Roy. Sci. Liège*, 6th Ser. **3**, 197.
Greenberg, J. M. and Yencha, A. J.: 1973, in J. M. Greenberg and H. C. van de Hulst (eds.), 'Interstellar Dust and Related Topics', *IAU Symp.* **52**, 309.

Greenberg, J. M., Yencha, A. J., Corbett, J. W., and Frisch, H. L.: 1972, *Mem. Soc. Roy. Sci. Liège*, 6th Ser. **3**, 425.

Habing, H. J.: 1968, *Bull. Astron. Inst. Neth.* **19**, 421.

Jackson, J. L.: 1959a, *J. Chem. Phys.* **31**, 154.

Jackson, J. L.: 1959b, *J. Chem. Phys.* **31**, 722.

Kittel, C.: 1956, *Introduction to Solid State Physics*, 2nd Edition, John Wiley and Sons, New York, p. 64.

Knacke, R. F., Cudaback, D. D., and Gaustad, J. E.: 1969, *Astrophys. J.* **158**, 151.

Merrill, K. M. and Soifer, B. T.: 1974, *Astrophys. J.* **189**, L27.

Minn, Y. K. and Greenberg, J. M.: 1975, *Astron. Astrophys.*, in press.

Morawetz, H.: 1960, in A. M. Bass and H. P. Broida (eds.), *Formation and Trapping of Free Radicals*, Academic Press, New York and London, Chapter 12, p. 363.

Morton, D. C.: 1974, *Astrophys. J.* **193**, L35.

Oort, J. H.: 1963, in *The Moon, Meteorites and Comets*, (The Solar System, Vol. IV), The University of Chicago Press, Chicago, p. 665.

Pimentel, G. C.: 1960, in A. M. Bass and H. P. Broida (eds.), *Formation and Trapping of Free Radicals*, Academic Press, New York and London, Chapter 4, p. 69.

Stief, L. J., De Carlo, V. J., and Hillman, J. J.: 1965, *J. Chem. Phys.* **43**, 2490.

Wall, L. A., Brown, D. W., and Florin, R. E.: 1959, *J. Chem. Phys.* **63**, 1762.

York, D., Drake, J., Jenkins, E., Morton, D., Rogerson, J., and Spitzer, L.: 1973, *Astrophys. J.* **182**, L1.

EFFECTS OF SUPRATHERMAL GRAINS

S. P. TARAFDAR and N. C. WICKRAMASINGHE

Dept. of Applied Mathematics and Astronomy, University College, Cardiff, Wales, U.K.

Abstract. Grains ejected from stars at velocities of $\sim 10^7$ cm s^{-1} and/or grains accelerated by the pressure of starlight in the intercloud medium to velocities in the range 2×10^6–10^7 cm s^{-1} are slowed to velocities of about 2×10^5 cm s^{-1} in a typical interstellar cloud. The interaction of fast grains with gas atoms as they are slowed in clouds could provide (a) the dominant heat source for interstellar clouds; (b) sites for molecule formation; and (c) a mechanism of providing a pressure balance between clouds and the intercloud medium.

1. Introduction

The possibility that interstellar grains originate in stellar atmospheres implies that they must be expelled into the interstellar medium with fairly high velocities. If grains are not accompanied by gas the final grain velocity reached is

$$V_\infty = \left\{ 2 \frac{P}{G} \gamma \frac{M}{R_S} \right\}^{1/2}, \tag{1}$$

where P/G is the ratio of radiation pressure force to gravity, γ is the gravitational constant, M is the mass of the star and R_S is the radial distance at which acceleration begins. It was argued by one of us (Wickramasinghe, 1972) that this velocity could exceed 3×10^8 cm s^{-1} for sufficiently luminous stars. Gilman (1973) has however pointed out that grain flow is likely to be coupled to gas flow in most cases and that final grain velocity reached in a star such as α-Ori could not exceed $\sim 10^7$ cm s^{-1}.

This controversy with regard to grain ejection velocities appears to have been resolved in a recent study by Salpeter (1974a, b). There are three critical luminosities involved in this analysis designated $L_{cr, p}$, $L_{cr, i}$, $L_{cr, z}$ which are defined by

$$L_{cr, p} = z^{2/3} L_{cr, i} = z L_{cr, z} \simeq L_\odot, \tag{2}$$

where z is the mass fraction of material condensable into grains and $L_\odot$ refers to the solar luminosity. For $z \simeq 10^{-3}$ this gives

$$L_{cr, p} = L_\odot, \qquad L_{cr, i} = 10^2 L_\odot, \qquad L_{cr, z} = 10^3 L_\odot. \tag{3}$$

For stars with $L > L_{cr, z}$ both gas and grains are expelled and the final velocity of grains relative to gas is likely to be below $\sim 5 \times 10^6$ cm s^{-1}. If $L_{cr, p} \lesssim L \ll L_{cr, z}$, pure grains can be expelled leaving behind most of the gas. Grain velocities are not suppressed in this case with asymptotic speeds of decoupled grains given by

$$V_{dg}(\infty) \sim (L/L_{cr, p})^{1/2} V_0, \tag{4}$$

where $V_0=(2GM/R)^{1/2}$ is the velocity of escape from the star. The upper limit for the luminosity in this case may be taken as $L_{cr,i}\approx10^2L_\odot$. A star with this luminosity is a giant with $V_0\sim100$–300 km s^{-1}, so that $V_{dg}(\infty)$ could be in the range 10^8–3×10^8 cm s^{-1}. Giant stars with luminosities in the range ~30–$100L_\odot$ provide, however, significantly lower rates of grain production than the more luminous stars. The grain flow rate is given (cf. Salpeter, 1974b) by

$$\phi_d \sim \frac{zQL}{uc} \simeq 4 \times 10^{-10} \frac{Q(L/100L_\odot)}{(u/5 \text{ km s}^{-1})} M_\odot \text{ yr}^{-1}, \tag{5}$$

where Q is the mean efficiency factor for radiation pressure on a grain, u is the thermal velocity, and z is set equal to 10^{-3}. For $L\sim L_{cr,i}$, $Q\sim0.25$, $u\sim5$ km s^{-1} we obtain $\phi_d\sim10^{-10}M_\odot$ yr^{-1}. This is 10 times less than the grain flow likely to be associated with supergiants such as α-Ori with an observed mass loss $\phi_{gas}\sim10^{-6}$ $M_\odot$ yr^{-1} (Deutsch, 1960; Weymann, 1963) if we assume a mass fraction of grains $\sim10^{-3}$. A determination of the mean injection speed of grains requires a knowledge of the luminosity distribution function amongst giants and supergiants which is at present uncertain. However, the analysis presented by Salpeter indicates grain injection velocities from giants and supergiants in the range 3×10^8 cm s^{-1} to $\sim3\times10^6$ cm s^{-1} for $L\sim10^2$–10^3 $L_\odot$, the more luminous stars producing grains at a more copious rate than the less luminous ones. A mean injection velocity $V_\infty\simeq10^7$ cm s^{-1} would appear likely and will be assumed in the ensuing discussion. We shall refer to such grains as 'suprathermal grains'.

Suprathermal grains may also be expelled from the environs of young O and B stars. There is evidence that circumstellar dust shells are associated with such stars and it is believed that these are remnants of primordial protostellar clouds (Reddish, 1967). The ejection of circumstellar shells from pre-Main-Sequence stars into the interstellar medium could constitute a source of fast grains. Fast grains may also be produced in material ejected from very massive stars (Hoyle *et al.*, 1973). Furthermore, as we shall see in Section 2, grain velocities $\sim2\times10^6$ cm s^{-1} could easily arise in the intercloud medium due to the low ambient gas densities and the unattenuated anisotropic component of starlight flux present in these regions.

The above considerations imply that a more or less steady and isotropic flux of energetic dust grains could be maintained by stellar sources in the intercloud medium. Significant slowing and trapping of these grains occur within clouds. In the present paper we investigate several possible astrophysical implications of the existence of suprathermal grains. Aspects of such grains which have already been discussed in some detail include ionization of the interstellar medium (Hayakawa, 1974, 1975), and excitation of rotational levels of H$_2$ (Tarafdar and Wickramasinghe, 1975), We confine our attention here to the role of suprathermal grains in (a) heating of gas clouds, (b) molecule formation and (c) contributing to the pressure balance in cloud/intercloud boundaries and thus to the accentuation of cloud structure of the interstellar medium.

2. Velocity Spectrum of Grains

We first note that the thermal velocity of grains in optically thin clouds will not be controlled by gas collisions, but rather by the asymmetric component of the interstellar radiation field. This asymetry could arise stochastically due to proximity of a grain (or cloud) to a 'nearest' star or due to excess radiation from the galactic bulge. If ζ is the fractional anisotropy in a typical cloud, the radiation pressure force acting on a grain of radius a, with efficiency factor for radiation pressure Q_{pr} is

$$F_{rad} \simeq Q_{pr}\pi a^2 U\zeta, \tag{6}$$

where U is the energy density of the stellar radiation field. The drag force on a grain moving at a marginally suprathermal speed v through a cloud of hydrogen density n_H is

$$F_{res} \simeq \pi a^2 m_H n_H v^2; \tag{7}$$

and equating (6) and (7) we obtain a terminal velocity

$$v_{min} = \left\{ \frac{Q_{pr}U\zeta}{m_H n_H} \right\}^{1/2}. \tag{8}$$

Setting $Q_{pr}=2$, $U=1$ eV cm^{-3}, $\zeta \simeq 0.2$, as representative values we obtain

$$v_{min} = 2 \times 10^5 \left(\frac{n_H}{10 \text{ cm}^{-3}} \right)^{-1/2} \text{cm s}^{-1}. \tag{9}$$

A similar argument was used by Salpeter and Wickramasinghe (1969) in the context of a grain alignment theory. In the ensuing discussion we adopt 2×10^5 cm s^{-1} as a reasonable value for the velocity of 'thermal' grains in a typical interstellar cloud, with $n_H \simeq 10$ cm^{-3}, and a dependence on n_H as approximately $n_H^{-1/2}$. Assuming a diameter of $D \simeq 10$ pc for such a cloud the lifetime of thermal grains against dispersal from the cloud is

$$t_d \simeq \frac{D}{v} \simeq 5 \times 10^6 \left(\frac{n_H}{10 \text{ cm}^{-3}} \right)^{1/2} \text{yr}. \tag{10}$$

Grains escaping from clouds are likely to be accelerated to much higher velocities in the intercloud medium. The intercloud gas density being $n_H \sim 0.1$ cm^{-3} Equation (9) already implies a terminal grain velocity $\sim 2.5 \times 10^6$ cm s^{-1}. Velocities in excess of 10^7 cm s^{-1} could arise due to stochastic variations of U during encounters with stars as are bound to occur. We thus assume that grains in the intercloud medium, irrespective of their mode of origin, have suprathermal speeds imparted to them by interaction with starlight.

On entry or re-entry into clouds these grains are slowed to velocities given by Equation (9) in a time-scale

$$t_s = \frac{4as}{3m_H n_H} \left\{ \frac{1}{v} - \frac{1}{v_i} \right\}, \tag{11}$$

where n_H refers to the number density of hydrogen in the cloud, a is the grain radius, s is the specific gravity of the grain material and v_i is the initial grain velocity (Wickramasinghe, 1972). In the approximation $v \ll v_i$ this gives

$$t_s = \frac{4as}{3m_H n_H v}.$$

(11a)

The ratio of suprathermal grains to thermal (or slow) grains in a cloud at any instant is

$$\xi = t_s/t_d \simeq \frac{t_s}{D/v} \simeq 0.65 \left(\frac{a}{10^{-5}\,\text{cm}}\right)\left(\frac{n_H}{10\,\text{cm}^{-3}}\right)^{-1}\left(\frac{D}{1\,\text{pc}}\right)^{-1}.$$

(12)

If we use Equation (11a) and set, as before, $s = 2.4$.

The equation of continuity of grain flow taken together with the equation of motion of a decelerating grain

$$\frac{dv}{dx} = -Av$$

gives us an expression for the velocity spectrum of grains. The number density of grains $f(v)\,dv$ with velocities between v and $v+dv$ is

$$f(v) = \alpha/v^2, \qquad v_{max} \geqslant v > v_{min}$$
$$= \beta\,\delta(v - v_{min}), \qquad v \approx v_{min},$$

(13)

where α, β are constants.

The total number density of grains is thus

$$n_g = \alpha \int_{v_{min}}^{v_{max}} \frac{dv}{v^2} + \beta = \alpha\left\{\frac{1}{v_{min}} - \frac{1}{v_{max}}\right\} + \beta = \xi\beta + \beta,$$

(14)

with ξ given by Equation (12) and where v_{max}, v_{min} are the maximum and minimum grain velocities. Equation (14) gives

$$\beta = \frac{1}{1+\xi}n_g, \qquad \alpha = \frac{n_g}{v_{min}^{-1} - v_{max}^{-1}}\frac{\xi}{1+\xi}.$$

(15)

Interstellar extinction data indicate that the average mass density of grains in a cloud is $\sim 10^{-2}$ the density of interstellar hydrogen. Thus

$$\tfrac{4}{3}\pi a^3 s n_g \simeq 10^{-2} n_H m_H \quad \text{and} \quad \frac{n_g}{n_H} \simeq \frac{4}{s} \times 10^{-12}\left(\frac{a}{10^{-5}\,\text{cm}}\right)^{-3}.$$

(16)

The velocity spectrum of grains is then

$$f(v) = 2.1 \times 10^{-6}\left(\frac{1}{1+\xi}\right)\left(\frac{a}{10^{-5}\,\text{cm}}\right)^{-2}\left(\frac{n_H}{10\,\text{cm}^{-3}}\right)^{-1/2}\left(\frac{D}{1\,\text{pc}}\right)^{-1} \times$$

$$\frac{1}{1 - v_{min}/v_{max}}\frac{1}{v^2}, \qquad v > v_{min},$$

$$= 1.6 \times 10^{-11}\left(\frac{1}{1+\xi}\right)\left(\frac{a}{10^{-5}\,\text{cm}}\right)^{-3}\left(\frac{n_H}{10\,\text{cm}^{-3}}\right)\delta(v - v_{min}),$$

$$v \approx v_{min}, \qquad (17)$$

if we use Equations (12), (13), (15) and (16) with $s = 2.5$, $f(v)\,dv$ now referring to the number density of grains with velocities in the range v, $v + dv$.

3. Heating of Gas Clouds by Suprathermal grains

Gas atoms impinging on suprathermal grains, as they are being slowed down in clouds, gain energy at the expense of the grains. This mechanism provides an efficient way of converting the energy of starlight photons into gas kinetic energy thus providing a heat source for interstellar clouds. The collision of an H atom with a charged suprathermal grain may be assumed to be quasi-elastic, the energy transfer per collision being $\sim \frac{1}{2} m_H v^2$ where v is the grain velocity. Since the grain is charged the interaction cross-section could be greater than πa^2 by a factor slightly in excess of unity. We assume it to be $\gamma^2 \pi a^2$, with $\gamma^2 \simeq 3$. The steady-state rate of energy transfer to the gas by this process is

$$\Gamma = \int_{v_{min}}^{v_{max}} (\pi \gamma^2 a^2 v n_H)(\tfrac{1}{2} m_H v^2) f(v)\,dv \ \text{erg cm}^{-3}\,\text{s}^{-1}. \tag{18}$$

Using Equations (12) and (17) and assuming $v_{min}/v_{max} < 1$ we may express Equation (18) in the form

$$\Gamma \cong 2.73 \times 10^{-23} \frac{\gamma^2}{1 + \xi} \frac{(n_H/10\ \text{cm}^{-3})^{1/2}}{(D/1\ \text{pc})} \left(\frac{v_{max}}{10^8\ \text{cm s}^{-1}}\right)^2 \times$$

$$\left[1 - \frac{v_{min}}{v_{max}} + \frac{2}{\xi}\left(\frac{v_{min}}{v_{max}}\right)^2\right] \text{erg cm}^{-3}\,\text{s}^{-1}. \tag{19}$$

Setting $\gamma^2 = 3$, $D = 10$ pc as typical values and assuming $v_{min}/v_{max} \ll 1$, $\xi < 1$ we obtain

$$\Gamma \cong 8.2 \times 10^{-24} \left(\frac{n_H}{10\ \text{cm}^{-3}}\right)^{1/2} \left(\frac{v_{max}}{10^8\ \text{cm s}^{-1}}\right)^2 \text{erg cm}^{-3}\,\text{s}^{-1}, \tag{20}$$

which gives

$$\Gamma \simeq 8.2 \times 10^{-26} \left(\frac{n_H}{10\ \text{cm}^{-3}}\right)^{1/2} \text{erg cm}^{-3}\,\text{s}^{-1}, \quad \text{if} \quad v_{max} \simeq 10^7\ \text{cm s}^{-1};$$

$$\tag{21}$$

$$\simeq 7.4 \times 10^{-27} \left(\frac{n_H}{10\ \text{cm}^{-3}}\right)^{1/2} \text{erg cm}^{-3}\,\text{s}^{-1}, \quad \text{if} \quad v_{max} \simeq 3 \times 10^6\ \text{cm s}^{-1}.$$

Thus, even for the lowest likely value of v_{max} we find a heating rate by the present process which is greater than or comparable with those appropriate to other processes which have thus far been proposed. We summarize this data in Table I.

TABLE I

Heating rates in erg cm^{-3} s^{-1} for processes other than suprathermal grains

Processes	Heating rates	Reference
(1) Photoionization	$\dfrac{6.4 \times 10^{-27}}{T^{1/2}} n_H^2$	Dalgarno and McCray (1972)
(2) 100 eV X-rays	$9.77 \times 10^{-27} n_H(\zeta/10^{-16})$ (with $(1+\varphi)E_n = 61$ eV)	Dalgarno and McCray (1972)
(3) 2 MeV Cosmic Rays	$4 \times 10^{-27}(1+12x)n_H(\zeta/10^{-16})$ (with $(1+\varphi)E_n = 25$ eV)	Dalgarno and McCray (1972)
(4) Photoelectric effect on dust grains	$(2.5-5) \times 10^{-27} n_H$	Watson (1972), cf. Field (1974)
(5) Chemical Reactions	$3.0 \times 10^{-29} n_{H_2}(n_H + 2n_{H_2})$	Dalgarno and Oppenheimer (1974), cf. Barlow and Silk (1975)
(6) H$_2$ Photodesorption	$1 \times 10^{-27} n_{H_2}\left(\dfrac{n_H + 2n_{H_2}}{n_H + n_{H_2}}\right)$	Barlow and Silk (1975)
(7) H$_2$ Recombination	$2.2 \times 10^{-28} n_H(n_H + 2n_{H_2})$	Barlow and Silk (1975)

4. Molecule Formation

The surfaces of suprathermal grains provide ideal sites for the operation of the Stecher-Williams theory of interstellar molecule formation (Stecher and Williams, 1966). The process envisaged by them involves the basic chemical exchange reaction

$$\text{Grain} - H + Y \rightarrow \text{Grain YH}, \tag{22}$$

where H is a chemisorbed hydrogen atom on the grain and Y ($\equiv$ H, C$^+$, O, N, etc.) is a second impinging atom. A large fraction of available valence sites on a grain composed of graphite may be assumed to be filled by hydrogen atoms at any instant and reaction (22) will proceed with an efficiency close to 100% provided the incoming atom Y has a relative kinetic energy greater than the activation energy for the reaction. Stecher and Williams (1966) points out that typical values of the activation energy may range from 0.2–0.3 eV. Although these authors discuss the possibility of molecule formation on fast grains they consider this only in the restricted context of the vicinity of hot stars. The same process will clearly be applicable throughout interstellar clouds where suprathermal grains are continually being slowed down. Grain speeds of $\gtrsim 10^5$ cm s^{-1} are ideally suited to overcoming the required activation barriers and these speeds are clearly present in the velocity spectrum discussed in Section 2.

Molecule formation on a fast grain and the release of such molecules into the gas will be effective for grains whose velocity satisfy

$$E_A < \tfrac{1}{2}mv^2 < E_e, \tag{23}$$

where E_A is the activation energy for a particular reaction and E_e is escape energy from the grain ~ 2 eV. Expressing the activation energy in terms of a temperature T_A this gives a velocity range

$$v_\alpha < v < v_\beta, \tag{24}$$

with

$$v_\alpha = \left(\frac{2kT_A}{m}\right)^{1/2}, \tag{25}$$

$$v_\beta = \left(\frac{2E_e}{m}\right)^{1/2} \simeq 2 \times 10^6 \left(\frac{m}{m_H}\right)^{-1/2}, \tag{26}$$

where m is the mass of the molecule.

The formation rate of a molecule YH is now given by

$$\gamma_{YH} = \int_{v_\alpha}^{v_\beta} \pi a^2 v n_Y f(v) \, dv \text{ cm}^{-3} \text{ s}^{-1}, \quad \text{if} \quad v_\alpha > v_{\min}, \tag{27a}$$

$$= \int_{v_{\min}}^{v_\beta} \pi a^2 v n_Y f(v) \, dv \text{ cm}^{-3} \text{ s}^{-1}, \quad \text{if} \quad v_\alpha \leqslant v_{\min}, \tag{27b}$$

where v_α, v_β are appropriately taken from Equations (25) and (26), $f(v)$ from (17) and n_Y refers to the ambient number density of species Y. Using Equations (12) and (17) we obtain

$$\gamma_{YH} = 2.1 \times 10^{-15} \frac{n_H^{-3/2}}{(1+\xi)} \left(\frac{D}{1 \text{ pc}}\right)^{-1} \frac{\ln(v_\beta/v_\alpha)}{1 - v_{\min}/v_{\max}} \text{ cm}^3 \text{ s}^{-1}$$

$$\text{if} \quad v_\alpha > v_{\min}, \tag{28a}$$

$$= 3.2 \times 10^{-16} \frac{n_H^{-1/2}}{(1+\xi)} \left(\frac{a}{10^{-5} \text{ cm}}\right)^{-1} \left(1 + \frac{\xi \ln(v_\beta/v_{\min})}{1 - v_{\min}/v_{\max}}\right) \text{ cm}^3 \text{ s}^{-1}$$

$$\text{if} \quad v_\alpha \leqslant v_{\min}. \tag{28b}$$

If we adopt $Y \equiv C^+$ and $kT_A \simeq 0.2$ eV (Stecher and Williams, 1966), $v_\alpha \simeq 1.7 \times 10^5$ cm s^{-1} from Equation (25). For the case of a 'canonical cloud' $n_H = 10$ cm^{-3} Equation (9) gives $v_{\min} = 2 \times 10^5$ cm s^{-1} which is at least marginally greater than v_α. Using $n_c/n_H = 3 \times 10^{-4}$ (Cameron, 1968), $E_e = 2$ eV, $a = 10^{-5}$ cm Equation (28b) with the help of (12), (17) and (26) gives

$$\gamma_{CH^+} = 1.0 \times 10^{-16} \text{ cm}^3 \text{ s}^{-1}. \tag{29}$$

This is at least one order of magnitude larger than the corresponding radiative association rate (Smith et al., 1973). If n_H is sufficiently large for $v_\alpha > v_{\min}$ to be satisfied then γ_{YH} is given by Equation (28a). For $n_H = 20$, $D = 10$ pc the Equation (28a) gives $\gamma_{CH^+} = 2.9 \times 10^{-18}$ cm^3 s^{-1} which is comparable to the radiative association rate given by Smith et al. (1973).

4.1 THE RADICALS CH, CH+

Estimates of the radiative recombination rates of CH, CH+ formed by two-body interactions have undergone several revisions since the early work of Kramers and ter Haar (1946). For a gas at kinetic temperature 100 K the recombination rates computed by several authors are set out in the first two rows of Table II.

TABLE II

Rates for formation by radiative recombination and destruction of CH and CH+

Process ($T=100$ K)	Smith et al.	Bates and Spitzer	Solomon and Klemperer	Designation
$C+H \rightarrow CH+h\nu$	1.5×10^{-17}	6×10^{-18}	3×10^{-18}	$\gamma_1(\text{cm}^3 \text{ s}^{-1})$
$C^+ +H \rightarrow CH^+ +h\nu$	5×10^{-18}	2×10^{-18}	3.5×10^{-17}	$\gamma_2(\text{cm}^3 \text{ s}^{-1})$
$CH+h\nu \rightarrow CH^+ +e^-$	$4 \times 10^{-11 a}$	8×10^{-12}	2.7×10^{-11}	$\beta_1(\text{s}^{-1})$
$CH+h\nu \rightarrow C+H$	11×10^{-11}	1.5×10^{-11}	1.7×10^{-11}	$\beta_2(\text{s}^{-1})$
$CH^+ +h\nu \rightarrow C^+ +H$	1×10^{-12}	5×10^{-13}	1×10^{-12}	$\beta_3(\text{s}^{-1})$
$CH^+ +e^- \rightarrow C+H$	$2.4 \times 10^{-9 b}$	~ 0	2.3×10^{-10}	$\alpha_1(\text{cm}^3 \text{ s}^{-1})$
$CH^+ +e^- \rightarrow CH+h\nu$	$\sim 0^b$	7×10^{-12}	2.3×10^{-10}	$\alpha_2(\text{cm}^3 \text{ s}^{-1})$
$CH^+ +H_2 \rightarrow CH_2^+ +H$	–	–	–	$\alpha_3(\text{cm}^3 \text{ s}^{-1})$
$CH_2^+ +H_2 \rightarrow CH_3^+ +H$	–	–	–	$\alpha_4(\text{cm}^3 \text{ s}^{-1})$
$CH_3^+ +e \rightarrow CH+H_2$	–	–	–	$\alpha_5(\text{cm}^3 \text{ s}^{-1})$

[a] Walker and Kelly (1972).
[b] Bardsley (1972); Krauss and Julienne (1973).

The subsequent rows give the main reactions proposed involving the destruction of CH and CH+ in a 'standard' interstellar cloud in the presence of an unattenuated interstellar radiation field. The last three reactions are from Watson (1974) and the values of α_3, α_4 and α_5 used are respectively 10^{-9} cm^3 s^{-1}, 10^{-9} cm^3 s^{-1} and 10^{-7} cm^3 s^{-1}.

For a standard cloud with $n_H \approx 10$ cm^{-3}, carbon is mainly in the form of C+. The grains are also probably positively charged. Therefore collision of C+ with a grain most probably results in the formation of CH+. However, the production of CH on the surface of grains may not always be negligible. Therefore, assuming a fraction x of the total formation rate as given by Equations (28a) and (28b) to produce CH+ and considering all reactions in Table II except the first two and setting $\alpha_2 = 0$, the equilibrium ratios of CH+ and CH to H can be written as

$$\frac{N(CH^+)}{N(H)} = \frac{\gamma_{CH} n(C)((\beta_1 + \beta_2)x + \beta_1(1 - x))}{(\beta_1 + \beta_2)\{\beta_3 + \alpha_1 n_e + \alpha_3 n(H_2)\} - \alpha_3 \beta_1 n(H_2)},$$

(30)

$$\frac{N(CH)}{N(H)} = \frac{\gamma_{CH} n(C)((1 - x)(\beta_3 + \alpha_1 n_e) + \alpha_3 n(H_2))}{(\beta_1 + \beta_2)\{\beta_3 + \alpha_1 n_e + \alpha_3 n(H_2)\} - \alpha_3 \beta_1 n(H_2)},$$

where γ_{CH} is the grain recombination rate given by Equations (28a) and (28b), α_1, β_1, β_2, β_3 are taken from Smith *et al.* (1973) (see Table II), α_3 is from Watson (1974) and $n(C)$, $n(H_2)$ and n_e are the densities of carbon, H_2 and electrons, respectively.

Substituting for γ_{CH} from Equation (28b) with $a = 10^{-6}$ cm, and $n(C)/n(H) = 3 \times 10^{-3}$ using $\alpha_3 = 10^{-10}$ cm^3 s^{-1} and other rates from Table II Equations (30) yield

$$\frac{N(CH^+)}{N(H)} = \frac{6.4 \times 10^{-10}n_H^{1/2}(11x + 4)}{0.01 + 24n_e + 0.73n(H_2)},$$

$$\frac{N(CH)}{n(H)} = \frac{6.4 \times 10^{-9}n_H^{1/2}((0.01 + 24n_e)(1 - x) + n(H_2))}{0.01 + 24n_e + 0.73n(H_2)}.$$

(31)

With $n_e/n_H = 3 \times 10^{-3}$, $n(H_2)/n_H = 1/3$ and $n_H = 10$ cm^{-3}, the Equation (31) gives $N(CH^+)/N(H) = 1.5 \times 10^{-8}$ and $N(CH)/N(H) = 3 \times 10^{-8}$, if $x = 1$; and $N(CH^+)/N(H) = 4 \times 10^{-9}$ and $N(CH)/N(H) = 3 \times 10^{-8}$, if $x = 0$. These values are not inconsistent with observations (Morton, 1975). However, we note that the rates α_1 and α_3 adopted in the calculation are quite uncertain and our conclusion regarding the ratios $N(CH^+)/N(H)$ and $N(CH)/N(H)$ must remain tentative.

4.2. OTHER DIATOMIC SPECIES

Molecular H_2 is likely to form most efficiently on the surfaces of thermal grains in the interiors of fairly dense clouds (Solomon and Wickramasinghe, 1969; Hollenbach *et al.*, 1971). The penetration of suprathermal grains into the interiors of such clouds being low, the fast grain formation process is unlikely to be important in this case. The rates of formation of OH and NH by this present process as given by Equation (28b) with $n_0/n_H \simeq 10^{-3}$, $n_N/n_H \simeq 10^{-4}$, $a = 10^{-5}$ cm taken in conjunction with dissociation rates $\beta_{OH} \simeq 10^{-10}$ s^{-1}, $\beta_{NH} \simeq 10^{-10}$ s^{-1} give equilibrium abundance ratios

$$\frac{N_{OH}}{N_H} \simeq 3 \times 10^{-9}n_H^{1/2},$$

(33)

$$\frac{N_{NH}}{N_H} \simeq 3 \times 10^{-10}n_H^{1/2}.$$

(34)

These values are consistent with their observational limits of $N_{OH}/N_H < 10^{-7}$, $N_{NH}/N_H < 5 \times 10^{-8}$.

The rates of formation of non-hydrogenic diatomic molecules such as CO, N_2, NO, CN on fast grains may be shown to be too low to account for the observed data in most cases. Efficient formation of CO and CN could however occur by gas phase reactions

$$\begin{array}{ll} CH^+ + O \rightarrow CO + H^+, & \gamma_{CO} \simeq 10^{-9} \text{ cm}^3 \text{ s}^{-1}; \\ CH^+ + N \rightarrow CN + H^+, & \gamma_{CN} \simeq 10^{-9} \text{ cm}^3 \text{ s}^{-1}. \end{array}$$

(35)

4.3. COMPLEX MOLECULES

Fast dust grains can play an important role in the formation of more complex molecules. Complex organic molecules which are frozen on 'thermal' grains could be expelled into the gas phase as these are accelerated to suprathermal speeds in the inter-cloud medium and are re-injected into denser clouds. It has always remained a puzzle to understand how large concentrations of molecules such as CH_3OH, HCN, CH_2O, CO and H_2O could co-exist in the gas phase with solid particles whose temperatures were less than 20 K. Such molecules are expected to freeze down on 'thermal' grains and the equilibrium vapour pressures at the grain temperatures in question will be far below what is observed. Photo-desorption is usually invoked to produce sufficiently rapid detachment, but this may not be a viable model in the denser clouds where complex molecules are observed. Sputtering of mantles off fast grains may be a more likely mechanism in some cases.

Fast grains may also serve to overcome the endothermicity of reactions which will not occur in a cold gas. Thus the reaction

$$H_2 + CO \rightarrow H_2CO$$

is endothermic by ~ 0.2 eV and cannot occur under normal circumstances in the gas phase. Fast grains being slowed in a cloud of molecular H_2 and CO could provide the venue for this reaction provided grain velocities are $\sim 10^6$ cm s^{-1}.

If we assume that most of the H is molecular and a large fraction (say 90%) of C is CO the rate constant for the fast grain reaction is

$$3.2 \times 10^{-20} n_{H_2}^{3/2} \text{ cm}^{-3} \text{ s}^{-1}$$

by use of Equation (28b) with $n_H \simeq \frac{1}{2} n_{H_2}$, $n_Y \simeq n_c$. With a photodissociation rate constant $\beta_{H_2CO} = 3 \times 10^{-10}$ s^{-1} (Herbst and Klemperer, 1973) this gives

$$\frac{n_{H_2CO}}{n_{H_2}} \simeq 1.1 \times 10^{-10} n_{H_2}^{1/2}, \qquad (37)$$

which is within the observed range of 10^{-10}–10^{-9} for this ratio if n_{H_2} is in the range 10–50 cm^{-3} (Davies and Matthews, 1972).

The H_2CO molecules thus formed on the surfaces of fast grains will be released into the gas phase and could re-condense on the surfaces of thermal grains (Wickrama-singhe, 1974, 1975).

5. Contribution to Pressure Balance in Clouds

Finally we consider the momentum transport and pressure due to fast grains injected into clouds. The momentum transport cm^{-2} s^{-1} due to this process is

$$P_{gr} \simeq \int_{v_{min}}^{v_{max}} (m_g v) v f(v) \, dv. \qquad (38)$$

Using Equations (12) and (17) and setting $m_g = \frac{4}{3}\pi a^3 s$ with $s = 2.5$, we obtain

$$P_{gr} \simeq 10^{-12}(1 + \xi)^{-1} \frac{(a/10^{-5}\text{ cm})}{(D/1\text{ pc})} \left(\frac{v_{min}}{10^5\text{ cm s}^{-1}}\right) \left(\frac{v_{max}}{10^8\text{ cm s}^{-1}}\right) \times$$

$$\left(1 + \frac{1}{\xi}\frac{v_{min}}{v_{max}}\right). \tag{39}$$

The condition for pressure balance across a cloud is thus

$$n_H kT \simeq P_{gr}; \tag{40}$$

and from Equations (9), (39) and (40) we obtain

$$\frac{T}{100\text{ K}} = 14.0(1 + \xi)^{-1} \left(\frac{a}{10^{-5}\text{ cm}}\right) \left(\frac{v_{max}}{10^8\text{ cm s}^{-1}}\right) \left(\frac{n_H}{10\text{ cm}^{-3}}\right)^{-3/2} \times$$

$$\left(\frac{D}{1\text{ pc}}\right)^{-1} \left(1 + \frac{1}{\xi}\frac{v_{min}}{v_{max}}\right). \tag{41}$$

With $D = 5$ pc, $n_H = 10$ cm^{-3}, $a \simeq 10^{-5}$ cm, $\xi = 0.1$ (Equation (13)) and $v_{max} = 2 \times 10^7$ cm s^{-1} this gives $T \simeq 60$ K. High-speed grains could, therefore, play a role in maintaining pressure balance between clouds and an intercloud medium.

A pressure balance of the type considered here could have an important effect on the process of star formation. A local increase of the momentum flux due to high-speed grains could precipitate gravitational collapse of clouds and the onset of star formation.

Since high-speed grain ejection is associated with regions of protostellar shells, the process of star formation would thus have a positive feedback. Star formation whenever it begins would therefore tend to proceed towards completion in any given region of interstellar space.

Acknowledgements

The work was supported in part by a grant from the Science Research Council. One of us (SPT) is on leave of absence from the Tata Institute of Fundamental Research, Bombay.

References

Bardsley, J. N.: 1972, unpublished results.
Barlow, M. J. and Solk, J.: 1975, preprints.
Bates, D. R. and Spitzer, L.: 1951, *Astrophys. J.* **113**, 441.
Cameron, A. G. W.: 1968, in L. H. Ahrens (ed.), *Origin and Distribution of the Elements*, Pergamon Press.
Dalgarno, A. and McCray, R. A.: 1972, *Ann. Rev. Astron Astrophys.* **10**, 375.
Dalgarno, A. and Oppenheimer, M.: 1974, *Astrophys. J.* **192**, 597.
Davies, R. D. and Matthews, H. E.: 1972, *Monthly Notices Roy. Astron. Soc.* **156**, 253.
Deutsch, A. J.: 1960, *Stars and Stellar Systems* **6**, 543.
Field, G. B.: 1974, *Proc. Les Houches Summer School*.
Frisch, P.: 1972, *Astrophys. J.* **173**, 301.

Gilman, R. C.: 1973, *Monthly Notices Roy. Astron. Soc.* **161**, 3P.

Hayakawa, S.: 1974, *Astrophys. Space Sci.* **31**, L13.

Hayakawa, S.: 1975, in N. C. Wickramasinghe and D. J. Morgan (eds.), *Publication of Cardiff Symposium on Solid State Astrophysics*, D. Reidel Publ. Co., Dordrecht, Holland, p. 93.

Herbst, E. and Klemperer, W. B.: 1973, *Astrophys. J.* **185**, 505.

Hollenbach, D. J., Werner, M. W., and Salpeter, E. E.: 1971, *Astrophys. J.* **163**, 165.

Hoyle, F., Solomon, P., and Woolf, W. J.: 1973, *Astrophys. J. Letters* **185**, L89.

Kramers, H. A. and ter Haar, D.: 1946, *Bull. Astron. Soc. Neth.* **10**, 137.

Krauss, M. and Julienne, P.: 1973, *Astrophys. J.* **183**, L139.

Morton, D. C.: 1975, *Astrophys. J.* **197**.

Reddish, V. C.: 1967, *Monthly Notices Roy. Astron. Soc.* **135**, 251.

Salpeter, E. E.: 1974a, *Astrophys. J.* **193**, 579.

Salpeter, E. E.: 1974b, *Astrophys. J.* **193**, 585.

Salpeter, E. E. and Wickramasinghe, N. C.: 1969, *Nature* **222**, 442.

Stecher, T. P. and Williams, D. A.: 1966, *Astrophys. J.* **146**, 88.

Solomon, P. M. and Klemperer, W. B.: 1972, *Astrophys. J.* **178**, 389.

Smith, W. H., Liszt, H. S., and Lutz, B. L.: 1973, *Astrophys. J.* **183**, 69.

Tarafdar, S. P. and Wickramasinghe, N. C.: 1975, *Nature* **254**, 203.

Walker, T. and Kelly, H. P.: 1972, *Chem. Phys. Letters* **16**, 511.

Watson, W. D.: 1972, *Astrophys. J.* **176**, 103.

Watson, W. D.: 1974, *Astrophys. J.* **183**, L17.

Weymann, R.: 1963, *Ann. Rev. Astron. Astrophys.* **1**, 97.

Wickramasinghe, N. C.: 1972, *Monthly Notices Roy. Astron. Soc.* **159**, 269.

Wickramasinghe, N. C.: 1974, *Nature* **252**, 462.

Wickramasinghe, N. C.: 1975, *Monthly Notices Roy. Astron. Soc.* **170**, 11P.

PART II

NEUTRON STAR PHYSICS

SOLID STATE PHYSICS AND COOLING OF
NEUTRON STARS

SACHIKO TSURUTA

Astronomy Centre, University of Sussex, Falmer, Brighton, England

Abstract. First we show the possible effect of the 'magnetic' condensation on cooling of neutron stars. Its observational significance (especially for younger pulsars such as the Crab pulsar) is emphasized. Other effects of solid state physics on cooling are also discussed.

1. Introduction

For a simple gas model without magnetic fields, the surface temperature of a neutron star is about a million degrees at the age of about a million years (Tsuruta and Cameron, 1966). Through recent developments, however, it was shown that the cooling time of a neutron star as a pulsar is significantly reduced due to various new factors, among which the major effects are (a) the reduction of opacities in the presence of strong magnetic fields and (b) the effects of baryon superfluidity. Thus, the surface temperature is reduced to 10^3 to 10^5 K at the age of about a million years, which is a typical age of a pulsar (Tsuruta *et al.*, 1972).

In the present paper, we wish to show that the effect of the solid state on cooling may be even more significant. This is mainly because the surface matter is condensed to the hair-like 'magnetic metal' (Ruderman, 1971) as the star cools down. Then, the surface density is increased from the photospheric value of about 1 gm cm^{-3} to about 10^4 gm cm^{-3}, and the temperature difference between the central isothermal core and the surface may be drastically reduced. Once this stage is reached, therefore, the residual stellar heat may escape very quickly. The exact degree of this effect depends on the delicate problem of phase transitions between the gas and solid-liquid state in the intermediate stages of cooling, and the electron conductivity in the condensed state. For a neutron star at the age of a typical pulsar, it can be safely said that the surface layers consist of the 'magnetic metal' while the intermediate layers are in the form of the Coulomb lattice. (It may extend right to the centre if the mass of the star is sufficiently low (Baym *et al.*, 1971).) The core or shells of superfluid baryons are also expected in the interior for more massive stars (Tsuruta, 1974). For younger pulsars such as the Crab pulsar NP 0532 and the Vela pulsar 0833-45, the effect of the 'magnetic' condensation may play a major role, competing with or even more important than the effect of pion condensation (if pions are present). Their observational significance is discussed. Other effects of solid state physics on cooling are also discussed.

2. Outer Layers of a Magnetic Neutron Star

In order to see the properties of the outer layers of a magnetic neutron star, various characteristic energies are shown in Figure 1 in units of the equivalent temperatures. $E_B^H(\text{Fe})$ is the binding energy of the iron 'magnetic metal', which is about 30 keV for $H \simeq 5 \times 10^{12}$ G (Ruderman, 1971, 1974), E_F^e is the electron Fermi energy, $E_c(\text{Fe})$ and $E_c(p)$ are the Coulomb energy of an iron and a proton, respectively, and E_H is the magnetic energy, where

$$E_c = kT_m = \frac{1}{\Gamma} \frac{(Ze)^2}{r_z} \exp(-r_z/r_{sc}),$$

$$\Gamma \simeq 50\text{–}100, \tag{1}$$

$$E_H = \frac{H^2}{4\pi n}.$$

More elaborate computer results for E_c (which corresponds to the melting temperature of the Coulomb lattice) are given by N. Itoh (private communication), but the above expression is adequate for the present purpose, and it qualitatively agrees with their

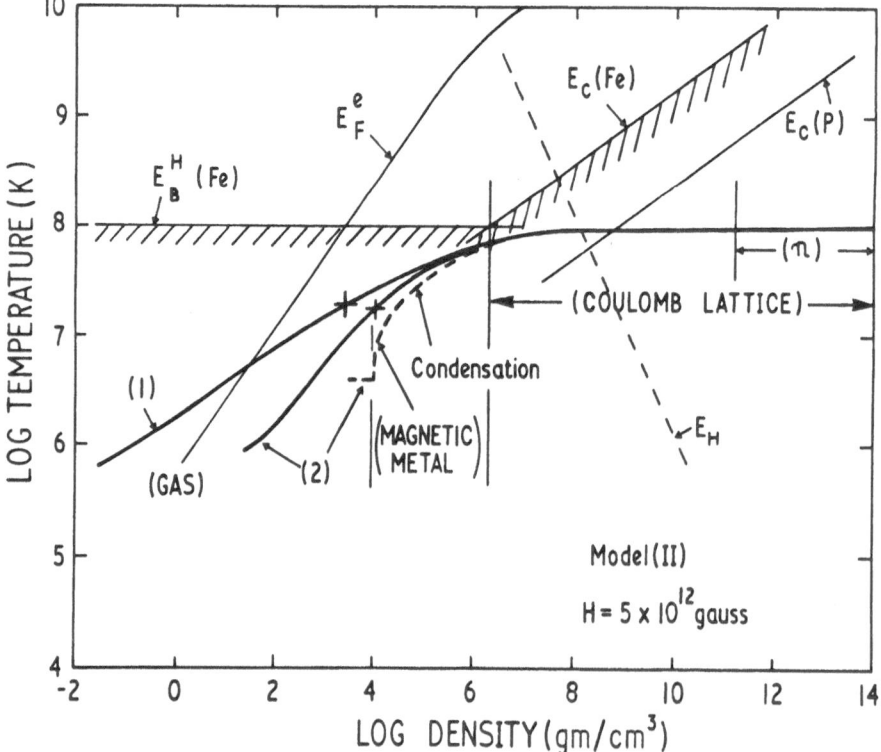

Fig. 1.   Properties of the outer layers of a neutron star (Model (II)) for (1) $H=0$ and (2) $H=5 \times 10^{12}$ G. The heavy dashed curve shows the condensed state.

results. The dashed line E_H (which corresponds to the magnetic field $H = 5 \times 10^{12}$ G) is to show the measure of importance of the magnetic effects.

The overall picture for the zero temperature matter, then, is (1) the 'magnetic metal' extends from the surface (where the density $\varrho_s = \sim 10^4$ gm cm^{-3}) up to the boundary of the crust with the Coulomb lattice where $\varrho \simeq 10^6$ gm cm^{-3}. Neutrons are present for $\varrho > \sim 3 \times 10^{11}$ gm cm^{-3}. The Coulomb lattice of heavy ions disappears near the nuclear density where the neutron core starts (Baym et al., 1971). In the present paper, we shall concentrate on the outer layers where most of the temperature changes take place.

Figure 1 also shows the temperature distributions in the outer layers as a function of densities (the heavy curves), where the branches (1) and (2) correspond to $H = 0$ and 5×10^{12} G, respectively. Here we used Model (II) of Tsuruta (1974) (hereafter called Paper I). This is a medium weight neutron star with the following properties: the mass $M = 0.476\ M_\odot$, the radius $R = 10.9$ km, and the central density $\varrho^c = 8 \times 10^{14}$ gm cm^{-3}. For each curve, the conductive opacity κ_c overtakes the radiative opacity κ_R beyond the point indicated by the cross.

The magnetic reduction of the opacities can be expressed as

$$\kappa_I(H) = a_I \kappa_I(0), \qquad\qquad (2)$$

where the subscript $I = R$ for the radiative opacity and $I = C$ for the conductive opacity. $\kappa(H)$ and $\kappa(0)$ are the opacities with and without the magnetic fields, respectively. The reduction factor for the radiative opacity a_R can be expressed as $(\omega/\omega_H)^2$, where ω is the radiation frequency and ω_H is the cyclotron frequency which depends on H as $\omega_H = eH/m_e c$ (Lodenquai et al., 1974; Tsuruta et al., 1972). The reduction factor a_C for the conductive opacity is a rather complicated function of the magnetic field, density and temperature (Canuto and Chiu, 1969). The typical values for a_C of a degenerate gas in the most important range of densities $\varrho = \sim 10^4$–10^6 gm cm^{-3}, are about 0.01 to 0.1 for $H = \sim 10^{12}$–10^{13} G (Paper I). They increase and approach 1 as the density increases. These values are the maximum reduction factors along certain spatial directions, but the cooling rates of the star as a whole are determined mainly by these factors. In the present calculations we used the values of a_C obtained in Paper I, which specifically apply to a gas. The effect of solidification can be treated either as lattice vibrations or as impurity. In the higher density and temperature regions and for zero magnetic fields, it was calculated by some authors (Canuto, 1970; Solinger, 1970). In the presence of the magnetic fields, it was reported that the effect of solidification is a further reduction of κ_C by a factor of 2 when $\varrho \simeq 10^6$ gm cm^{-3} (Canuto, private communication). But I am not aware of any work done on a_C for the 'magnetic metal' in the most critical region of $\varrho = \sim 10^4$–10^6 gm cm^{-3}. This is one of the major causes of uncertainties in the present results, and it is strongly recommended that someone should make better theoretical estimates of the electron conductivity in the 'magnetic metal'. However, according to Ruderman (private communication), the electron conductivity may be significantly increased in the condensed state.

We can say that at least in the denser regions the matter is in the solid state if the heavy solid curve (the stellar temperature distribution) is lower than the melting temperature lines expressed as E_c (the Coulomb lattice) or E_B^H (the 'magnetic metal'). Thus, at a particular stage of cooling when the internal temperature is about 10^8 K, the star is already in the solid state (the Coulomb lattice) at least down to the layer where the density is $\sim 10^6$ gm cm^{-3}. We see that the magnetic effects are not important in these inner layers, except at the boundary regions ($\sim 10^6$–10^8 gm cm^{-3}).

In the outermost layers, the problem is more complicated because the condensation depends on pressures as well as temperatures. We can, however, make rough estimates of the state of matter through the application of the theory of phase changes. According-ing to the Clausius-Clapeyron equation, phase changes take place along the curve in the P-T (pressure-temperature) plane determined by

$$\frac{dP_{cc}}{dT} = \frac{\Delta S}{\Delta V};$$

(3)

i.e., the slope is equal to the entropy change divided by the volume change. Taking into account that the volume of a solid is negligible as compared with that of a gas, it reduces to

$$P_{cc} = K \exp\left(-\frac{L}{RT}\right),$$

(4)

where K is some proportionality constant, L is the latent heat (of sublimation) and R is the universal gas constant. Using microscopic considerations, the above equation can be expressed as

$$P_{cc} = n_s kT \exp\left[-\frac{E_B^H(\text{Fe})}{kT}\right],$$

(5)

where n_s is the number density of matter in the solid state. (Note that the latent heat in this particular case is related to the binding energy of the 'magnetic metal' by the relation $L = N_0 E_B^H(\text{Fe})$, where N_0 is the Avogadro's number.) Then, at a given tem-perature, T, the condensation to the 'magnetic metal' takes place if the density of the particular layer is sufficiently high so that the pressure there exceeds P_{cc}. Applying this line of analysis, it was concluded that the solidification to the 'magnetic metal' should take place when the core temperature is about 10^8 K. In this case, the surface starts rather abruptly at $\varrho_s \simeq 10^4$ gm cm^{-3} and the temperature distribution is roughly as shown in Figure 1 by the heavy dashed curve, where $H = 5 \times 10^{12}$ G. Here it is essential to have the accurate values of a_c. It was assumed in the present calculations that κ_C for the 'magnetic metal' is reduced by a factor of ~ 50 as compared with the corresponding value for a magnetized gas. The density of atmospheres (gases) at the stellar surface ϱ_s^G can be found from the equation

$$\varrho_s^G = \varrho_s^s \exp\left(-E_B^H/kT_s\right),$$

(6)

which can be easily derived from the Clausius-Clapeyron equation; ϱ_s^s being the density

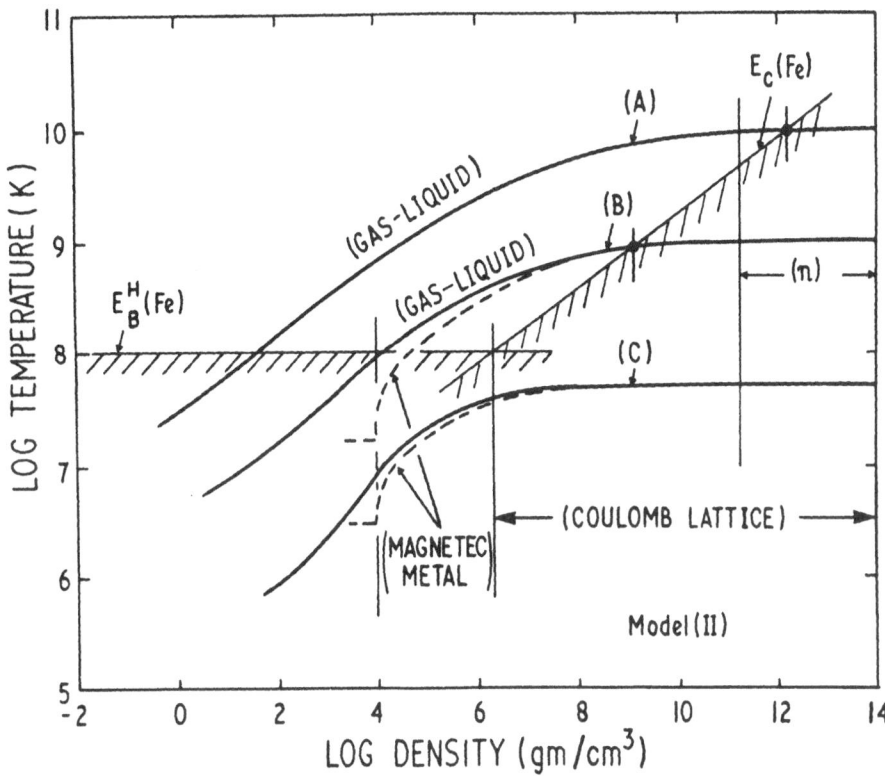

Fig. 2. Temperature distributions (vs density) in the outer layers of a neutron star (Model (II)), at different stages of cooling. The dashed curves show the effect of the 'magnetic' condensation. $H = 5 \times 10^{12}$ G.

of the solid surface. The gas density at the surface is less than 0.1% of ϱ_s^s already at $T_s = 10^7$ K.

In Figure 2, the temperature distributions in the outer envelopes are shown as a function of densities at the three typical points during the cooling: when (A) $T_c = 10^{10}$ K, (B) $T_c = 10^9$ K and (C) $T_c = 5 \times 10^7$ K. (T_c is the temperature of the isothermal core.) In case (A), temperatures are sufficiently high all through the outer envelopes so that they stay in the gas-liquid phase all the way up to the boundary to the isothermal core at about 10^{12} gm cm^{-3}.* Even though the temperatures are lower than the melting temperature of the 'magnetic metal' near the photosphere, the gas pressure is too low to allow for the condensation. At much higher temperatures, the present method may break down, because then the radiation pressure will play the major role. For instance, it may blow away the surface particles (E. E. Salpeter, private communication). In case (B), the star has cooled down sufficiently to allow for the 'magnetic' condensation of the atmospheric particles, according to the rough estimates through the above analysis. Thus, it seems that the surface of a neutron star is a particularly hostile place where, at certain stages of cooling, magnetic dusts (or rather needles) are pouring down.

* At such high temperatures, the effect of superfluidity is not important.

It may be noted that in stage (B) there is a possibility that the temperatures are still high enough to keep some liquid layers trapped between the 'magnetic metal' surface layers and the internal (Coulomb lattice) crust layers (or core). In stage (C), however, the whole outer envelopes are safely in the solid state right up to the boundary to the neutron core. In Figure 2, the temperature distributions for the condensed surface layers are shown by the dashed curves for cases (B) and (C). Here again, it was assumed that $H = 5 \times 10^{12}$ G and that κ_C for the 'magnetic metal' is reduced by a factor of ~ 50. We note that the temperature difference between the interior and the surface is significantly reduced for the condensed state.

3. Cooling Rates and Their Observational Significance

The cooling rates can be calculated by the standard method. The details are explained in Paper I. The rates depend on the total thermal energy and luminosity of the star. The total energy is the sum over the thermal tails of degenerate particles and the energy of non-degenerate heavy ions. The total luminosity includes both the neutrino and photon luminosities. The total energy and neutrino luminosity depend on the internal temperature while the photon luminosity is determined by the surface temperature. It may be noted that both the 'magnetic' condensation of the surface matter and the magnetic reductions of the opacities enter the problem mainly through the significant reduction of the temperature difference between the interior and the surface (Section 2).

Another major effect on cooling comes from baryon superfluidity (and superconductivity). This is because the specific heat of a superfluid particle is reduced by the factor

$$Y_s \propto \exp\left(-1.44 T_0/T\right) \frac{T_0}{T}, \quad \text{for} \quad T \ll T_0, \tag{7}$$

where the critical temperature T_0 is proportional to the superfluid energy gap. The effect is uncertain to the extent of the uncertainty of the energy gap. The maximum effects are given by Tsuruta *et al.* (1972). In Paper I, I used the latest information on the energy gap which, according to the Kyoto group, is possibly the best estimate available. Another effect of superfluidity is to suppress the neutrino URCA rates.

The solidification of heavy ions also affects the cooling rates by reducing the specific heat by the Debye factor $\mathscr{D}(\theta/T)$ sufficiently below the melting temperature. This effect is not significant for the medium weight and heavy stars if superfluidity is neglected, because then the major contributions to the total energy come from neutrons and other baryons. When the energies of these baryons are drastically reduced in the presence of superfluid, however, the effect of the state of heavy ions cannot be neglected. For low mass stars where heavy nuclei may be present right to the centre, its effect becomes significant. The star cools faster after the solidification to the Coulomb lattice. The further details are found in Paper I. The equations of state are also affected by the solidification of nuclei, but its effect on the cooling problems is relatively minor. (However, it may be noted that the presence of strong magnetic fields may somewhat reduce the pressure near the solid surface (for $\varrho = \sim 10^4$ gm cm^{-3}) at sufficiently low temperatures ($< \sim 10^6$ K)).

The star may cool very fast if pions are present in the dense interior, due to the increased neutrino escaping rates. The possibility of the presence of pions in neutron stars and the question of where (at which critical density) they should appear are still controversial. In this paper, we shall not dwell on these questions because the other authors in this volume are expected to cover the subject. Assuming that pions are indeed present, I am not aware of any derivation of the exact equations or any exact calculations of the cooling rates of pion stars. However, some crude preliminary estimates seem to indicate that the pion cooling may be very important, competing with the effect of the 'magnetic' condensation of the surface matter. In the central core there is also a possibility for the solidification of neutrons, which may have some observational significance, but it seems that its effect on cooling is relatively minor.

The cooling curves (the surface temperatures T_s as a function of time) are shown in Figure 3 for Model (II) (of Paper I). The curve (I) shows the zero field result. In the

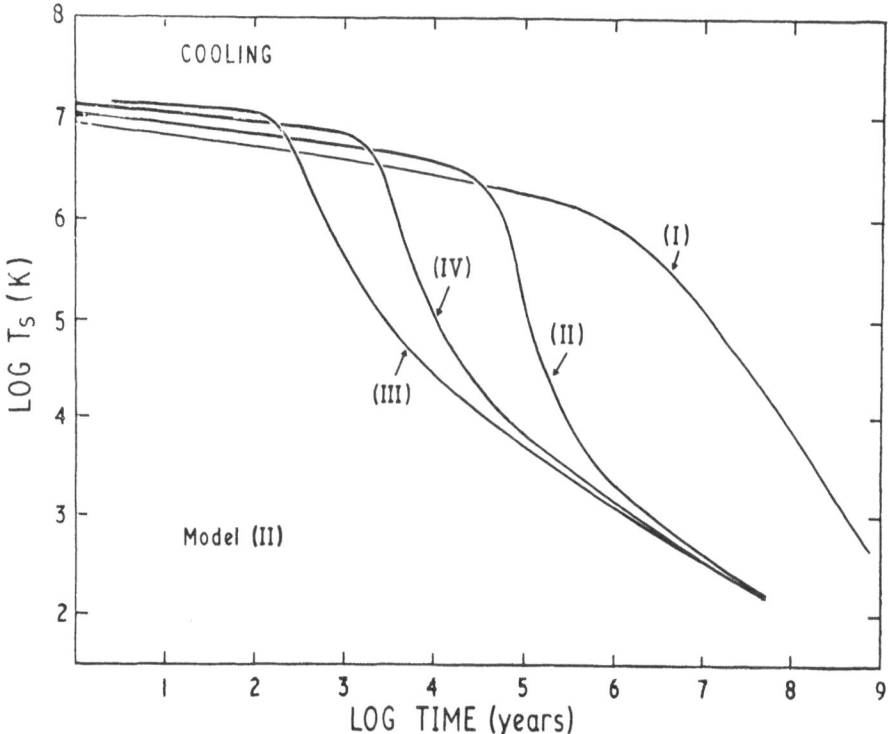

Fig. 3.   Cooling curves (surface temperature T_s vs time) for a neutron star (Model (II)) for (I) $H=0$, (II) with the effects of magnetic fields ($H=5\times10^{12}$ G), superfluidity and crystallization of heavy nuclei, (III) with the additional effect of the 'magnetic' condensation, and (IV) with the effect of the increased magnetic field strength ($H=5\times10^{13}$ G).

curve (II), $H = 5 \times 10^{12}$ G, and the effects of the presence of the magnetic field, super-fluidity and crystallization of heavy nuclei are included in the manner described in Paper I. The curve (III) shows the additional effect of the 'magnetic' condensation of the surface layers. All other conditions are the same as in the curve (II). Here, it was assumed that the conductive opacity is further reduced by a factor of ~ 50. The curve (IV) shows the effect of the increase of the magnetic field strength, by a factor of 10. Other conditions are the same as in the curve (II). Since the main purpose of the present report is to compare the effects of the different factors, a definite assumption (which is hoped to be the best estimate in Paper I) was made, in order to take into account the effect of superfluidity.

Finally, the following conclusions may be drawn. (1) A typical pulsar (the age $\sim 10^6$ yr and $H = 10^{12} - 10^{13}$ G) is dead cold ($\sim 10^3 - 10^5$ K), independent of the effects of the different factors mentioned above. (This conclusion does not change if the mass of the pulsar is increased or decreased.) In the absence of the magnetic fields, however, the star at the same age is as hot as about a million degrees.

(2) The situation is less certain for a neutron star at the age of $\sim 10^5$ yr (which may apply to some of the younger pulsars and possibly some of the binary X-ray sources). First, let us forget about the 'magnetic' condensation. Then, when $H = 5 \times 10^{12}$ G, $T_s \simeq 10^5$ K (Figure 3). However, the temperature goes up to nearly 10^6 K if $H \simeq 10^{12}$ G. It goes up further if the star is more massive or if the effect of superfluidity is less than the present estimates (see Tsuruta et al., 1972). With the further increase of the electron conductivity by a factor of 50 due to the condensation to the 'magnetic metal', or with the further increase of the magnetic field strength by a factor of 10, however, the surface temperature of the star at the same age goes down to $\lesssim 10^4$ K. According to Greenstein (1971), the stellar surface may be kept hot at the level of $\gtrsim 10^6$ K for younger pulsars (with the age of $\lesssim 10^5$ yr) due to a turbulent heating mechanism, and soft X-rays should be detectable from some of these younger pulsars. So far no observations of such X-rays have been reported. The continued failure of such detections in the future observations may be used against Greenstein's theory. That may also mean several things; (i) the significant reduction of the conductive opacity in the 'magnetic metal', (ii) the field strength of $H \gtrsim 5 \times 10^{12}$ G, (iii) the greater effect of superfluidity than the estimates used in the present calculations, (iv) the pion condensation, and (or) (v) very small stellar mass ($< 0.2 \, M_\odot$).

(3) An important test may come from what happens to very young neutron stars, especially to the Crab pulsar at the age of about one thousand years. This is because the star should be hot enough to emit the detectable amount of soft X-rays, unless the conductive opacity is drastically reduced in the 'magnetically condensed' state, or pions are present. It is so even if the maximum possible effect of superfluidity is included (for $H < 5 \times 10^{13}$ G). The stellar radiation may go below the detectable range if the pulsar is a very light neutron star ($M < 0.2 \, M_\odot$), but that seems to contradict with other theoretical and observational information on the Crab pulsar (Pines et al., 1974).[*]

* Also a star with $M < 0.2 \, M_\odot$ may be unstable (J. M. Cohen, private communication).

If future observations (e.g. the occultation experiment of the UK5 satellite) fail to detect a compact X-ray component in the Crab, for photons of ~ 0.5 to 2 keV with the intensity somewhere at the level of $\sim 10^{33}$–10^{36} ergs s^{-1}, that may strongly favour the presence of either (or both) of the above two phenomena, namely the 'magnetic' condensation in the surface layers and (or) the pion condensation in the interior (assuming that $M > \sim 0.2\ M_\odot$). If such a compact X-ray component is detected, that may mean that the pulsar is still hot enough, due to the absence of pions and the lack of the drastic increase of the electron conductivity in the 'magnetic metal'. In this case, this component may or may not be pulsed, depending on the magnetic field strength. If the polar spots are heated to sufficiently high temperatures ($\gtrsim 10^8$ K) and if the electron conductivity is significantly increased in the 'magnetic metal' (at least by a factor of 10), a cold neutron star can be heated to the level where the soft X-rays from the stellar surface should be detectable (Tsuruta and Rees, 1975). In this case, the radiation from the stellar surface should be pulsed, because the isotropy of the opacities should be lost under the strong magnetic fields which are implied by the efficient heat conduction. However, the pulse shape may be quite different from that of the main pulsar component. Furthermore, some resourceful persons may be able to think of some other heat sources, even a plastic flow (J. Shaham, private communication). Heating of a neutron star by the polar accretion and the radiation of X-rays as a consequence are interesting problems especially for some of the binary X-ray sources too, but I shall confine the present discussions only to problems directly related to cooling, because the following speaker, Professor Rees, is expected to give a more general review.

It is clear from the above discussions that at the present stage there are too many unknown factors to allow us to say anything definite. But as an optimist, I shall venture to suggest that the combination of different tests together (some of which are mentioned above) within the next several years may give tremendous insight to various problems, including the problems of solid state physics.

Acknowledgements

I wish to thank Professor Ruderman for invaluable discussions and suggestions, especially on the problem of the 'magnetic' condensation. I also with to thank the other participants of the Symposium on Solid State Astrophysics, especially Professors Salpeter and Rees, for valuable discussions and comments.

References

Baym, G., Pethick, C. J., and Sutherland, P.: 1971, *Astrophys. J.* **170**, 299.
Canuto, V.: 1970, *Astrophys. J.* **159**, 641.
Canuto, V. and Chiu, H.-Y.: 1969, *Phys. Rev.* **188**, 2446.
Greenstein, G.: 1971, *Nature Phys. Sci.* **232**, 117.
Lodenquai, J., Canuto, V., Ruderman, M., and Tsuruta, S.: 1974, *Astrophys. J.* **190**, 140.

Pines, D., Shaham, J., and Ruderman, M.: 1974, in C. J. Hansen (ed.), *Physics of Dense Matter*, Reidel Publ. Co., Dordrecht, p. 189.

Ruderman, M.: 1971, *Phys. Rev. Letters* **27**, 1306.

Ruderman, M.: 1974, in C. J. Hansen (ed.), *Physics of Dense Matter*, Reidel Publ. Co., Dordrecht, p. 117.

Solinger, A. B.: 1970, *Astrophys. J.* **161**, 553.

Tsuruta, S.: 1974, in C. J. Hansen (ed.), *Physics of Dense Matter*, Reidel Publ. Co., Dordrecht, p. 209.

Tsuruta, S. and Cameron, A. G .W.: 1966, *Can. J. Phys.* **44**, 1863.

Tsuruta, S. and Rees, M. J.: 1975, to be published.

Tsuruta, S., Canuto, V., Lodenquai, J., and Ruderman, M.: 1972, *Astrophys. J.* **176**, 739.

EVIDENCE ON NEUTRON STAR STRUCTURE FROM PULSARS AND RELATED OBJECTS

M. J. REES

Institute of Astronomy, Cambridge, England

Abstract. Properties of pulsars and binary X-ray sources which seem particularly relevant to theories of neutron star structure and strong magnetic fields are reviewed and discussed.

1. Introduction

Studies of pulsars and binary X-ray sources provide many opportunities for testing theories of dense matter against observations. I shall concentrate on the aspects of the data which seem most directly relevant to the actual structure of neutron stars, and the effects of ultra-strong magnetic fields, omitting those which are discussed by other speakers at this meeting. Since my remarks are mainly of review nature, only a brief summary is presented.

2. Masses, etc.

2.1. MASS DETERMINATIONS

Two well-studied X-ray sources Her X-1 and Cen X-3 are believed to be spinning neutron stars in close binary systems, the emission being caused by material accreted from their companions (see Giacconi, 1974; and Rees, 1974a, b, for reviews of data and interpretation). The prospects of reliable mass determinations are obviously much better in these cases than for isolated pulsars.

The mass function $M_c^3 \sin^3 i/(M_c + M_X)^2$ can be accurately deduced from the X-ray timing data (M_X and M_c being the masses of the neutron star and its companion, respectively). However a second relationship is required before M_X itself can be determined. Even though HZ Her (the visible companion of Her X-1) is brighter than 13th mag., its radial velocity cannot be directly inferred from its optical spectrum. This is largely because X-ray irradiation heats one side of the star so severely that the spectral lines give a complicated 'weighted average' of the velocity over the surface of HZ Her, and the velocity of its centre of mass cannot be readily unravelled. Very recently, however, Middledich and Nelson (1974) claim to have estimated the mass of Her X-1 with 10% precision by an ingenious technique. Weak optical pulsations are sometimes observed from the Her X-1–HZ Her system, and are believed to be caused by gas, heated by the X-rays, whose cooling time is $\lesssim$ the ~ 1.24 s pulse period. By comparing the Doppler effects for the optical and X-ray pulses, one can infer where in the system they originate. It is argued that HZ Her is overflowing its Roche lobe, and that these optical pulses come from the vicinity of the inner Lagrangian point. This then provides a second relationship between M_X and M_c; and, if $i \simeq 90°$, yields the result that $M_X \simeq 1.3\ M_\odot$. This estimate is important because (as discussed below), some interpretations of the complex long term variability of Her X-1 constrain the mass to be in a certain range.

The mass of the recently discovered binary pulsar (Hulse and Taylor, 1975) can, in principle, be determined even if its companion remains undetectable. This however will need precise observations spread over a long enough interval to allow the orbital position angle ω (advancing at $\sim 4.2°$ per year) to change significantly; and will also need a better understanding of the other effects – both Newtonian and relativistic – affecting the orbit.

The maximum permitted mass for a slowly rotating neutron star is estimated (on the basis of general relativity) to be $\sim 1.7\ M_\odot$ (Pandharipande, 1974), and is of course sensitive to the equation of state at high densities. It would plainly be important to know more masses, in order to decide whether this upper limit is tenable. This could be regarded as a test either of general relativity, or of the physics of dense matter – the emphasis depending, presumably, on one's relative confidence in the validity of these two theories! It is possible that several other X-ray sources in the UHURU catalogue involve neutron stars in binary systems, in which case further mass estimates may be made in the next few years.

2.2. Moments of Inertia

If pulsars derive their power from rotational kinetic energy, then one can in principle set a lower limit to the moment of inertia I, if Ω and $\dot{\Omega}$ are known. It was indeed the order-of-magnitude agreement between $|I\Omega\dot{\Omega}|$ for the Crab pulsar and the estimated power supply into the Nebula which helped to clinch the case for the rotating neutron star pulsar model. However it is not possible to derive any firm constraints on the pulsar parameters. The distance to the Crab Nebula is in any case uncertain by a factor ~ 1.5. The only convincing lower limit to the present-day power input into the Nebula is that associated with the short-lived electrons responsible for the X-ray and optical continuum. (Even this inference is not watertight – these electrons could conceivably be accelerated in the Nebula itself, their power deriving from magnetic fields built up in the first ~ 100 yr after the supernova.) The optical and X-ray energy could be supplied by a neutron star of almost any mass. If one makes allowances for 'inefficiency factors' in the electron acceleration, or assumes that energy is being supplied in other forms – magnetic fields, bulk kinetic energy, etc. – at the mean input rate required over the lifetime of the Nebula (Trimble and Rees, 1970) then the implied lower limit of $\sim 4 \times 10^{44}$ gm cm^2 on I would exclude low-mass neutron stars. But the uncertainties are so great that this constraint perhaps cannot be taken seriously.

In the case of old pulsars, the only power output for which we have any evidence is that associated with the radio pulses themselves. Moreover the distances are uncertain, and we do not know what allowances to make for possible beaming. The only pulsar for which an interesting limit on I can conceivably be set is JP 1953, which has $|\Omega/\dot{\Omega}| \gtrsim 10^{10}$ yr.

3. Magnetic Fields

Surface magnetic fields of $\sim 10^{12}$ G are a feature of almost all pulsar models, and are also invoked in models of Her X-1 and Cen X-3 which involve accretion onto neutron stars. It is important to note that these estimates refer to a dipole component: there is no direct evidence against a *smaller-scale* surface field which is even stronger, nor against a stronger toroidal field *inside* neutron stars.

Fields of this strength cause the surface layers to form an anisotropic lattice, in which the ions (e.g., Fe^{56}) and inner electrons form themselves into chains, surrounded by an outer sheath of electrons. The chains attract laterally, and in a field 2×10^{12} G the lattice spacing would be $\sim 10^{-9}$ cm (implying densities $\sim 10^4$ gm cm^{-3}); the binding energy per ion would be ~ 14 keV and the Fermi energy of the sheath electrons ~ 750 eV.

Ruderman and Sutherland (1975) suggest that the entire pulsar phenomenon depends crucially on the fact that the surface electric fields predicted by the 'unipolar inductor' mechanism cannot attain the $\sim 10^{12}$ V cm^{-1} needed in order to extract ions from this lattice. (The maximum field available, even if the potential gradients are concentrated near the star rather than in the outer magnetosphere, is $\sim 10^{11}(\Omega/2\pi)$ V cm^{-1}, Ω being the pulsar's angular velocity.) Nor can the surface of pulsars (except, perhaps, the Crab pulsar) be hot enough for thermal emission of ions. For this reason, the current circuit is instead completed by an *inward*-flowing stream of *electrons*, electron-positron pairs being created in the magnetosphere by interaction of γ-rays with the strong magnetic fields. The positrons stream outward and escape from the pulsar. The γ-rays themselves result from curvature radiation by the electrons and positrons. Cascades of electron-positron pairs are produced in localized 'sparks', in terms of whose properties Ruderman and Sutherland (following earlier ideas of Sturrock, 1971) explain the 'marching sub-pulse' phenomenon. The coherent radio emission is attributed to plasma oscillations excited by two-stream instabilities. (If these ideas are correct, they incidentally imply that pulsars could be a source of primary cosmic ray positrons: indeed the dominance of electrons in the cosmic radiation – and the fact that the observed positrons can all be explained as secondaries – may set some constraints on the quantitative details of the Ruderman-Sutherland scheme.)

The value of $\dot{\Omega}$ has now been measured for about 80 pulsars. The value of $|\dot{\Omega}/\Omega^3|$ is a measure of the surface magnetic field. It might eventually be possible to decide whether pulsars eventually 'die' because their fields decay, because electrons cannot be extracted from the crust when Ω falls below some critical value, or because core-crust coupling breaks down. At the moment, however, one cannot reliably decide between these (and many other) possibilities.

Strong magnetic fields enhance the thermal cooling rate for neutron stars. Moreover, a magnetized neutron star radiating its internal energy would be hotter at the magnetic poles, and would thus appear periodic if it were rotating. Even though the binary X-ray sources are deriving their energy from accretion, the emission probably occurs near the magnetic poles. The Larmor energy is $h\nu_L \simeq 11.6(B/10^{12}\ \text{G})$ keV, and so the magnetic field is likely to be strong enough to cause high polarization (modulated each rotation period) of the observed X-ray pulses. Both linear and circular polarization may occur.

4. Rigidity, etc.

The rigidity of a neutron star's crust (and the much higher rigidity of the solid core, if this exists) have been invoked to explain various peculiarities revealed by the observations. For a detailed discussion of these effects, see Pines *et al.* (1973) or Lamb (1975).

4.1. SEISMIC ACTIVITY

The 'restless' behaviour of the Crab pulsar has been attributed to 'starquakes', and a detailed theory of this phenomenon has been developed by the Illinois group in particular. The 'glitch' observed in 1969, for which $\Delta\Omega/\Omega \simeq 4 \times 10^{-9}$, is interpreted as a sudden change in the oblateness; and the smaller continual random 'jitter' in period is considered to result from 'microquakes', in which elastic strain associated with a misalignment of the rotation axis and axis of symmetry is released. The Crab timing data can all apparently be explained in terms of 'crustquakes'; and it is then possible to infer the crust rigidity, the relative moments of inertia of the crust and the superfluid core, etc. The elastic strain is rebuilt after each macroquake primarily by the reduction in centrifugal effects as the pulsar slows down. However, there are other possibilities – for instance, an initial small-scale surface field of $\geqslant 10^{12}$ G would decay owing to the imperfect conductivity, causing strain to develop. However it does not seem that alternative theories (e.g., those involving imperfect coupling between crust and superfluid, or even plasma effects in the magnetosphere) can yet be excluded, though these have not received such close theoretical scrutiny. It would be valuable to have some independent evidence for seismic activity: for example, might there not be a burst of thermal X-rays or γ-rays if a seismic event dissipated energy near the stellar surface?

The large ($\Delta\Omega/\Omega \simeq 10^{-6}$) glitch observed in the Vela pulsar cannot plausibly be explained as a crustquake, except possibly in the case of a neutron star of very low mass, large radius, and thick crust. But it could be due to sudden release of energy in a solid core (which would require, in most theories, a mass $\gtrsim 1\ M_\odot$).

If the Crab pulsar had a solid core, its deviation from axisymmetry could be large enough for gravitational radiation to be the dominant energy loss, which would invalidate any *upper* limit that might be set on I from estimates of the energy budget in the Crab nebula. If such gravitational waves could be detected and their polarization measured (as might be possible by exploiting resonance techniques) this would permit the orientation of the rotation axis to be determined – information unobtainable by other methods.

4.2. WOBBLE

No evidence has been found for free precession in pulsars. However, the 35-day periodicity of the 'extended lows' in Her X-1 (see Giacconi, 1974) has been attributed to a 'wobble' which either causes a beam to precess out of our line of sight, or else modulates the accretion rate. (In the former case an amplitude $\gtrsim 30°$ would be required; but a smaller effect might be sufficient to control an 'accretion gate'.) The free precession frequency depends both on oblateness and on rigidity. If Her X-1 has a solid core, a wobble period of 35 days can readily be explained; but if there is no solid core, then a period as short as 35 days can be obtained only for a low mass (thick crust) neutron star unless the critical strain angle for fracture is unexpectedly high. The relevant theory of this process is reviewed by Lamb (1975), who also discusses how the accretion torque, modulated over the 35 day period, could maintain a big wobble amplitude despite the damping associated with friction between the solid and liquid portions of the star.

References

Giacconi, R.: 1974, in 'Astrophysics and Gravitation', *Proc. 16th Solvay Conference*, Editions de l'Université de Bruxelles, Brussels, p. 27.

Hulse, R. A. and Taylor, J. H.: 1975, *Astrophys. J. (Letters)* **195**, L51.

Lamb, F. K.: 1975, *Proc. 7th Texas Conference*, (in press).

Middledich, J. and Nelson, J.: 1975, *Proc. 7th Texas Conference*, (in press).

Pandharipande, V. R.: 1974, in 'Astrophysics and Gravitation', *Proc. 16th Solvay Conference*, Editions de l'Université de Bruxelles, Brussels, p. 177.

Pines, D., Shaham, J., and Ruderman, M. A.: 1973, in C. J. Hansen (ed.), *Physics of Condensed Matter*, D. Reidel Publ. Co., Dordrecht-Holland, p. 89.

Rees, M. J.: 1974a, in 'Astrophysics and Gravitation', *Proc. 16th Solvay Conference*, Editions de l'Université de Bruxelles, Brussels, p. 97.

Rees, M. J.: 1974b, in G. Contopoulos (ed.), *Highlights of Astronomy*, Vol. 3, D. Reidel Publ. Co., Dordrecht-Holland, p. 89.

Ruderman, M. A. and Sutherland, P. S.: 1975, *Astrophys. J.* **196**, 51.

Sturrock, P. A.: 1971, *Astrophys. J.* **164**, 529.

Trimble, V. L. and Rees, M. J.: 1970, *Astrophys. Letters* **5**, 93.

NEUTRON STAR CORES

R. G. PALMER*

Physics Dept., Princeton University, Princeton, N.J., U.S.A.

Abstract. Current theories, and the astrophysical implications, of the nature of high density neutron star matter are reviewed. Suggestions are made for a compromise between the alternatives of neutron crystallization and pion condensation.

1. Introduction

When neutron stars were first conceived in the 1930's, they were envisaged as a neutron-proton-electron Fermi liquid, it being realized that the nucleon Fermi pressure – and internucleon forces – could withstand the gravitational pressure of a suitably collapsed body. Our present picture of neutron stars is in essence unchanged, but much detailed structure and dynamics has emerged, and neutron star models are well on the way towards explaining the observed behaviour of pulsars. There is by now satisfactory agreement on the structure of neutron stars up to, and somewhat beyond, the density of ordinary nuclear matter ($\sim 3 \times 10^{14}$ gm cc^{-1}), but there is still considerable controversy as to the nature of the matter at higher densities appropriate to the core of large neutron stars. It is this core region that we discuss here.

For orientation, Figure 1 shows schematically the major structural regions of a typical neutron star. The surface starts abruptly with a thin layer of atoms, greatly modified by the intense magnetic field, before giving way to a crust region in which electrons are no longer bound to individual nuclei, but form a nearly uniform degenerate gas bathing a lattice of nuclei. The nuclei are at first Fe^{56}, but with increasing density the higher electron Fermi pressure shifts β-equilibrium more and more towards neutrons, leading to increasingly neutron-rich nucleides. Since $n-n$ forces are less attractive than $n-p$ forces, the additional neutrons are progressively less strongly bound in nuclei, and are eventually not bound at all. The excess neutrons then form a delocalized neutron fluid, in the inner crust region of Figure 1. As the density is further increased the fluid and nucleide densities converge, finally blending into a uniform neutron-proton-electron fluid. Both the neutrons and the protons are probably superfluid in this region. Further details of the outer regions of neutron stars may be found in the reviews of Ruderman (1972) and Canuto (1974).

The boundary between crust and superfluid represents a density of about 2.5×10^{14} gm cc^{-1}. The fluid may extend to the centre of the star, or a further high density

* Supported in part by NSF Contract GH 40474.

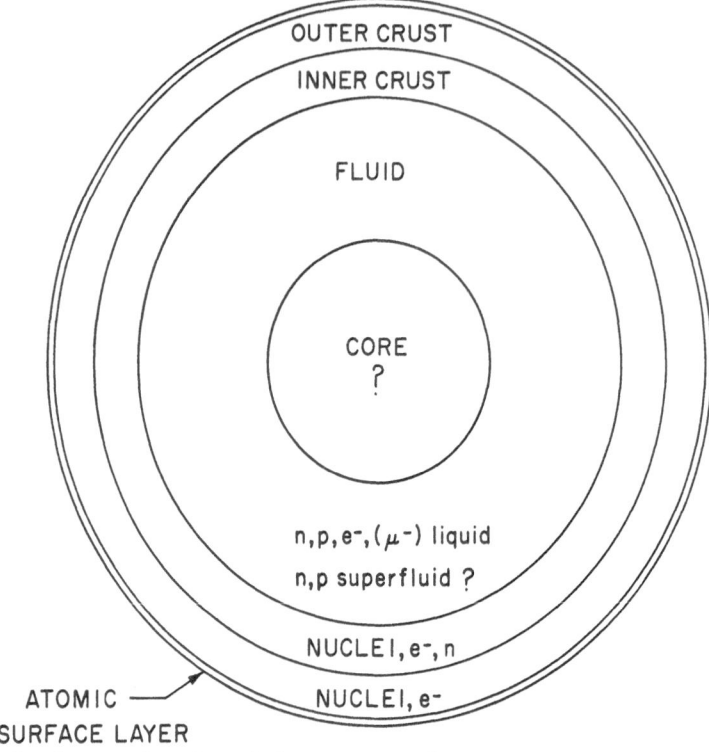

Fig. 1.   Cross section of a typical neutron star, of mass $\sim M_\odot$ and radius ~ 10 km.

transition may occur, leading to a separate 'core' region. In this paper we discuss the various models for this region and attempt to resolve some of the controversy between the alternatives. We first point out in Section 2 the relevance of the core to experimental observations and to astrophysics in general. In Section 3 we review the crystallization theories, which propose a solid core. The alternative of pion condensation is discussed in Section 4, in which we also suggest the possibility of isospin-wave condensation. This may lead to a unified view of crystallization and pion condensation.

2. Implications of Core Structure

Knowledge of the structure of the core region of neutron stars is surprisingly important in understanding the dynamics of pulsars, and for evaluating other phenomena attributed to neutron stars. In this section we review those aspects of neutron star behavior that depend on core structure.

2.1. MAXIMUM MASS

The mass-radius relationship for a neutron star can be obtained by integrating the Tolman-Oppenheimer-Volkoff hydrostatic support equations

$$\frac{dP}{dr} = -\frac{G(\varrho + P/c^2)(m + 4\pi r^3 P/c^2)}{r(r - 2Gm/c^2)},$$

$$\frac{dm}{dr} = 4\pi r^2 \varrho.$$

Solution of the equations requires knowledge of the equation of state $P(\varrho)$. Stability is not ensured by the above equations, but can be tested by an ancillary condition (Chandrasekhar, 1964). This procedure leads to a maximum allowable mass M_{max}, which is quite sensitive to the high density character of $P(\varrho)$, corresponding to the core region. Estimates of M_{max} generally lie in the range $0.5–1.5 \times M_\odot$, and M_{max} is almost certainly less than $3.2\ M_\odot$ (Rhoades and Ruffini, 1974). Models involving high density transitions to a solid state, or especially to a pion condensed state, will have somewhat softer equations of state, implying a smaller maximum mass. A collapsed object with $M > M_{max}$ must be a black hole, and thus knowledge of M_{max} is important in observationally distinguishing neutron stars and black holes. In the X-ray binary Cyg X-1 there is a collapsed component of mass $> 3\ M_\odot$. This is in all probability a black hole, but a better understanding of neutron star cores could clinch this judgement and possibly aid in resolving more borderline cases yet to be discovered.

2.2. COOLING RATE

Neutron stars can cool by both photon and neutrino emission. In the earlier, hotter, stages of their evolution ($> 10^8$ K) neutrino emission is the most important effect, occurring through URCA processes, plasmon decay, and neutrino bremsstrahlung, typified by

$$n + n \rightarrow n + p + e^- + \bar{\nu},$$
$$\text{plasmon} \rightarrow \nu + \bar{\nu},$$
$$e^- + n \rightarrow e^- + n + \nu + \bar{\nu},$$

respectively. When lower temperatures are reached the surface photon opacity is reduced and X-ray emission becomes the dominant process.

The occurrence of pion condensation in neutron star cores could provide additional neutrino loss processes, such as

$$n + \pi^- \rightarrow n + e^- + \bar{\nu},$$

which are estimated (Bahcall and Wolf, 1965) to be very efficient in cooling the star, probably to $< 10^6$ K within a few weeks after formation for an isolated neutron star. The high efficiency is essentially due to the boson nature of pions, which are thus less restricted by the exclusion principle than fermions. However, condensed state pions are in chemical equilibrium with nucleons via $n \leftrightarrow p + \pi^-$, and the pionic cooling rate is probably considerably less than the original estimates (Baym and Flowers, 1974).

The temperature has little influence on the structure of a neutron star, it being essentially zero on a nuclear energy scale (1 MeV $\sim 10^{10}$ K), but is of importance in some dynamical processes (such as the coupling to superfluid neutrons, mentioned below) and in soft X-ray emission.

2.3. GLITCH RECOVERY

The Crab and Vela pulsars have both shown sudden decreases in period, or *glitches*, followed by partial recovery over a period of days (Crab) or years (Vela). Figure 2 shows the gross features of a typical glitch. There are a number of models for the origin of the glitch, but however the initial spin-up AC is produced the relaxation CD may be explained by a simple two component theory. The neutron star is envisaged as consisting of two rotating components which are only very weakly coupled to each other. One component is the neutron superfluid, while all other particles form the second component. Assuming that both neutrons and protons are superfluid (in the 'fluid' regime of Figure 1), the two components are only coupled by the magnetic interaction of electrons with neutrons in vortex cores, and neutron-vortex/proton-vortex interactions (Baym *et al.*, 1969). If the original spin-up occurs only in the second component, and the two components then relax to a common angular velocity, simple dynamics shows that $Q = I_n/I$, where Q is defined by Figure 2 and $I_n(I)$

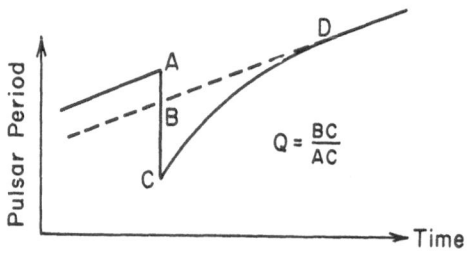

Fig. 2.   Idealized glitch.

is the moment of inertia of the neutron superfluid (entire star). Observations give $Q \sim 0.95$ for Crab and $Q \sim 0.15$ for Vela. The value for Crab is in satisfactory agreement with theory for a fairly light neutron star ($\sim 0.3\ M_\odot$) in which the fluid region occupies most of the star and extends to the centre. However, Vela is not so obliging; in the absence of any high density transition $Q \sim 0.15$ implies an extremely light neutron star ($\sim 0.1\ M_\odot$), consisting almost entirely of crust, whose genesis is very hard to explain in a supernova event. Pines *et al.* (1972) have suggested the alternative of Vela being a heavy neutron star ($\sim 0.7\ M_\odot$) consisting largely of a solid core with only a thin superfluid layer near the surface. This theory appears to fit the available data, but reliable estimates of the fluid-core transition density are still needed.

2.4. COREQUAKES

Many suggestions have been made for the source of the glitches in Crab and Vela, including planetary perturbations, accretion of massive objects, plasma instabilities in the magnetosphere, hydrodynamic instabilities in the fluid region, and starquakes. A review of these theories may be found in Shaham *et al.* (1973). The most promising explanation seems to be the starquake theory, which postulates a sudden cracking of the rigid crust to decrease the overall oblateness as the rotation rate decreases. For Crab the fractional decrease in period is $\sim 10^{-9}$, which is consistent with the expected strength of the crust, but Vela shows much larger glitches (Δ period/period $\sim 10^{-6}$) which imply an improbably large accumulation of strain in the crust. This problem can be avoided if Vela has a solid core, which could also undergo quakes (Pines *et al.*, 1972). The core can easily resist the necessary stresses, particularly if it retains much of its original oblateness and releases only a little in each quake. This hypothesis is also consistent with the observed frequency of glitches in Vela (~ 2 years between events), whereas crustquakes alone predict an average interval several orders of magnitude too large.

2.5. WOBBLES

If large neutron stars possess solid cores, oblate due to rotation, some new dynamical modes are to be expected. In particular a precession, or wobble, of frequency $\Omega_P \sim \frac{3}{2}\varepsilon\Omega$ (where ε is the core oblateness, Ω is the rotation frequency) is predicted by Pines and Shaham (1972). Such a wobble has been postulated as the cause of the observed 35 day high-low cycle (~ 12 days on, ~ 23 days off) of the Her X-1 X-ray binary pulsar (Pines *et al.*, 1973). The X-rays are thought to originate in a hot spot near the magnetic pole of neutron star, which is heated by accretion of matter from the binary companion. The accretion mechanism could conceivably be sufficiently marginal to be turned on and off by a neutron star wobble. No simpler theory seems to fit the facts. To obtain a wobble period of 35 days an oblate solid core is essential; the free precession frequency would be several orders of magnitude higher without such a core.

Wobbles might also be observable as periodic fluctuations in pulsar intensities. Lastly, Dyson (1972) has suggested that a wobbling oblate core would be a very efficient source of gravitational radiation.

3. Neutron Crystallization

The fluid region of a neutron star is a highly quantum interacting Fermi liquid, effectively at zero temperature. Another such system is He^3 near $T=0$ K. Helium solidifies under a reasonable pressure (~ 30 atm) and it is natural to ask whether neutron matter might also solidify under pressure, leading perhaps to a crystalline core in neutron stars. Controversy still rages over this question, and although most (though not all) authors agree that crystallization *does* finally occur, estimates of the transition density and pressure are in considerable disagreement.

One general argument is of interest. Consider the case of helium, in which the repulsive part of the potential may be represented as $V(r) \propto r^{-12}$. At sufficiently high densities, where the attractions may be ignored, the total potential energy therefore rises as ϱ^4, while the kinetic energy is roughly proportional to $\varrho^{2/3}$ ($\varrho^{1/3}$ relativistically). As the density ϱ is increased the potential energy must become dominant. Since the potential energy is a function of *position* coordinates only, it can be minimized in an ordered, or solid, phase. Hence helium *must* solidify at sufficiently high density (in fact it does so long before the foregoing argument becomes compelling). The same argument can be applied to any quantum system that possesses

(a) a sufficiently repulsive core ($V(r) \propto r^{-\alpha}$, $\alpha > 2$),

(b) a pair potential that is a function of particle separation only.

While nuclear matter almost certainly satisfies (a), there are tensor forces, angular-momentum (L) dependent forces, and three body forces that violate (b). Unless these can be considered as unessential details, we cannot definitely infer crystallization at any density. This reasoning shows the importance of including the full internucleon interaction in any realistic model.

We now review some of the detailed calculations aimed at computing the crystallization density, if any. The models and results are summarized in Table I.

TABLE I

Crystallization theories

Authors	Method	Approximate crystallization density (gm cc$^{-1} \times 10^{14}$)
Banerjee *et al.* (1970)	Debye model	–
Anderson and Palmer (1971)	Corresponding states	3.5
Clark and Chao (1972)	Corresponding states	4
Østgaard (1973)	Quantum crystal/Brueckner	None below 20
Canuto and Chitre (1974)	Quantum crystal/T-matrix	16
Coldwell (1972)	Variational (one particle functions)	5
Nosanow and Parish (1973)	Variational (Jastrow – Gaussian)	4.5
Pandharipande (1971, 1973)	Variational (Jastrow – Gaussian)	None below 60
Schiff (1973)	Expansion about Bose hard spheres	30
Clark (1972)	Hard spheres/Corresponding states	10
Iachello *et al.* (1974)	Quarks	6

The first suggestion of a neutron crystal was made by Banerjee *et al.* (1970), who used a harmonic Debye model. This approach is well known to fail in highly quantum systems, in which the anharmonic terms are essential (the same method gives imaginary phonon frequencies for solid helium). The predicted equation of state for the solid was not compared to a liquid equation of state, and so no crystallization density was computed.

Several authors (Anderson and Palmer, 1971; Palmer and Anderson, 1974; Clark and Chao, 1972) have attempted to exploit the known solidification properties of helium to estimate those of neutron matter. The method employed is the quantum theory of corresponding states (de Boer, 1948) which allows one to interpolate and extrapolate amongst systems having similarly shaped pair potentials, differing only in length and energy scales. Further allowance is made for the differences in shape between typical nuclear and rare gas potentials. Clark and Chao use a parameterized nuclear potential as starting point, choosing the parameters to give the same results as a full Reid potential in a variational calculation. Palmer and Anderson avoid assuming an effective potential from the outset, instead deriving one empirically by fitting corresponding states predictions to the known properties of nuclei. The two methods predict similar crystallization densities of about 3–4×10^{14} gm cc^{-1}. Neither approach is able to take fully into account the complexities of realistic nuclear interactions. There are also considerable uncertainties due to the lack of adequate extrapolation data from rare gas systems, especially with respect to the solidification properties themselves; only He^4 and He^3 have positive solidification pressures at absolute zero, and they differ in their statistics (Nosanow *et al.*, 1974, find a considerable Bose/Fermi difference in the solidification properties of otherwise similar systems).

Canuto and Chitre (1974) and Østgaard (1973) have applied quantum crystal theories to the problem of a neutron crystal. Østgaard uses a modified Brueckner theory with a simple central potential, and finds no crystallization below 2×10^{15} gm cc^{-1}. Canuto and Chitre apply a more elaborate T-matrix approach and are able to treat the full Reid potential. In their model neutron matter solidifies at a density of 1.6×10^{15} gm cc^{-1}.

Variational methods have been used by Coldwell (1972), Pandharipande (1973), and Nosanow and Parish (1973). Coldwell constructs orthogonal single particle wave functions that are capable of describing either a solid or a gas phase. Making partial allowance for the spin and L-dependence of the potential, he finds crystallization at about 5×10^{14} gm cc^{-1}. Nosanow and Parish take trial functions of the Jastrow-Gaussian form, which are able to model short range correlations. They analyze the properties of a neutron system having a simple two-component potential, using Monte-Carlo integration and cluster expansion, and find crystallization at about 4.5×10^{14} gm cc^{-1}. Pandharipande also uses a Jastrow-Gaussian wave function, but imposes constraints on the two-body correlation part to suppress the n-body $(n > 2)$ cluster contributions to the energy. This technique enables him to calculate – and minimize – the energy directly without resort to Monte-Carlo methods. Spin and L-dependence of the potential is fully allowed for, but tensor forces are omitted. No indication of crystallization is obtained below 6×10^{15} gm cc^{-1}.

It seems probable that crystallization – if it occurs – is governed mainly by the hard repulsive core of the internucleon interaction. Schiff (1973), and Clark (1972) have exploited this by basing their estimates on the properties of hard sphere systems. Schiff uses numerical computations of the equations of state of Bose hard sphere fluids and solids, and then treats an assumed attractive potential (central, one component) by perturbation theory. Lastly the Bose/Fermi difference is incorporated by Wu-Feenberg expansion. A result of 2.4–3.6×10^{15} gm cc^{-1} is found for the crystallization density. Clark applies corresponding states theory on the assumption that solidification in all substances may be related to an effective hard sphere system, and finds crystallization· at about 1×10^{15} gm cc^{-1}. Doubt is cast on both these procedures by the work of Cochran and Chester (1973), who show that repulsive Yukawa interactions are not at all well represented by 'equivalent' hard spheres.

Finally, Iachello *et al.* (1974) have proposed a crude quark model in which nucleons are built from one light and one heavy quark bound by exchange of a massless gluon. At a density of around 6×10^{14} gm cc^{-1} they find it energetically favourable to localize the heavy quarks (bosons) in a lattice while delocalizing the light quarks into a Fermi sea. This represents crystallization.

The great diversity of approaches described above is well matched by the spread of numerical results (Table I). One of the chief difficulties is the choice of an internucleon potential, a problem that is solved differently by each author, making comparison of methods all the more hazardous. It is not even clear what potential *should* be used if simplifications were unnecessary; even the full Reid potential, used only by Canuto and Chitre (1973), may be inadequate at the high densities considered. Further difficulties are appropriate to each method individually, and detailed criticisms of each other's work may be found in many of the papers cited above.

4. Pion Condensation

As the density of neutron star matter is increased several new particles may appear. For example muons will start to be formed when the electron chemical potential exceeds the muon rest mass, although this has little effect on nuclear properties other than slowing the trend towards a smaller proton fraction. Similarly, hyperons (Σ^-, Λ°, etc.) will appear when the neutron chemical potential becomes sufficiently high. This will certainly soften the equation of state, but it is difficult to treat in detail because of our ignorance of hyperonic interactions. Pandharipande (1971) has computed a liquid phase equation of state including hyperons (assuming a universal interaction between all baryons), and Canuto and Chitre hope to incorporate hyperons in their crystal model.

Pions may also be formed at high density, and this possibility has been the concern of much recent work. If pions could be treated as non-interacting particles they would be produced via $n \rightarrow p + \pi^-$ as soon as $\mu_n - \mu_p > m_\pi c^2 \approx 140$ MeV (the μ's being chemical potentials). From β-equilibrium $\mu_n - \mu_p = \mu_e$, and the electron chemical potential μ_e exceeds 140 MeV at about 6×10^{14} gm cc^{-1} in typical fluid models. But pions *do* interact strongly with nuclear matter. The S-wave interaction alone would probably suppress pion formation for ever; the energy of a π^- interacting only in the S-wave may be approximated (Bethe, 1971) by

$$\omega_{\pi^-} \sim 140 + 219(\varrho_n - \varrho_p) \text{ MeV} \qquad (\varrho\text{'s in nucleons fm}^{-3}),$$

and this quantity is always greater than μ_e in present calculations. The pion-nucleon P-wave interaction can, however, lower the pion energy again at finite momentum, reopening the question of pion production. If pions do appear it is advantageous for them, being bosons, to occupy a single macroscopic mode, thus forming a finite-k condensate.

Two basic approaches have been used to study the possibility of pion condensation. One, adopted by Sawyer and Scalapino (1973), Sawyer and Yao (1973), and Kogut and Manassah (1972), is to write down a many-body Hamiltonian for an interacting pion-nucleon system, and attempt to find the state which minimizes the energy at a given density. The problem may be made tractable by replacing the pion annihilation and creation operators by their mean field values ($a_k \to \sqrt{n_k}$, etc.), and by restricting attention to only one pion mode. Constraints are needed for conservation of charge, baryon number, and total momentum. In their simplest model Sawyer and Scalapino find a second order phase transition at 4.5×10^{14} gm cc^{-1}, beyond which pions start to occupy a single mode at $k = 0.9$ fm^{-1}. Corrections due to nucleon-nucleon interactions, S-wave pion-nucleon interactions, pion-pion interactions, and the presence of non-condensed pion modes can be included, but do not modify the basic conclusion; taking all effects into account, Sawyer and Scalapino predict a pion condensation threshold between 0.8 and 1.7×10^{15} gm cc^{-1}. Kogut and Manassah follow a very similar approach. Sawyer and Yao study a standing wave pion condensate, finding considerable dependence on nucleon-nucleon interactions.

The other basic approach is to look for the elementary excitations of neutron star matter. Poles in the interacting pion propagator define the pion energy-momentum relation $\omega(k)$, and we look for a density at which $\omega(k)$ goes soft, implying pion condensation. This is the viewpoint of Migdal (1973a, b), Barshay et al. (1973), and Palmer (1973). Baym and Flowers (1974) use an approach intermediate between those described here, and pay particular attention to the condensation condition; $\omega(k) = \mu_e$ is necessary but not always sufficient for condensation (see also Au et al., 1974).

We now describe an example of the second approach, as developed by the author, P. W. Anderson, E. Tosatti, N. Itoh, and M. A. Alpar; full details may be found in Anderson et al. (1975). The work of Migdal and of Barshay et al. is very similar. The calculation shows some new features, suggesting in particular a modified form of pion condensation which we call *isospin wave* condensation. We begin by writing the full pion propagator $D(\omega, k)$ in terms of the free propagator $D_0(\omega, k)$ and a polarization function $\Pi(\omega, k)$ as

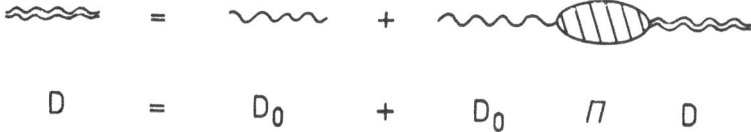

The poles of $D(\omega, k)$ locate the real pion, whose dispersion is, therefore, given by

$$\omega^2(k) = m_\pi^2 + k^2 + \Pi(\omega, k)$$

if we absorb some normalization factors into Π. We have considered various approximations for the polarization function Π, which represents all possible irreducible scattering processes that can be undergone by a pion. The first approximation (Palmer, 1973) was to consider only excitation of $\Delta(1238)$ states, which form the dominant resonances in low energy pion-nucleon scattering,

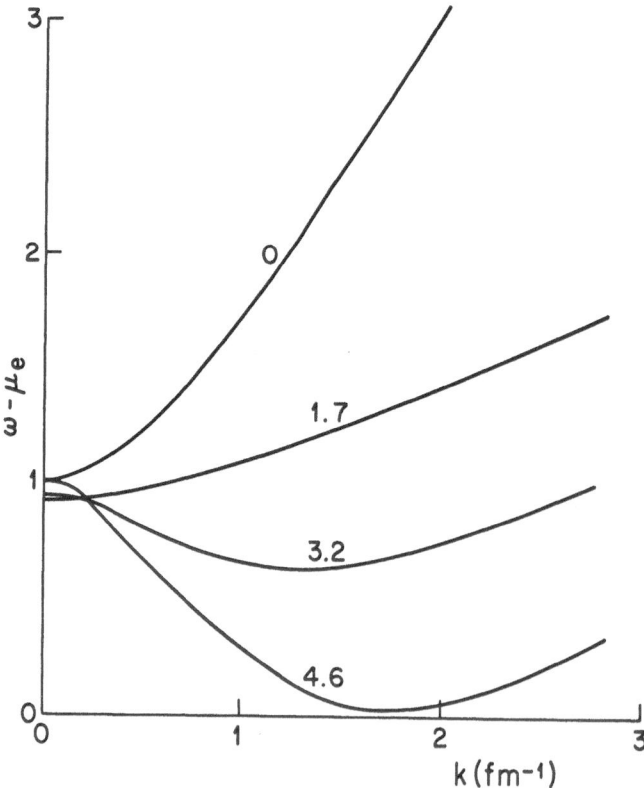

The factors of $\frac{1}{3}$ are for isospin projection. This expression was evaluated by standard methods, using $V_k = f\boldsymbol{\sigma}\cdot\mathbf{k}/\sqrt{\omega_k^0 \Omega}$ for the vertices and ignoring the finite width of the Δ-resonance. Typical results for several densities are shown in Figure 3 for a proton fraction of 5%. In this approximation we see a pion condensation threshold (for π^-) at a density of about 5×10^{14} gm cc^{-1}, with a condensate k of 1.75 fm^{-1}. The con-

Fig. 3. The π^- energy-momentum relation at several densities, including only the Δ-resonance channel and S-wave scattering. $\omega - \mu_e$ is in units of $m_\pi c^2$, densities are in units of gm cc$^{-1}\times 10^{14}$. The proton fraction is 5%.

densation condition for π^- is not $\omega(k)=0$, but $\omega(k)=\mu_e$, since one electron must be eliminated on introduction of a π^- in order to maintain charge conservation. In fact all energies should be referred to a ground state μ_e above the neutral ground state for consistent treatment of negative excitations. For π^+ the sign of this shift is reversed.

A more realistic approximation for Π includes the possibility of neutron-proton polarization

Following Migdal (1973a) we allow for nucleon-nucleon interactions by renormalizing the vertex as

where the interaction Γ is obtained from Fermi-liquid theory. Unfortunately, the Fermi-liquid parameters are not very accurately known, and this is reflected in a considerable quantitative uncertainty in our results, but does not affect our qualitative conclusions. It also pinpoints the strong dependence of pion condensation on the short range nucleon-nucleon interaction, which has been noted by several authors. The question of pion condensation will probably not be settled quantitatively until we have a more complete understanding of this interaction.

The improved model shows a surprising new feature; additional branches of the pion spectrum appear. In Figure 4 we plot our calculated spectrum for π^+ at a density of 1.7×10^{14} gm cc^{-1}. The central loop first appears at lower density entirely within the $p\bar{n}$ particle-hole pair spectrum (where it is heavily damped; the imaginary part of Π was ignored in the calculation), and grows until it emerges to cross the $\omega(k)=\mu_e$ axis as shown. A new soft mode is indicated, occurring considerably before the π^- branch nears condensation. The new mode must have the same quantum numbers as the pion, and we identify it with an isospin wave (ISW), analogous to a conventional spin wave.

The model is crude, and many details call for study, but the condensation of a positively charged ISW suggests the formation of a proton lattice in neutron star cores. As the amplitude of the positive instability grows, negative charge must also be supplied, and it seems that most of this charge must come by removing protons from the Fermi sea. It is likely that this process runs to completion and that all of the positive charge resides in the ISW; one simply has a proton lattice with no free proton Fermi surface. Crystallization of the protons may also trigger crystallization

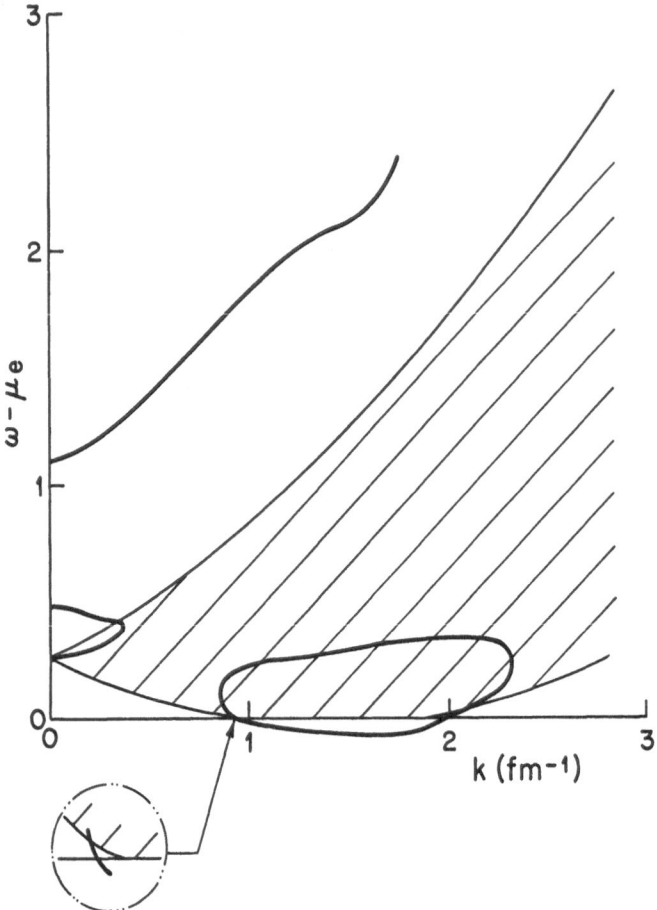

Fig. 4.   The π^+ energy-momentum relation at a density of 1.7×10^{14} gm cc^{-1} (with 5% protons), including nuclear polarization as well as Δ-resonance and S-wave scattering. The parameters used in the calculation were those of Migdal (1973a). $\omega - \mu_e$ is in units of $m_\pi c^2$. Also shown is the proton/neutron-hole pair excitation region (shaded).

of the neutrons, leading to a completely solid phase such as discussed in the last section. Migdal (1974) has come to somewhat similar conclusions, predicting a laminated structure for nuclear matter in the presence of a pion condensate.

5. Conclusions

This review has indicated that understanding of high density neutron star matter is still far from complete. Neutron crystallization and pion condensation both seem very serious contenders for such matter. A possible compromise between these effects is indicated by some detailed studies of pion condensation, as outlined above, which suggest that the first pion-like mode to condense is a positively charged isospin wave. Condensation of this mode may well imply crystallization, thus unifying the two possibilities.

Two further considerations (Palmer *et al.*, 1973) suggest that pion condensation and neutron crystallization may be mutually advantageous. Firstly, if a pion standing wave appeared in the system it would set up a periodic potential for nucleons. The density and spin assignments seem compatible with crystallization with this periodicity, though perhaps only in one dimension, forming a laminated structure. Secondly, formation of such a commensurate lattice would enable saving of some recoil kinetic energy in pion-nucleon interactions by utilizing Umklapp processes to remove unwanted momentum.

Much further work is needed to decide whether this compromise – or any other – is really viable. Neutron star cores should continue to provide stimulating problems in diverse fields for a considerable time hence.

References

Anderson, P. W., Itoh, N., Tosatti, E., Alpar, M. A., and Palmer, R. G.: 1975, to be published.
Anderson, P. W. and Palmer, R. G.: 1971, *Nature Phys. Sci.* **231**, 145.
Au, C. K., Baym, G., and Flowers, E.: 1974, preprint.
Bahcall, J. N. and Wolf, R. A.: 1965, *Phys. Rev.* **140**, B1452.
Banerjee, B., Chitre, S. M., and Garde, V. K.: 1970, *Phys. Rev. Letters* **25**, 1125.
Barshay, S., Vagradov, G., and Brown, G. E.: 1973, *Phys. Letters* **43B**, 359.
Baym, G. and Flowers, E.: 1974, *Nucl. Phys.* **A222**, 29.
Baym, G., Pethick, C. J., and Pines, D.: 1969, *Nature* **223**, 673 and 674.
Bethe, H. A.: 1971, *Ann. Rev. Nucl. Sci.* **21**, 93.
de Boer, J.: 1948, *Physics* **14**, 139.
Canuto, V.: 1974, *Ann. Rev. Astron. Astrophys.* **12**, 167.
Canuto, V. and Chitre, S. M.: 1974, *Phys. Rev.* **D9**, 1587.
Chandrasekhar, S.: 1964, *Astrophys. J.* **140**, 417.
Clark, J. W.: 1972, *Proceedings of the Symposium on the Many-Body Problem*, Rome, 19–23 Sept.
Clark, J. W. and Chao, N.-C.: 1972, *Nature Phys. Sci.* **236**, 37.
Cochran, S. and Chester, G. V.: 1973, Materials Science Center Report #2127, Cornell University.
Coldwell, R. L.: 1972, *Phys. Rev.* **D5**, 1273.
Dyson, F.: 1972, *Sixth Texas Symposium on Relativistic Astrophysics*, New York.
Iachello, F., Langer, W. D., and Lande, A.: 1974, preprint.
Kogut, J. and Manassah, J. T.: 1972, *Physics Letters* **41A**, 129.
Migdal, A. B.: 1973a, *Phys. Rev. Letters* **31**, 257.
Migdal, A. B.: 1973b, *Nucl. Phys.* **A210**, 421.
Migdal, A. B.: 1974, preprint.
Nosanow, L. H. and Parish, L. J.: 1973, *Ann. NY Acad. Sci.* **224**, 226.
Nosanow, L. H., Parish, L. J., and Pinski, F. J.: 1974, preprint.
Østgaard, E.: 1973, *Phys. Letters* **47B**, 303.
Palmer, R. G.: 1973, Thesis, Cambridge Univ., unpublished.
Palmer, R. G. and Anderson, P. W.: 1974, *Phys. Rev.* **D9**, 3281.
Palmer, R. G., Tosatti, E., and Anderson, P. W.: 1973, *Nature Phys. Sci.* **245**, 119.
Pandharipande, V. R.: 1971, *Nucl. Phys.* **A178**, 123.
Pandharipande, V. R.: 1973, *Nucl. Phys.* **A217**, 1.
Pines, D., Pethick, C. J., and Lamb, F. K.: 1973, *Ann. NY Acad. Sci.* **224**, 237.
Pines, D. and Shaham, J.: 1972, *Phys. Earth Planetary Interiors* **6**, 103.
Pines, D., Shaham, J., and Ruderman, M. A.: 1972, *Nature Phys. Sci.* **237**, 83.
Rhoades, C. E., Jr. and Ruffini, R.: 1974, *Phys. Rev. Letters* **32**, 324.
Ruderman, M. A.: 1972, *Ann. Rev. Astron. Astrophys.* **10**, 427.

Sawyer, R. F. and Scalapino, D. J.: 1973, *Phys. Rev.* **D7**, 953.
Sawyer, R. F. and Yao, A. C.: 1973, *Phys. Rev.* **D7**, 1579.
Schiff, D.: 1973, *Nature Phys. Sci.* **243**, 130.
Shaham, J., Pines, D., and Ruderman, M. A.: 1973, *Ann. NY Acad. Sci.* **224**, 190.

STRUCTURE OF NEUTRON STAR CORES

V. CANUTO*

Institute for Space Studies, Goddard Space Flight Center, NASA, New York, N.Y., U.S.A.

B. DATTA

Dept. of Physics, City College of the City University of New York, New York, N.Y., U.S.A.

and

J. LODENQUAI

Dept. of Physics, University of the West Indies, Kingston, Jamaica

Abstract. After reviewing the outer and central regions of a neutron star, we discuss the central region and the possibility that the core has a solid structure. We present the work of different groups on the solidification problem, suggesting that the neutron star-cores are indeed solid.

1. Introduction

Neutron stars are believed to be formed in the gravitational implosion of aged, massive stars which have used up their thermonuclear energy. Matter in such stars is so dense that the gravitational attraction is balanced by the pressure of the highly-degenerate neutrons. This requires densities of the order $\sim 10^{14}$ g cm^{-3}. Since stellar masses are comparable to the solar mass ($= 2 \times 10^{33}$ g), the radius of a neutron star may be expected to be of the order ~ 10 km. The cross-section of a typical medium-weight neutron star is shown in Figure 1.

It has been suggested that matter at high densities, such as found in neutron star cores, has a crystalline structure. After briefly describing the crust and central region of a neutron star, we discuss the core region and the possibility for a crystalline structure to exist.

2. The Crusts and the Central Region

The main constituents of the outer crust are Fe^{56} nuclei (the end point of thermonuclear burning) and a gas of free degenerate electrons, produced because of pressure ionization. These nuclei, embedded in a gas of freely-moving and relativistic electrons, are only weakly screened by the electrons, and as Ruderman (1968) first suggested, they arrange themselves in a crystalline structure.

As the density increases, the electron chemical potential also increases so that the process of inverse beta-decay becomes energetically favorable, i.e.,

$$(Z, A) + e^- \rightarrow (Z - 1, A) + \nu.$$

* Also with the Department of Physics, City College of New York.

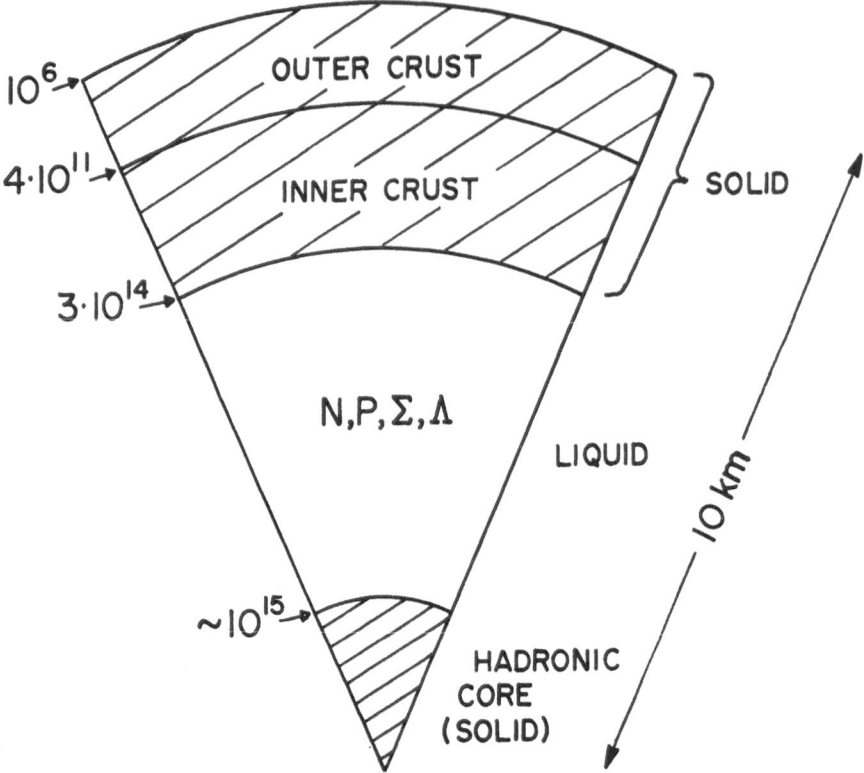

Fig. 1.   Schematic representation of a neutron star interior. The numbers on the left are the densities (in g cm⁻³) at which the several phase-transitions occur.

This process produces nuclei which are more and more neutron-rich, and at the same time stable against beta-decay, because of the presence of the degenerate electron gas. At a density $\sim 4 \times 10^{11}$ g cm^{-3}, the neutron excess reaches a point where the last neutron is no longer bound. The neutrons then start to drip out of the nuclei and the system is now composed of nuclei, electrons and free neutrons. This transition, known as the Neutron Drip Point, marks the separation of the inner crust from the outer crust. As the density increases, the protons within the nuclei find it energetically more favorable to spread around uniformly rather than be clumped together in the nuclei. The nuclei become less and less localized in space, and finally dissolve in the surrounding medium. The region beneath the inner crust, the central region, is thus composed of a mixture of three degenerate quantum liquids: neutrons, protons and electrons. It is expected that the neutrons and the protons can be in superfluid states, while the electrons remain a normal, weakly interacting relativistic plasma.

3. The Core Region and the Solidification Problem

As the density exceeds the nuclear matter density, one reaches the so-called hadronic core wherein mesons and hyperons are expected to be present in addition to neutrons, protons and electrons. The calculation of the equation of state for this quantum system of baryons is a formidable task because it needs a knowledge of the hyperon-nucleon and hyperon-hyperon interactions as well as that of a reliable many-body technique.

An important question is whether the neutron matter inside the neutron star core will turn into a quantum crystal. The problem has been investigated by several groups over the last few years, and most of them suggest that solidification of neutron matter will take place at such high densities. There is, however, no consensus as to the exact value of the solidification density which varies from 4.2×10^{14} g cm^{-3} to $(2.9 \pm 0.5) \times 10^{15}$ g cm^{-3}. The problem is complicated by the fact that one does not know, as yet, the behaviour of N-N interaction near the origin. We discuss below the work that has so far been done on this solidification problem.

The old hard-core N-N potential almost by definition leads to a solid structure at particle densities of the order of $n_B = (\frac{4}{3}\pi r_c^3)^{-1}$, because at these densities every particle feels an infinite repulsion due to every other particle around it. This leads to a localization of the particles. With soft-core N-N potentials, however, one cannot make such definitive conclusions without first performing a microscopic calculation.

The first workers who pointed out that physically the solid state could be a more preferable configuration for neutron matter than a liquid state were Cazzola et al. (1966), who obtained the equation of state by solving the Dirac equation for each nucleon, assumed to be moving in a square-well potential. Their work, however, was inconclusive in the sense that they did not actually show that a crystal structure indeed occurred. Anderson and Palmer (1971) suggested that one could draw a possible analogy between a neutron liquid and He^3 liquid (see also Palmer and Anderson, 1974), and calculate the solidification density by applying the so-called law of corresponding states. Their calculations predicted the solidification density to be 5×10^{14} g cm^{-3}. Although the analogy between neutron liquid and He^3 can be instructive, it must be pointed out that it is not entirely correct, because the stiffness of the Lennard-Jones 6–12 potential, due to the term r^{-12}, has no analog in the presently known nucleon-nucleon potentials, whose forms tend to be somewhat softer. Moreover, the Lennard-Jones potential is spherically symmetric whereas the N-N potential has important non-central components like spin and angular momentum dependences, which cannot be ignored in realistic calculations. An attempt employing the Hartree-Fock variational method was tried by Coldwell (1972) who used the Reid soft-core potential for the N-N interaction and found that nucleons become localized at densities higher than $\sim 7 \times 10^{14}$ g cm^{-3}. However his work is incomplete in the sense that the trial wave function he used does not take into account the distortion of the two-body wave-function when the relative separation of the two particles becomes small. Pandhari-pande (1973) tackled the solidification problem by expanding the ground state energy in clusters up to second order and then minimizing the energy. The correlated 2-body wave-function was, in turn, found by solving the homogeneous Bethe-Goldstone equation. The basic idea of his computation, which is called lowest order constrained variation (LOCV), is to put restrictions on the correlation function (which is the ratio of the correlated 2-body wave function to the uncorrelated 2-body wave function) so as to make the truncated cluster expansion a reasonable one. Pandharipande's treatment of the Bethe-Goldstone equation is, however, not fully convincing. Firstly, we should note that the interaction Hamiltonian employed in the study of quantum crystals is not invariant under the transformation $\mathbf{r} \rightarrow -\mathbf{r}$. This implies that the wavefunction does not possess spatial symmetry as regards the angular-momentum decomposition. Consequently, all the angular momentum components of the wavefunction are coupled. This makes the problem rather complicated because at these high densities one has to include interactions up to 5 or 6 partial waves. This, in turn, gives rise to about 20 to 24 coupled differential equations. The coupling terms are not trivial and the simple decoupling of all the waves, as performed by Pandharipande in his work, leaves out many interesting features of the problem. His results indicate that the neutron matter will not solidify until a very high density ($\sim 8 \times 10^{15}$ g cm^{-3}).

Canuto and Chitre (1973, 1974) developed an extensive and thorough way of handling the Bethe-Goldstone equation, without ignoring the coupling of all partial waves and using the most general form of the nucleon-nucleon potential. Once the wave functions are obtained by solving the Bethe-Goldstone equation, they can be used in either the variational approach or the t-matrix method. Canuto and Chitre used the t-matrix approach. The variational method has the shortcoming that it cannot handle the spin and angular momentum dependences of nuclear forces in a straight-forward manner. It has the advantage that it enables one to take into account the higher-order many-body contributions by putting certain restrictions on the correlation function. However, if the effects of higher-order correlations are not important, then the t-matrix method is better because it can fully handle, to any degree of accuracy, the state dependence of the nuclear forces. Considering a system of nucleons, the hamiltonian for the system is

$$H = -\frac{\hbar^2}{2m} \sum_i \nabla_i^2 + \frac{1}{2} \sum_{i<j} V_{ij}. \tag{1}$$

The Slater determinant for the system was built up of single-particle wave functions of the gaussian form

$$\phi(i) = \frac{\alpha^{3/2}}{\pi^{3/4}} e^{-\alpha^2/2|r_i - R_i|^2}, \quad \alpha^2 = \frac{m\omega}{\hbar}, \tag{2}$$

where R_i is the ith lattice site around which the particle performs an oscillatory motion due to the influence of the remaining $(N-1)$ particles. The t-matrix expansion gives the following expression for the energy per particle (up to and including 2-body clusters)

$$\frac{E}{N} = \tfrac{3}{4}\hbar\omega + \sum_{i,j} \frac{\int \psi_{ij}^* V_{ij} \phi_{ij} \, \mathrm{d}^3 r_i \, \mathrm{d}^3 r_j}{2N \int \psi_{ij}^* \phi_{ij} \, \mathrm{d}^3 r_i \, \mathrm{d}^3 r_j}$$

$$= \tfrac{3}{4}\hbar\omega + \tfrac{1}{2} \sum_k n_k \varepsilon_k, \tag{3}$$

where ϕ_{ij} = uncorrelated 2-body wave function

         = $\phi(i)\phi(j)$

      ψ_{ij} = correlated 2-body wave function

ψ_{ij} is found by solving the Bethe-Goldstone equation

$$\left\{ -\frac{\hbar^2}{m}\nabla_r^2 + \tfrac{1}{4}m\omega^2(\mathbf{r} - \Delta)^2 + V(r) \right\} \psi(\mathbf{r}) = \{-\tfrac{3}{2}\hbar\omega - 2U(0)\}\psi(\mathbf{r}) \qquad (4)$$

where $\Delta = \mathbf{R}_1 - \mathbf{R}_2$.

Equation (4) is difficult to solve because it contains the term $\mathbf{r}\cdot\Delta$ which couples even waves with odd waves. If an angular momentum expansion is made of $\psi(\mathbf{r})$, then an infinite set of coupled differential equations results. All previous work that dealt with such an equation invariably used some average over $\mathbf{r}\cdot\Delta$.

To judge the t-matrix method and the handling of Equation (4), Canuto *et al.* (1974) tested it for the case of solid He^3. The results are shown in Figure 2. For the

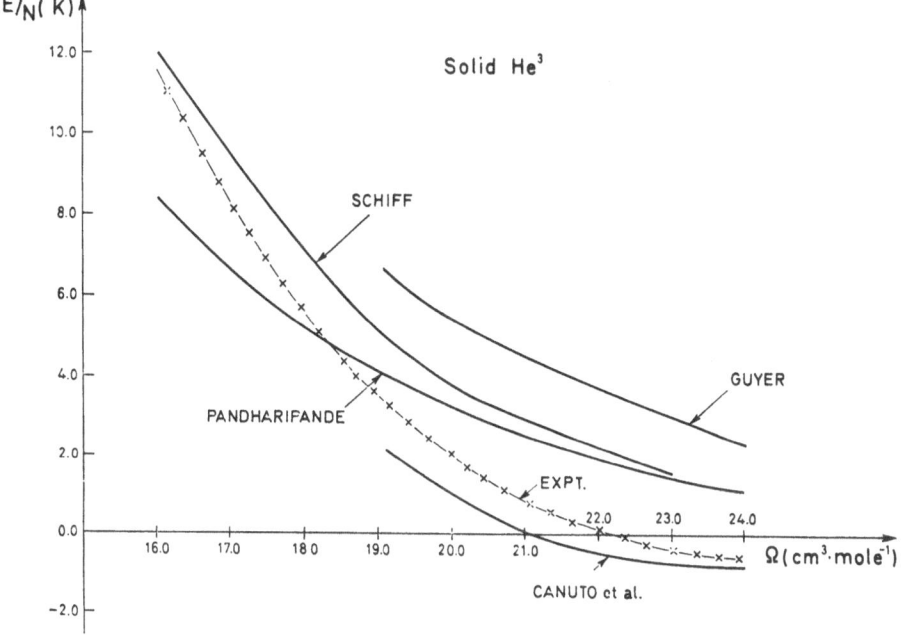

Fig. 2.   Ground-state energy versus molar volume for solid He^3 as obtained by Canuto *et al.* (1974). The results of Guyer (1969), Schiff (1973) and Pandharipande (1973) are reported for comparison.

time being, significant comparison should be made between the results of Guyer (1969) who used the same method with the approximation $\mathbf{r}\cdot\Delta \simeq r\Delta$ and the results of Canuto *et al.* who expanded the wave function into its partial waves, and solved a resulting set of 25 coupled differential equations. The results of Canuto *et al.* show a significant improvement over the results of Guyer. This shows that the treatment of the Bethe-Goldstone equation as done by Canuto and Chitre is more reliable.

Now, for a complete description of the neutron matter the spin variables should be taken into account. Equation (4) then reduces to three sets of equations corresponding to $S=0$, $M_s=0$ and $S=1$ and $M_s=\pm 1$. Canuto and Chitre solved the three sets of 7, 13 and 18 coupled differential equations by considering six partial waves which they found to be large enough for the system to be stable. Their results indicate that the neutron matter will solidify at a density of about 1.6×10^{15} g cm^{-3} and that an FCC configuration is energetically more favorable than a BCC one.

Schiff (1973) has studied the solidification problem using the idea that for a high-density system of nucleons, strong repulsion is more important than the Fermi statistics, so that one can consider the Pauli principle to be a perturbation to a Bose system, for which it is easier to calculate the ground state energy. The result of his computations yields a solidification density of $(2.9\pm 0.5)\times 10^{15}$ g cm^{-3}. More recently, Nosanow and Parish (1974) used the variational approach together with Monte-Carlo techniques (to take into account the many-body effects of the short-range correlations), and obtained a solidification density at about 4.2×10^{14} g cm^{-3}, which corresponds to the region just below the superfluid region of the neutron star. However, in their investigation the spin and angular momentum dependences were treated only approximately, and they do not indicate how sensitive their results are to the choice of potentials.

The results of the calculations on the solidification problem as performed by the different groups are shown in Figure 3.

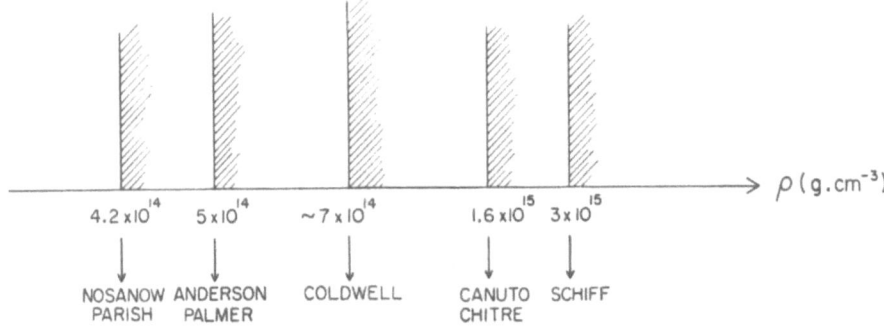

Fig. 3.   Present status of the results on the solidification problem.

4. Discussion

From Figure 3, it is clear that the spread in the value of the solidification density is rather large. Hence there is considerable room for controversy about the reliability of the different calculations. From our discussions on the solidification problem, it is clear that a microscopic calculation should be the most desirable one. This makes the results of Anderson and Palmer and of Coldwell look somewhat unconvincing. The result of Noshanow and Parish, based as it is on the Monte Carlo methods, is difficult to judge in terms of a microscopic description. Besides, they do not take into account the spin and angular momentum dependences of nuclear forces explicitly. As for the computations of Schiff, one has the feeling that it does not describe a real system of neutrons in that the actual short-range repulsion is not hard-core type, as assumed by Schiff, but rather of a soft-core type. This leaves us with the only two microscopic computations by Canuto and Chitre on one hand and Pandharipande on the other hand. Canuto and Chitre's reason for adopting the t-matrix approach is based on the fact that such an approach gives good results for the test case of solid He^3 and also that it enables one to deal with the state-dependence of N-N interaction in an exact manner. On the other hand, Pandharipande's computation, though microscopic in its scope, is incomplete in that it does not treat the solution of the Bethe-Goldstone equation in a convincing manner. Perhaps the only unsatisfactory feature of Canuto and Chitre's calculations is existence of the parity-violating term $r \cdot \Delta$, whose implications are yet to be fully understood.

Finally, although the theoretical computation suggests that there is a very good chance that the neutron star cores will be in a solid state, the real proof has to come from astrophysical considerations. There is one piece of observational evidence which makes plausible the existence of solid cores in neutron stars. This is in reference to speed-up of the Vela pulsar. The star-quake theory (Ruderman, 1969), which quite successfully explains the speed-up of the Crab pulsar, can also explain the observational features of the Vela pulsar if the latter is assumed to have a solid core (Pines *et al.*, 1972).

Acknowledgement

Two of us (B.D. and J.L.) would like to thank Dr R. Jastrow for his hospitality at the Institute for Space Studies, New York.

References

Anderson, P. W. and Palmer, R. G.: 1971, *Nature Phys. Sci.* **231**, 145.
Canuto, V. and Chitre, S. M.: 1973, *Phys. Rev. Letters* **30**, 999.
Canuto, V. and Chitre, S. M.: 1973, *Nat. Phys. Sci.* **243**, 63.
Canuto, V. and Chitre, S. M.: 1974, *Phys. Rev.* **D9**, 1587.
Canuto, V., Lodenquai, J., Parish, L., and Chitre, S. M.: 1974, *J. Low Temp. Phys.* **17**, 179.
Cazzola, P., Lucaroni, L., and Scaringi, C.: 1966, *Il Nuovo Cimento* **43**, 250.
Coldwell, R. L.: 1972, *Phys. Rev.* **D5**, 1273.
Guyer, R. A.: 1969, *Solid State Commun.* **7**, 315.
Nosanow, L. H. and Parish, L. J.: 1974, *6th Texas Symposium on Relativistic Astrophysics*, published by the New York Academy of Sciences, **224**, 226.
Palmer, R. G. and Anderson, P. W.: 1974, *Phys. Rev.* **D9**, 3281.
Pandharipande, V. R.: 1973, *Nucl. Phys.* **A217**, 1.
Pines, D., Shaham, J., and Ruderman, M. A.: 1972, *Nature* **237**, 83.
Ruderman, M. A.: 1968, *Nature* **218**, 1128.
Ruderman, M. A.: 1969, *Nature* **223**, 597.
Schiff, D.: 1973, *Nature Phys. Sci.* **243**, 130.

LIST OF PARTICIPANTS

Alpar, M. A., New Cavendish Laboratory (T.C.M.), Madingley Road, Cambridge CB3 0HE

Ausburn, K. J., c/o Department of Physics, University of Exeter, Exeter EX4 4QL

Barlow, M. J., Astronomy Centre, University of Sussex, Falmer, Brighton BN1 9QH

Bonetti, A. M., Space Physics University, Via S. Bonaventura 13, 50145 Firenze, Florence, Italy

Bussoletti, E., University of Lecce, Istituto di Fisica, Via Arnesano, Lecce, Italy

Canuto, V., Nordisk Institut for Teoretisk Atomfysik, Blegdamsvej 17, DK-2100, Copenhagen, Denmark

Code, A. D., Astronomy Department, 475 N. Charter Street, University of Wisconsin, Madison, Wisconsin 53706, U.S.A.

Cole, E. A. B., School of Mathematics, University of Leeds, Leeds 2

Dempsey, M. J., Department of Applied Mathematics and Astronomy, University College, Cardiff

Dopita, M. A., Department of Astronomy, Manchester University, Manchester M13 9PL

Dorschner, J., Universitaets-Sternwarte, DDR – 69 Jena, East Germany

Edmunds, M. G., Department of Applied Mathematics and Astronomy, University College, Cardiff

Evans, D. A., Department of Applied Mathematics and Astronomy, University College, Cardiff

Feuerbacher, B., Estec, Domeinweg, Noordwijk, Holland

Fitton, B., Estec, Domeinweg, Noordwijk, Holland

Greenberg, J. M., Department of Astronomy, State University of New York, Albany, New York, U.S.A.

Hayakawa, S., Department of Physics, Nagoya University, Chikusa-Ku, Nagoya, Japan

Huffman, D. R., Physics Department, University of Arizona, Tucson, Arizona 85721, U.S.A.

Itoh, N., Cavendish Laboratory, Madingley Road, Cambridge CB3 0HE

John, T. L., Department of Applied Mathematics and Astronomy, University College, Cardiff

Kegel, W. H., Lehrstuhl Fur Theor. Astrophys., 69 Heidelberg, IM Neuenheimer Feld 294, West Germany

Lee, T. J., Royal Observatory, Edinburgh, Blackford Hill, Edinburgh EH9 2HJ

Lukes, T., Department of Applied Mathematics and Astronomy, University College, Cardiff

Manning, P. G., c/o Chemistry Department, Edward Davies Chemical Laboratories, University College of Wales, Aberystwyth

Martin, D. H., Department of Physics, Queen Mary College, University of London, Mile End Road, London E1 4NS

Moorwood, A. F. M., Estec, Domeinweg, Noordwijk, Holland

Morgan, D. H., Royal Observatory, Edinburgh, Blackford Hill, Edinburgh EH9 3HJ

Morgan, D. J., Department of Applied Mathematics and Astronomy, University College, Cardiff

Nandy, K., Royal Observatory, Edinburgh, Blackford Hill, Edinburgh EH9 3HJ

Nelson, A. H., Department of Applied Mathematics and Astronomy, University College, Cardiff

Novotny, E., Department of Applied Mathematics and Astronomy, University College, Cardiff

Palmer, R. G., Physics Department, Princeton University, Jadwin Hall, P.O. Box 708, Princeton, New Jersey 08540, U.S.A.

Pim Fitzgerald, M., Astronomisches Institut der Ruhr-Universitat, D-4630 Bochum, West Germany

Pounds, K. A., Department of Physics, University of Leicester, University Road, Leicester LE1 7RH

Rees, M., University of Cambridge, Institute of Astronomy, Madingley Road, Cambridge CB3 0HA

Salpeter, E. E., 308 Newan L.N.S., Cornell University, Ithaca, New York 14850, U.S.A.

Schmidt, K. H., Zentral Institut fur Astrophysik, Sternwarte Babelsberg, Rosa-Luxemburg-Str. 17A, DDR-1502, Potsdam-Babelsberg, West Germany

Sedlmayr, E., University of Heidelberg, Lehrstuhl Fur Theor. Astrophysik, 69 Heidelberg 1, Neuenheimerfeld, West Germany

Sherwood, V. E., Astronomisches Institut Der Rhur-Universitat, 463 Bochum-Querenburg, Postfach 2148, West Germany

Smith, L. F., Max Planck Institut Fur Radio Astronomie, 53 Bonn 1, Auf Dem Hugel 69, West Germany

Snow, T. P., Princeton University Observatory, Peyton Hall, Princeton, New Jersey 08540, U.S.A.

Suggett, G. J., Department of Applied Mathematics and Astronomy, University College, Cardiff

Tarafdar, S. P., Department of Applied Mathematics and Astronomy, University College, Cardiff

Thomas, B., Department of Applied Physics, University of Wales Institute of Science and Technology, Cardiff

Tsuruta, S., Astronomy Centre, Physics Building, University of Sussex, Falmer, Brighton BN1 9QH

Vanýsek, V., Department of Astronomy, Charles University, 15000 Praha 5 – Smichov, Svedska 8, Czechoslovakia

Whitworth, A. P., Department of Applied Mathematics and Astronomy, University College, Cardiff

Wickramasinghe, N. C., Department of Applied Mathematics and Astronomy, University College, Cardiff

Williams, R. J., Department of Applied Mathematics and Astronomy, University College, Cardiff

Willis, R. F., Estec, Domeinweg, Noordwijk, Holland

INDEX OF NAMES

INDEX OF SUBJECTS

ASTROPHYSICS AND SPACE SCIENCE LIBRARY

Edited by

J. E. Blamont, R. L. F. Boyd, L. Goldberg, C. de Jager, Z. Kopal, G. H. Ludwig, R. Lüst,
B. M. McCormac, H. E. Newell, L. I. Sedov, Z. Švestka, and W. de Graaff

23. A. Muller (ed.), *The Magellanic Clouds. A European Southern Observatory Presentation: Principal Prospects, Current Observational and Theoretical Approaches, and Prospects for Future Research. Based on the Symposium on the Magellanic Clouds, held in Santiago de Chile, March 1969, on the Occasion of the Dedication of the European Southern Observatory.* 1971, XII + 189 pp.

24. B. M. McCormac (ed.), *The Radiating Atmosphere. Proceedings of a Symposium Organized by the Summer Advanced Study Institute, held at Queen's University, Kingston, Ontario, August 3–14, 1970.* 1971, XI + 455 pp.

25. G. Fiocco (ed.), *Mesospheric Models and Related Experiments. Proceedings of the 4th ESRIN-ESLAB Symposium, held at Frascati, Italy, July 6–10, 1970.* 1971, VIII + 298 pp.

26. I. Atanasijević, *Selected Exercises in Galactic Astronomy.* 1971, XII + 144 pp.

27. C. J. Macris (ed.), *Physics of the Solar Corona. Proceedings of the NATO Advanced Study Institute on Physics of the Solar Corona, held at Cavouri-Vouliagmeni, Athens, Greece, 6–17 September 1970.* 1971, XII + 345 pp.

28. F. Delobeau, *The Environment of the Earth.* 1971, IX + 113 pp.

29. E. R. Dyer (general ed.), *Solar-Terrestrial Physics/1970. Proceedings of the International Symposium on Solar-Terrestrial Physics, held in Leningrad, U.S.S.R., 12–19 May 1970.* 1972, VIII + 938 pp.

30. V. Manno and J. Ring (eds.), *Infrared Detection Techniques for Space Research, Proceedings of the 5th ESLAB-ESRIN Symposium, held in Noordwijk, The Netherlands, June 8–11, 1971.* 1972, XII + 344 pp.

31. M. Lecar (ed.), *Gravitational N-Body Problem. Proceedings of IAU Colloquium No. 10, held in Cambridge, England, August 12–15, 1970.* 1972, XI + 441 pp.

32. B. M. McCormac (ed.), *Earth's Magnetospheric Processes. Proceedings of a Symposium Organized by the Summer Advanced Study Institute and Ninth ESRO Summer School, held in Cortina, Italy, August 30–September 10, 1971.* 1972, VIII + 417 pp.

33. Antonin Rükl, *Maps of Lunar Hemispheres.* 1972, V + 24 pp.

34. V. Kourganoff, *Introduction to the Physics of Stellar Interiors.* 1973, XI + 115 pp.

35. B. M. McCormac (ed.), *Physics and Chemistry of Upper Atmospheres. Proceedings of a Symposium Organized by the Summer Advanced Study Institute, held at the University of Orléans, France, July 31–August 11, 1972.* 1973, VIII + 389 pp.

36. J. D. Fernie (ed.), *Variable Stars in Globular Clusters and in Related Systems. Proceedings of the IAU Colloquium No. 21, held at the University of Toronto, Toronto, Canada, August 29–31, 1972.* 1973, IX + 234 pp.

37. R. J. L. Grard (ed.), *Photon and Particle Interaction with Surfaces in Space. Proceedings of the 6th ESLAB Symposium, held at Noordwijk, The Netherlands, 26–29 September, 1972.* 1973, XV + 577 pp.

38. Werner Israel (ed.), *Relativity, Astrophysics and Cosmology. Proceedings of the Summer School, held 14–26 August, 1972, at the Banff Centre, Banff, Alberta, Canada.* 1973, IX + 323 pp.

39. B. D. Tapley and V. Szebehely (eds.), *Recent Advances in Dynamical Astronomy, Proceedings of the NATO Advanced Study Institute in Dynamical Astronomy, held in Cortina d'Ampezzo, Italy, August 9–12, 1972.* 1973, XIII + 468 pp.

40. A. G. W. Cameron (ed.), *Cosmochemistry. Proceedings of the Symposium on Cosmochemistry, held at the Smithsonian Astrophysical Observatory, Cambridge, Mass., August 14–16, 1972.* 1973, X + 173 pp.

41. M. Golay, *Introduction to Astronomical Photometry.* 1974, IX + 364 pp.

42. D. E. Page (ed.), *Correlated Interplanetary and Magnetospheric Observations. Proceedings of the 7th ESLAB Symposium, held at Saulgau, W. Germany, 22–25 May, 1973.* 1974, XIV + 662 pp.

43. Riccardo Giacconi and Herbert Gursky (eds.), *X-Ray Astronomy.* 1974, X + 450 pp.

44. B. M. McCormac (ed.), *Magnetospheric Physics. Proceedings of the Advanced Summer Institute, held in Sheffield, U.K., August 1973.* 1974, VII + 399 pp.

45. C. B. Cosmovici (ed.), *Supernovae and Supernova Remnants. Proceedings of the International Conference on Supernovae, held in Lecce, Italy, May 7–11, 1973.* 1974, XVII + 387 pp.

46. A. P. Mitra, *Ionospheric Effects of Solar Flares.* 1974, XI + 294 pp.

50. Zdeněk Kopal and Robert W. Carder, *Mapping of the Moon.* 1974, VIII + 237 pp.

52. V. Formisano (ed.), *The Magnetospheres of the Earth and Jupiter. Proceedings of the Neil Brice Memorial Symposium, held in Frascati, May 28–June 1, 1974.* 1975, XI + 485 pp.